T0360396

STOCHASTIC INTEREST RATE MODELING WITH FIXED INCOME DERIVATIVE PRICING

3rd Edition

ADVANCED SERIES ON STATISTICAL SCIENCE & APPLIED PROBABILITY

Editor: Ole E. Barndorff-Nielsen

This Series comprises accounts, by leading experts, from those areas of statistical science and applied probability where advanced mathematical tools are essential. Both research monographs and textbooks will be included, and the series promotes purely theoretical works as well as studies that combine theory and applications.

Published*

*To view the complete list of the published volumes in the series, please visit:
http://www.worldscientific.com/series/asssap

Advanced Series on

Statistical Science &

Applied Probability

Vol. 22

STOCHASTIC INTEREST RATE MODELING WITH FIXED INCOME DERIVATIVE PRICING

3rd Edition

Nicolas Privault

Nanyang Technological University, Singapore

World Scientific

NEW JERSEY · LONDON · SINGAPORE · BEIJING · SHANGHAI · HONG KONG · TAIPEI · CHENNAI · TOKYO

Published by

World Scientific Publishing Co. Pte. Ltd.

5 Toh Tuck Link, Singapore 596224

USA office: 27 Warren Street, Suite 401-402, Hackensack, NJ 07601

UK office: 57 Shelton Street, Covent Garden, London WC2H 9HE

Library of Congress Control Number: 2021931518

British Library Cataloguing-in-Publication Data
A catalogue record for this book is available from the British Library.

Advanced Series on Statistical Science and Applied Probability — Vol. 22
STOCHASTIC INTEREST RATE MODELING WITH FIXED INCOME
DERIVATIVE PRICING
Third Edition

ISBN 978-981-122-660-1 (hardcover)
ISBN 978-981-122-661-8 (ebook for institutions)
ISBN 978-981-122-662-5 (ebook for individuals)

For any available supplementary material, please visit
https://www.worldscientific.com/worldscibooks/10.1142/12000#t=suppl

Printed in Singapore

Preface

The goal of this book is two-fold. On the one hand, it is to serve as an introduction to stochastic interest rate models, ranging from short rate to forward rate models such as the Heath-Jarrow-Morton (HJM) and the Brace-Gatarek-Musiela (BGM) models. On the other hand, it is to treat the pricing of interest rate and fixed income derivatives such as bond options, caps, and swaptions, using forward measure techniques. An introduction to default bond pricing and an outlook on model calibration are also included as additional topics. The book is targeted at advanced undergraduate and beginning graduate students, assuming that they have already received an introduction to the fundamentals of probability theory and stochastic calculus. The focus is placed on a step-by-step introduction of concepts with explicit calculations.

This third edition represents a significant update on the 2012 second edition titled *An Elementary Introduction to Stochastic Interest Rate Modeling*. Most chapters have been reorganized and largely rewritten with additional details and supplementary solved exercises. New graphs and simulations based on market data have been included, together with the corresponding R codes. This new edition also contains 75 exercises and 4 problems, with their complete solutions.

This text grew from lecture notes of a course on stochastic interest rate models given in the Master of Science in Mathematics for Finance and Actuarial Science at City University of Hong Kong, based on the research work undertaken in the MathFi project at INRIA Paris-Rocquencourt, France. The material in this third edition has also been developed through teaching in the Master of Science in Financial Engineering of the Nanyang Business School, at the Nanyang Technological University in Singapore. I thank those institutions for excellent working conditions and for the possibility to facilitate these courses. Thanks are also due to participating students, as well as to Ming Gao (J.P. Morgan), Kazuhiro Kojima, Sijian Lin, Xiao Wei (Central University of Finance and Economics, Beijing), and Ubbo Wiersema (University of Reading), for suggestions and corrections that have led to substantial improvements.

Introduction

Let us shortly describe the main objectives of interest rate modeling. According to the rules of continuous-time compounding of interest, the value V_t at time $t \geqslant 0$ of a bank account earning interest at a fixed rate $r > 0$ is given by

$$V_t = V_0 e^{rt}, \qquad t \geqslant 0.$$

This relation can be restated in differential form as

$$\frac{dV_t}{V_t} = rdt,$$

to yield the dynamics of $(V_t)_{t\geqslant 0}$. The reality of the financial world is however more complex, as it allows interest rates to become functions of time that can be subject to random fluctuations. In this case, the value of V_t becomes

$$V_t = V_0 \exp\left(\int_0^t r_s ds \right),$$

where $(r_s)_{s\geqslant 0}$ is a time-dependent random process, called here a short-term interest rate process. The dynamics of this type of processes can be modeled in various ways using stochastic differential equations.

Short-term interest rate models are still not sufficient for the needs of financial institutions, who may require the possibility to agree at a present time t on a loan to be implemented over a future period of time $[T, S]$, at a rate $r(t, T, S)$ depending on $t \leqslant T \leqslant S$. This adds another level of sophistication to the modeling of interest rates, introducing the need for *forward* interest rate processes $r(t, T, S)$, which now depend on three time indexes. The instantaneous forward rates, defined as $T \mapsto r(t, T) :=$

$r(t, T, T)$, can be viewed at a fixed time t as functions of the single variable T, the maturity date.

Forward rate processes $r(t, T, S)$ are of special interest from a functional analytic point of view, because they can be reinterpreted as processes $t \mapsto r(t, \cdot, \cdot)$ taking values in a space of functions of two variables. Thus the modeling of forward rates makes a heavy use of stochastic processes taking values in (infinite-dimensional) function spaces, adding another level of technical difficulty in comparison with standard equity models.

Let us turn to the contents of this text. The first two chapters are devoted to reviews of stochastic calculus and of classical Black-Scholes pricing and hedging for options on equities. Indeed, the Black-Scholes framework is a fundamental tool for the pricing of interest rate derivatives via the Black pricing formula, and it can also be used for approximations.

Next, after a rapid presentation of mean-reverting short-term interest rate models in Chapter 3, we turn to the pricing of coupon and zero-coupon bonds in Chapter 4. Bond prices are constructed from short-term interest rate processes, and they will be used for the definition of forward rate processes.

The constructions of forward rates, instantaneous forward rates, LIBOR rates and swap rates, are presented in Chapter 5. The problems of estimating and fitting interest rate curves in the Nelson-Siegel and Svensson spaces are considered in Chapter 6, which also includes an introduction to two-factor models. Although LIBOR rates are to be phased out by the end of year 2021, we will still use this terminology when referring to simple or linear compounded forward rates.

The Heath-Jarrow-Morton (HJM) and Brace-Gatarek-Musiela (BGM) models of forward rates are described in Chapter 7, along with the HJM condition for absence of arbitrage. The construction of forward measures and its consequences on the pricing of interest rate derivatives are given in Chapter 8, with application to the pricing of bond options.

Chapter 9 is devoted to the pricing of caps and swaptions on LIBOR rates, including in the BGM model, with an outlook on its calibration. The pricing of default bonds is considered in Chapter 10, together with associated derivatives (credit default swaps) that are designed as a protection against default. We also refer the reader to James and Webber (2001), Björk (2004), Cairns (2004), Brigo and Mercurio (2006), Schoenmakers (2005), Carmona and Tehranchi (2006), Filipović (2009), Wu (2019), for related expositions of the theory and applications of interest rate modeling, and to Bielecki and Rutkowski (2002) for additional material on credit risk modeling.

The book is completed by an appendix on mathematical prerequisites, and by 75 exercises and 4 problems. Complete solutions to the exercises proposed in each chapter are provided at the end of the book.

Contents

List of Figures

List of Tables

Chapter 1

A Review of Stochastic Calculus

We start with a review of the Brownian motion and stochastic integrals, which are key tools for the modeling of interest rate processes. For simplicity of exposition, our presentation of stochastic integrals is focused on square-integrable processes, and we refer the reader to more advanced texts such as *e.g.* Protter (2004) for a comprehensive introduction.

1.1 Brownian Motion

We work on a probability space $(\Omega, \mathcal{F}, \mathbb{P})$. The modeling of random assets in finance is mainly based on stochastic processes, which are families $(X_t)_{t \in I}$ of random variables indexed by a time interval I.

First of all we recall the definition of Brownian motion, which is a fundamental example of a stochastic process.

Definition 1.1. *The standard Brownian motion is a stochastic process* $(B_t)_{t \in \mathbb{R}_+}$ *such that*

1. $B_0 = 0$ *almost surely,*
2. *The sample paths* $t \mapsto B_t$ *are (almost surely) continuous.*
3. *For any finite sequence of times* $t_0 < t_1 < \cdots < t_n$, *the increments*
$$B_{t_1} - B_{t_0}, B_{t_2} - B_{t_1}, \ldots, B_{t_n} - B_{t_{n-1}}$$
   are independent.
4. *For any times* $0 \leqslant s < t$, $B_t - B_s$ *is normally distributed with mean zero and variance* $t - s$.

For convenience we will sometimes interpret a Brownian motion as a random walk over infinitesimal time intervals of length dt, with increments

ΔB_t over $[t, t+dt]$ given by

$$\Delta B_t = \pm\sqrt{dt} \qquad (1.1)$$

with equal probabilities $1/2$. In the sequel, we consider the filtration[1] $(\mathcal{F}_t)_{t\in\mathbb{R}_+}$ generated by $(B_t)_{t\in\mathbb{R}_+}$, defined as

$$\mathcal{F}_t = \sigma(B_s \ : \ 0 \leqslant s \leqslant t), \qquad t \in \mathbb{R}_+.$$

The n-dimensional Brownian motion can be constructed as

$$\left(B_t^1, \ldots, B_t^n\right)_{t\in\mathbb{R}_+}$$

where $\left(B_t^1\right)_{t\in\mathbb{R}_+}, \ \ldots, \ \left(B_t^n\right)_{t\in\mathbb{R}_+}$ are independent copies of $(B_t)_{t\in\mathbb{R}_+}$.

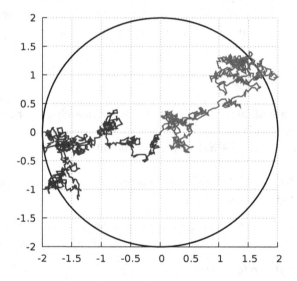

Fig. 1.1: Sample paths of a two-dimensional Brownian motion.

Next, we turn to simulations of two-dimensional, resp. three-dimensional, Brownian motion, cf. Figure 1.1, resp. 1.2. Recall that the Brownian motion was originally observed in an experimental setting from the 2-dimensional the movement of pollen particles by R. Brown in 1827.

---

[1]The filtration $(\mathcal{F}_t)_{t\in\mathbb{R}_+}$ is a nondecreasing family of sub σ-algebras of $\mathcal{F}$, see the appendix for details.

1.2 Stochastic Integration

In this section we construct the Itô stochastic integral of square-integrable adapted processes with respect to the Brownian motion. The main use of stochastic integrals in finance is to model the wealth of a portfolio driven by a (random) risky asset price process.

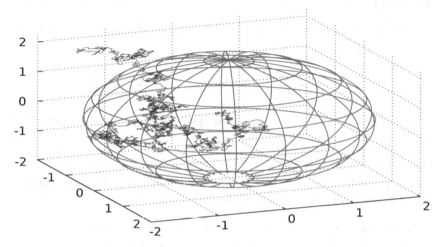

Fig. 1.2: Sample path of a three-dimensional Brownian motion.

Definition 1.2. *A process $(X_t)_{t\in\mathbb{R}_+}$ is said to be $(\mathcal{F}_t)_{t\in\mathbb{R}_+}$-adapted if X_t is $\mathcal{F}_t$-measurable for all $t \in \mathbb{R}_+$.*

In other words, $(X_t)_{t\in\mathbb{R}_+}$ is $(\mathcal{F}_t)_{t\in\mathbb{R}_+}$-adapted when the value of X_t at time t only depends on information contained in the Brownian path up to time t.

Definition 1.3.

(1) Let $L^p(\Omega\times\mathbb{R}_+)$ denote the space of p-integrable processes, i.e. the space of stochastic processes $u : \Omega \times \mathbb{R}_+ \to \mathbb{R}$ such that

$$\mathbb{E}\left[\int_0^\infty |u_t|^p dt\right] < \infty.$$

(2) Let $L^p_{\mathrm{ad}}(\Omega \times \mathbb{R}_+)$, $p \in [1,\infty]$, denote the space of $(\mathcal{F}_t)_{t\in\mathbb{R}_+}$-adapted processes in $L^p(\Omega \times \mathbb{R}_+)$.

A naive definition of the stochastic integral with respect to the Brownian motion would consist in writing

$$\int_0^\infty f(t)dB_t = \int_0^\infty f(t)\frac{dB_t}{dt}dt,$$

however this definition fails because the paths of Brownian motion are not differentiable:

$$\frac{dB_t}{dt} = \frac{\pm\sqrt{dt}}{dt} = \pm\frac{1}{\sqrt{dt}} \simeq \pm\infty.$$

Instead, stochastic integrals will be constructed as integrals of simple predictable processes.

Definition 1.4. *Let $\mathcal{P}$ denote the space of simple predictable processes $(u_t)_{t\in\mathbb{R}_+}$ of the form*

$$u_t = \sum_{i=1}^{n} F_i \mathbb{1}_{(t_{i-1}^n, t_i^n]}(t), \qquad t \in \mathbb{R}_+, \tag{1.2}$$

where $F_i \in L^2(\Omega, \mathcal{F}_{t_{i-1}}, \mathbb{P})$ is $\mathcal{F}_{t_{i-1}}$-measurable, $i = 1, 2, \ldots, n$, and $0 = t_0 < t_1 < \cdots < t_{n-1} < t_n = T$.

One easily checks that the set $\mathcal{P}$ of simple predictable processes forms a linear space. In the next proposition, we construct the stochastic integral with respect to the Brownian motion.

Proposition 1.1. *The stochastic integral with respect to the Brownian motion $(B_t)_{t\in\mathbb{R}_+}$, defined on simple predictable processes $(u_t)_{t\in\mathbb{R}_+}$ of the form (1.2) by*

$$\int_0^\infty u_t dB_t := \sum_{i=1}^{n} F_i(B_{t_i} - B_{t_{i-1}}), \tag{1.3}$$

extends to $u \in L_{ad}^2(\Omega \times \mathbb{R}_+)$ via the isometry formula

$$\mathbb{E}\left[\left(\int_0^\infty u_t dB_t\right)^2\right] = \mathbb{E}\left[\int_0^\infty |u_t|^2 dt\right], \quad u \in L_{ad}^2(\Omega \times \mathbb{R}_+). \tag{1.4}$$

In addition, the random variable $\int_0^T u_t dB_t$ is centered, i.e., we have

$$\mathbb{E}\left[\int_0^T u_t dB_t\right] = 0. \tag{1.5}$$

Proof. We start by showing that the Itô isometry (1.4) holds for the simple predictable process u of the form (1.2). We have

$$\mathbb{E}\left[\left(\int_0^T u_t dB_t\right)^2\right] = \mathbb{E}\left[\left(\sum_{i=1}^{n} F_i(B_{t_i} - B_{t_{i-1}})\right)^2\right]$$

$$= \mathbb{E}\left[\sum_{i,j=1}^{n} F_i F_j (B_{t_i} - B_{t_{i-1}})(B_{t_j} - B_{t_{j-1}})\right]$$

$$= \mathbb{E}\left[\sum_{i=1}^{n} |F_i|^2 (B_{t_i} - B_{t_{i-1}})^2\right]$$

$$+ 2\,\mathbb{E}\left[\sum_{1 \leqslant i < j \leqslant n} F_i F_j (B_{t_i} - B_{t_{i-1}})(B_{t_j} - B_{t_{j-1}})\right]$$

$$= \sum_{i=1}^{n} \mathbb{E}\left[\mathbb{E}\left[|F_i|^2 (B_{t_i} - B_{t_{i-1}})^2 \mid \mathcal{F}_{t_{i-1}}\right]\right]$$

$$+ 2 \sum_{1 \leqslant i < j \leqslant n} \mathbb{E}[\mathbb{E}[F_i F_j (B_{t_i} - B_{t_{i-1}})(B_{t_j} - B_{t_{j-1}}) \mid \mathcal{F}_{t_{j-1}}]]$$

$$= \sum_{i=1}^{n} \mathbb{E}\left[|F_i|^2\,\mathbb{E}\left[(B_{t_i} - B_{t_{i-1}})^2 \mid \mathcal{F}_{t_{i-1}}\right]\right]$$

$$+ 2 \sum_{1 \leqslant i < j \leqslant n} \mathbb{E}[F_i F_j (B_{t_i} - B_{t_{i-1}})\,\mathbb{E}[(B_{t_j} - B_{t_{j-1}}) \mid \mathcal{F}_{t_{j-1}}]]$$

$$= \sum_{i=1}^{n} \mathbb{E}\left[|F_i|^2\,\mathbb{E}\left[(B_{t_i} - B_{t_{i-1}})^2\right]\right]$$

$$+ 2 \sum_{1 \leqslant i < j \leqslant n} \mathbb{E}[F_i F_j (B_{t_i} - B_{t_{i-1}})\,\mathbb{E}[B_{t_j} - B_{t_{j-1}}]]$$

$$= \sum_{i=1}^{n} \mathbb{E}\left[|F_i|^2 (t_i - t_{i-1})\right]$$

$$= \mathbb{E}\left[\sum_{i=1}^{n} |F_i|^2 (t_i - t_{i-1})\right]$$

$$= \mathbb{E}\left[\int_0^T |u_t|^2 dt\right],$$

where we applied the "tower property" (11.7) of conditional expectations and the facts that $B_{t_i} - B_{t_{i-1}}$ is independent of $\mathcal{F}_{t_{i-1}}$ and

$$\mathbb{E}[B_{t_i} - B_{t_{i-1}}] = 0, \quad \mathbb{E}\left[(B_{t_i} - B_{t_{i-1}})^2\right] = t_i - t_{i-1}, \quad i = 1, 2, \ldots, n.$$

The extension of the stochastic integral to square-integrable $(\mathcal{F}_t)_{t \in \mathbb{R}_+}$-adapted processes $(u_t)_{t \in \mathbb{R}_+}$ is obtained by a denseness and Cauchy sequence argument using the isometry (1.4). By Lemma 1.1 of Ikeda and Watanabe (1989), pages 22 and 46, or Proposition 2.5.3 in Privault (2009), the set of simple predictable processes forms a linear space which is dense in the subspace $L^2_{\mathrm{ad}}(\Omega \times \mathbb{R}_+)$ of square-integrable $(\mathcal{F}_t)_{t \in \mathbb{R}_+}$-adapted processes in $L^2(\Omega \times \mathbb{R}_+)$. In other words, given u a square-integrable $(\mathcal{F}_t)_{t \in \mathbb{R}_+}$-adapted process there exists a sequence $(u^{(n)})_{n \geqslant 1}$ of simple predictable processes converging to u in $L^2(\Omega \times \mathbb{R}_+)$, *i.e.*

$$\lim_{n \to \infty} \|u - u^{(n)}\|_{L^2(\Omega \times [0,T])} = \lim_{n \to \infty} \sqrt{\mathbb{E}\left[\int_0^T |u_t - u_t^{(n)}|^2 dt\right]} = 0.$$

Since the sequence $(u^{(n)})_{n \geqslant 1}$ converges, it is a Cauchy sequence in $L^2(\Omega \times \mathbb{R}_+)$ hence by the Itô isometry (1.4), the sequence $\left(\int_0^T u_t^{(n)} dB_t\right)_{n \geqslant 1}$ is itself a Cauchy sequence in $L^2(\Omega)$, therefore it admits a limit in the complete space $L^2(\Omega)$. In this case, we let

$$\int_0^T u_t dB_t := \lim_{n \to \infty} \int_0^T u_t^{(n)} dB_t$$

and the limit is unique from (1.4) and satisfies (1.4). The fact that the random variable $\int_0^T u_t dB_t$ is *centered* can be proved first on a simple predictable process u of the form (1.2), as

$$\mathbb{E}\left[\int_0^T u_t dB_t\right] = \mathbb{E}\left[\sum_{i=1}^n F_i(B_{t_i} - B_{t_{i-1}})\right]$$

$$= \mathbb{E}\left[\sum_{i=1}^n F_i(B_{t_i} - B_{t_{i-1}})\right]$$

$$= \sum_{i=1}^n \mathbb{E}[\mathbb{E}[F_i(B_{t_i} - B_{t_{i-1}}) \mid \mathcal{F}_{t_{i-1}}]]$$

$$= \sum_{i=1}^n \mathbb{E}[F_i \, \mathbb{E}[B_{t_i} - B_{t_{i-1}} \mid \mathcal{F}_{t_{i-1}}]]$$

$$= \sum_{i=1}^n \mathbb{E}[F_i \, \mathbb{E}[B_{t_i} - B_{t_{i-1}}]]$$

$$= 0,$$

and this identity can be extended as above from simple predictable processes to $(\mathcal{F}_t)_{t \in \mathbb{R}_+}$-adapted processes $(u_t)_{t \in \mathbb{R}_+}$ in $L^2(\Omega \times \mathbb{R}_+)$, which yields (1.5). □

In practice, the stochastic integral (1.3) will be used to express the value of a self-financing portfolio as in Relation (2.11) below.

Note also that by bilinearity, the Itô isometry (1.4) can be written as

$$\mathbb{E}\left[\int_0^T u_t dB_t \int_0^T v_t dB_t\right] = \mathbb{E}\left[\int_0^T u_t v_t dt\right],$$

for all square-integrable $(\mathcal{F}_t)_{t\in\mathbb{R}_+}$-adapted processes $(u_t)_{t\in\mathbb{R}_+}$, $(v_t)_{t\in\mathbb{R}_+}$. Note that when the integrand $(u_t)_{t\in\mathbb{R}_+}$ is not a deterministic function, the random variable $\int_0^T u_t dB_t$ no longer has a Gaussian distribution, except in some exceptional cases.

We close this section with a remark on the gaussianity of stochastic integrals of deterministic functions.

Proposition 1.2. *Let* $f \in L^2(\mathbb{R}_+)$. *The stochastic integral*

$$\int_0^\infty f(t)dB_t$$

is a Gaussian random variable with mean 0 and variance

$$\int_0^\infty |f(t)|^2 dt.$$

Proof. From the relation

$$\text{Var}[\alpha X] = \alpha^2 \text{Var}[X],$$

cf. (11.1) in the appendix, the stochastic integral

$$\int_0^\infty f(t)dB_t := \sum_{k=1}^n a_k(B_{t_k} - B_{t_{k-1}}),$$

of the simple function

$$f(t) = \sum_{k=1}^n a_k \mathbb{1}_{(t_k,t_{k-1}]}(t),$$

has a centered Gaussian distribution with variance

$$\text{Var}\left[\int_0^\infty f(t)dB_t\right] = \sum_{k=1}^n a_k \text{Var}[B_{t_k} - B_{t_{k-1}}]$$

$$= \sum_{k=1}^n |a_k|^2 (t_k - t_{k-1})$$

$$= \sum_{k=1}^n |a_k|^2 \int_{t_{k-1}}^{t_k} dt$$

$$= \int_0^\infty |f(t)|^2 dt.$$

The result is extended by the denseness of simple functions in $L^2(\mathbb{R}_+)$. $\square$

In particular, for any $f \in L^2(\mathbb{R}_+)$ the Itô isometry (1.4) reads

$$E\left[\left(\int_0^\infty f(t)dB_t\right)^2\right] = \int_0^\infty |f(t)|^2 dt.$$

1.3   Definite Stochastic Integral

The definite stochastic integral of an $(\mathcal{F}_t)_{t\in\mathbb{R}_+}$-adapted process $(u_t)_{t\in\mathbb{R}_+}$ over an interval $[a, b] \subset [0, T]$ is defined as

$$\int_a^b u_t dB_t := \int_0^T \mathbb{1}_{[a,b]}(t) u_t dB_t,$$

with in particular

$$\int_a^b dB_t = \int_0^T \mathbb{1}_{[a,b]}(t) dB_t = B_b - B_a, \quad 0 \leqslant a \leqslant b.$$

We also have the Chasles relation

$$\int_a^c u_t dB_t = \int_a^b u_t dB_t + \int_b^c u_t dB_t, \quad 0 \leqslant a \leqslant b \leqslant c,$$

and the stochastic integral has the following linearity property:

$$\int_0^T (u_t + v_t) dB_t = \int_0^T u_t dB_t + \int_0^T v_t dB_t, \quad u, v \in L^2(\mathbb{R}_+).$$

As an application of the Itô isometry (1.4) we note for example that

$$\mathbb{E}\left[\left(\int_0^T B_t dB_t\right)^2\right] = \mathbb{E}\left[\int_0^T |B_t|^2 dt\right] = \int_0^T \mathbb{E}\left[|B_t|^2\right] dt = \int_0^T t\,dt = \frac{T^2}{2}.$$

The next proposition shows how to compute the conditional expectation of a stochastic integral by truncation of the integration interval. It also shows that the Itô integral yields a martingale with respect to the Brownian filtration $(\mathcal{F}_t)_{t\in\mathbb{R}_+}$.

Proposition 1.3. *The stochastic integral process* $\left(\int_0^t u_s dB_s\right)_{t\in\mathbb{R}_+}$ *of a square-integrable* $(\mathcal{F}_t)_{t\in\mathbb{R}_+}$*-adapted process* $u \in L^2_{\mathrm{ad}}(\Omega \times \mathbb{R}_+)$ *is a martingale, i.e.:*

$$\mathbb{E}\left[\int_0^t u_\tau dB_\tau \,\Big|\, \mathcal{F}_s\right] = \int_0^s u_\tau dB_\tau, \quad 0 \leqslant s \leqslant t. \qquad (1.6)$$

Proof. The statement is first proved in case $(u_t)_{t\in\mathbb{R}_+}$ is a simple predictable process, and then extended to the general case, cf. *e.g.* Proposition 2.5.7 in Privault (2009). For example, for $(u_t)_{t\in\mathbb{R}_+}$ a simple process of the form $u_t := F\mathbb{1}_{[a,b]}(t)$, $t \in \mathbb{R}_+$, with F an $\mathcal{F}_a$-measurable random variable and $t \in [a,b]$, we have

$$\mathbb{E}\left[\int_0^\infty u_s dB_s \,\Big|\, \mathcal{F}_t\right] = \mathbb{E}\left[\int_0^\infty F\mathbb{1}_{[a,b]}(s)dB_s \,\Big|\, \mathcal{F}_t\right]$$
$$= \mathbb{E}[F(B_b - B_a) \mid \mathcal{F}_t]$$
$$= F\,\mathbb{E}[B_b - B_a \mid \mathcal{F}_t]$$
$$= F(B_t - B_a)$$
$$= \int_a^t F\mathbb{1}_{[a,b]}(s)dB_s$$
$$= \int_0^t F\mathbb{1}_{[a,b]}(s)dB_s$$
$$= \int_0^t u_s dB_s, \qquad a \leqslant t \leqslant b.$$

On the other hand, when $t \in [0,a]$ we have

$$\int_0^t u_s dB_s = \int_0^t F\mathbb{1}_{[a,b]}(s)dB_s = 0,$$

and we need to check that

$$\mathbb{E}\left[\int_0^\infty u_s dB_s \,\Big|\, \mathcal{F}_t\right] = \mathbb{E}\left[\int_0^\infty F\mathbb{1}_{[a,b]}(s)dB_s \,\Big|\, \mathcal{F}_t\right]$$
$$= \mathbb{E}[F(B_b - B_a) \mid \mathcal{F}_t]$$
$$= \mathbb{E}[\mathbb{E}[F(B_b - B_a) \mid \mathcal{F}_a] \mid \mathcal{F}_t]$$
$$= \mathbb{E}[F\,\mathbb{E}[B_b - B_a \mid \mathcal{F}_a] \mid \mathcal{F}_t]$$
$$= 0, \qquad 0 \leqslant t \leqslant a,$$

where we used the tower property (11.7) of conditional expectations and the fact that Brownian motion $(B_t)_{t\in\mathbb{R}_+}$ is a martingale, as

$$\mathbb{E}[B_b - B_a \mid \mathcal{F}_a] = \mathbb{E}[B_b \mid \mathcal{F}_a] - B_a = B_a - B_a = 0.$$

The extension from simple processes to square-integrable $(\mathcal{F}_t)_{t\in\mathbb{R}_+}$-adapted processes in $L^2_{\mathrm{ad}}(\Omega \times \mathbb{R}_+)$ can be proved similarly to Proposition 1.1. Indeed, given $(u^{(n)})_{n\geqslant 1}$ be a sequence of simple predictable processes converging to u in $L^2(\Omega \times [0,T])$ cf. Lemma 1.1 of Ikeda and Watanabe (1989), pages 22 and 46, by Fatou's Lemma and the continuity of the conditional expectation on L^2 we have:

$$\mathbb{E}\left[\left(\int_0^t u_s dM_s - \mathbb{E}\left[\int_0^\infty u_s dM_s \,\bigg|\, \mathcal{F}_t\right]\right)^2\right]$$

$$\leqslant \liminf_{n\to\infty} \mathbb{E}\left[\left(\int_0^t u_s^{(n)} dM_s - \mathbb{E}\left[\int_0^\infty u_s dM_s \,\bigg|\, \mathcal{F}_t\right]\right)^2\right]$$

$$= \lim_{n\to\infty} \mathbb{E}\left[\left(\mathbb{E}\left[\int_0^\infty u_s^{(n)} dM_s - \int_0^\infty u_s dM_s \,\bigg|\, \mathcal{F}_t\right]\right)^2\right]$$

$$\leqslant \lim_{n\to\infty} \mathbb{E}\left[\mathbb{E}\left[\left(\int_0^\infty u_s^{(n)} dM_s - \int_0^\infty u_s dM_s\right)^2 \,\bigg|\, \mathcal{F}_t\right]\right]$$

$$= \lim_{n\to\infty} \mathbb{E}\left[\left(\int_0^\infty (u_s^{(n)} - u_s) dM_s\right)^2\right]$$

$$= \lim_{n\to\infty} \mathbb{E}\left[\int_0^\infty |u_s^{(n)} - u_s|^2 ds\right]$$

$$= 0,$$

where we used the Itô isometry (1.4). $\square$

As a consequence of Proposition 1.3, we also have

$$\mathbb{E}\left[\int_t^\infty u_\tau dB_\tau \,\bigg|\, \mathcal{F}_t\right] = 0, \quad \text{and} \quad \mathbb{E}\left[\int_0^t u_\tau dB_\tau \,\bigg|\, \mathcal{F}_t\right] = \int_0^t u_\tau dB_\tau. \quad (1.7)$$

In particular, $\int_0^t u_\tau dB_\tau$, $t \in \mathbb{R}_+$, is $\mathcal{F}_t$-measurable for all u in the space $L^2_{\mathrm{ad}}(\Omega \times \mathbb{R}_+)$ of square-integrable $(\mathcal{F}_t)_{t\in\mathbb{R}_+}$-adapted processes. In addition, since $\mathcal{F}_0 = \{\emptyset, \Omega\}$, Relation (1.6) applied with $t = 0$ recovers the fact that the Itô integral is a centered random variable:

$$\mathbb{E}\left[\int_0^\infty u_s dB_s\right] = \mathbb{E}\left[\int_0^\infty u_s dB_s \,\bigg|\, \mathcal{F}_0\right] = \int_0^0 u_s dB_s = 0.$$

1.4   Quadratic Variation

We now introduce the notion of quadratic variation of the Brownian motion.

Definition 1.5. *The quadratic variation of $(B_t)_{t\in\mathbb{R}_+}$ is the process $([B,B]_t)_{t\in\mathbb{R}_+}$ defined as*

$$[B,B]_t = B_t^2 - 2\int_0^t B_s dB_s, \quad t \in \mathbb{R}_+.$$

Let now

$$\pi^n = \{0 = t_0^n < t_1^n < \cdots < t_{n-1}^n < t_n^n = t\}$$

denote a family of subdivisions of $[0, t]$, such that

$$|\pi^n| := \operatorname*{Max}_{i=1,2,\ldots,n} |t_i^n - t_{i-1}^n|$$

converges to 0 as n goes to infinity.

Proposition 1.4. *We have*

$$[B, B]_t = \lim_{n \to \infty} \sum_{i=1}^n \left(B_{t_i^n} - B_{t_{i-1}^n} \right)^2, \qquad t \geqslant 0,$$

where the limit exists in $L^2(\Omega)$ and is independent of the sequence $(\pi^n)_{n \geqslant 1}$ of subdivisions chosen.

Proof. As an immediate consequence of the Definition 1.3 of the stochastic integral we have

$$B_s(B_t - B_s) = \int_s^t B_s dB_\tau, \qquad 0 \leqslant s \leqslant t,$$

hence

$$[B, B]_{t_i^n} - [B, B]_{t_{i-1}^n} = B_{t_i^n}^2 - B_{t_{i-1}^n}^2 - 2\int_{t_{i-1}^n}^{t_i^n} B_s dB_s$$

$$= \left(B_{t_i^n} - B_{t_{i-1}^n} \right)^2 - 2\int_{t_{i-1}^n}^{t_i^n} B_s dB_s + 2\left(B_{t_{i-1}^n} - B_{t_i^n} \right) B_{t_{i-1}^n}$$

$$= \left(B_{t_i^n} - B_{t_{i-1}^n} \right)^2 + 2\int_{t_{i-1}^n}^{t_i^n} \left(B_{t_{i-1}^n} - B_s \right) dB_s,$$

hence

$$\mathbb{E}\left[\left([B, B]_t - \sum_{i=1}^n \left(B_{t_i^n} - B_{t_{i-1}^n} \right)^2 \right)^2 \right]$$

$$= \mathbb{E}\left[\left(\sum_{i=1}^n [B, B]_{t_i^n} - [B, B]_{t_{i-1}^n} - \left(B_{t_i^n} - B_{t_{i-1}^n} \right)^2 \right)^2 \right]$$

$$= 4\,\mathbb{E}\left[\left(\sum_{i=1}^n \int_0^t \mathbb{1}_{(t_{i-1}^n, t_i^n]}(s) \left(B_s - B_{t_{i-1}^n} \right) dB_s \right)^2 \right]$$

$$= 4 \, \mathbb{E} \left[\sum_{i=1}^{n} \int_{t_{i-1}^n}^{t_i^n} \left(B_s - B_{t_{i-1}^n} \right)^2 ds \right]$$

$$= 4 \, \mathbb{E} \left[\sum_{i=1}^{n} \int_{t_{i-1}^n}^{t_i^n} \left(s - t_{i-1}^n \right)^2 ds \right]$$

$$\leqslant 4t |\pi^n|.$$

$\square$

In view of the informal construction (1.1) of the Brownian motion as a random walk, the next proposition can be simply interpreted by writing $(\Delta B_t)^2 = dt$.

Proposition 1.5. *The quadratic variation of the Brownian motion* $(B_t)_{t \in \mathbb{R}_+}$ *is*

$$[B, B]_t = t, \qquad t \in \mathbb{R}_+.$$

Proof.   (See *e.g.* Protter (2004), Theorem I-28). For every subdivision

$$\{0 = t_0^n < \cdots < t_n^n = t\}$$

we have, by independence of the increments of the Brownian motion:

$$\mathbb{E} \left[\left(t - \sum_{i=1}^{n} \left(B_{t_i^n} - B_{t_{i-1}^n} \right)^2 \right)^2 \right]$$

$$= \mathbb{E} \left[\left(\sum_{i=1}^{n} \left(B_{t_i^n} - B_{t_{i-1}^n} \right)^2 - \left(t_i^n - t_{i-1}^n \right) \right)^2 \right]$$

$$= \sum_{i=1}^{n} \left(t_i^n - t_{i-1}^n \right)^2 \mathbb{E} \left[\left(\frac{\left(B_{t_i^n} - B_{t_{i-1}^n} \right)^2}{t_i^n - t_{i-1}^n} - 1 \right)^2 \right]$$

$$= \mathbb{E} \left[(Z^2 - 1)^2 \right] \sum_{i=0}^{n} \left(t_i^n - t_{i-1}^n \right)^2$$

$$\leqslant t |\pi^n| \, \mathbb{E} \left[g(Z^2 - 1)^2 \right],$$

where Z is a standard Gaussian random variable. We conclude from the vanishing of $\lim_{n \to \infty} |\pi^n| = 0$. $\square$

We note that by the Lévy characterization theorem, the Brownian motion is the only continuous martingale with quadratic variation $[B_t, B_t] = t$, $t \in \mathbb{R}_+$, see *e.g.* Theorem IV.3.6 in Revuz and Yor (1994).

1.5 Itô's Formula

Using the rule $(dB_t)^2 = (\pm\sqrt{dt})^2 = dt$, Taylor's formula reads informally

$$df(B_t) = f'(B_t)dB_t + \frac{1}{2}f''(B_t)(dB_t)^2$$

$$= f'(B_t)dB_t + \frac{1}{2}f''(B_t)dt.$$

The Itô formula provides a generalization of this identity to Itô processes $(X_t)_{t\in\mathbb{R}_+}$ of the form

$$X_t = X_0 + \int_0^t v_s ds + \int_0^t u_s dB_s, \qquad t \in \mathbb{R}_+, \tag{1.8}$$

where $(u_t)_{t\in\mathbb{R}_+}$, $(v_t)_{t\in\mathbb{R}_+}$, are $(\mathcal{F}_t)_{t\in\mathbb{R}_+}$-adapted and sufficiently integrable processes.

The Itô formula can be stated in integral form as

$$f(t, X_t) = f(0, X_0) + \int_0^t \frac{\partial f}{\partial s}(s, X_s)ds \tag{1.9}$$

$$+ \int_0^t v_s \frac{\partial f}{\partial x}(s, X_s)ds + \int_0^t u_s \frac{\partial f}{\partial x}(s, X_s)dB_s + \frac{1}{2}\int_0^t u_s^2 \frac{\partial^2 f}{\partial x^2}(s, X_s)ds,$$

for $f \in \mathcal{C}_b^{1,2}(\mathbb{R}_+ \times \mathbb{R})$ or, in differential form, as:

$$df(t, X_t) = \frac{\partial f}{\partial t}(t, X_t)dt + \frac{\partial f}{\partial x}(t, X_t)dX_t + \frac{1}{2}\frac{\partial^2 f}{\partial x^2}(t, X_t)dX_t \cdot dX_t$$

$$= \frac{\partial f}{\partial t}(t, X_t)dt + v_t \frac{\partial f}{\partial x}(t, X_t)dt + u_t \frac{\partial f}{\partial x}(t, X_t)dB_t + \frac{1}{2}u_t^2 \frac{\partial^2 f}{\partial x^2}(t, X_t)dt,$$

$$\tag{1.10}$$

with

$$dX_t \cdot dX_t = (v_t dt + u_t dB_t) \cdot (v_t dt + u_t dB_t) = u_t^2 dB_t \cdot dB_t = u_t^2 dt$$

according to the Itô multiplication table.

$\cdot$	dt	dB_t
dt	0	0
dB_t	0	dt

Table 1.1: Itô multiplication table.

Example

Applying Itô's formula (1.10) to B_t^2 with

$$B_t^2 = f(t, B_t) \quad \text{and} \quad f(t, x) = x^2,$$

and

$$\frac{\partial f}{\partial t}(t, x) = 0, \qquad \frac{\partial f}{\partial x}(t, x) = 2x, \qquad \frac{1}{2}\frac{\partial^2 f}{\partial x^2}(t, x) = 1,$$

we find

$$
\begin{aligned}
d(B_t^2) &= df(B_t) \\
&= \frac{\partial f}{\partial t}(t, B_t)dt + \frac{\partial f}{\partial x}(t, B_t)dB_t + \frac{1}{2}\frac{\partial^2 f}{\partial x^2}(t, B_t)dt \\
&= 2B_t dB_t + dt.
\end{aligned}
$$

Next, by integration in $t \in [0, T]$ we find

$$B_T^2 = B_0 + 2\int_0^T B_s dB_s + \int_0^T dt = 2\int_0^T B_s dB_s + T, \tag{1.11}$$

hence the relation

$$\int_0^T B_s dB_s = \frac{1}{2}\left(B_T^2 - T\right).$$

Similarly, we have

$$
\begin{cases}
d(B_t^3) = 3B_t^2 dB_t + 3B_t dt, \\[2mm]
d(\sin B_t) = \cos(B_t)dB_t - \dfrac{1}{2}\sin(B_t)dt, \\[2mm]
d\,\mathrm{e}^{B_t} = \mathrm{e}^{B_t}dB_t + \dfrac{1}{2}\mathrm{e}^{B_t}dt, \\[2mm]
d\log B_t = \dfrac{1}{B_t}dB_t - \dfrac{1}{2B_t^2}dt, \\[2mm]
d\,\mathrm{e}^{tB_t} = B_t\,\mathrm{e}^{tB_t}dt + \dfrac{t^2}{2}\mathrm{e}^{tB_t}dt + t\,\mathrm{e}^{tB_t}dB_t.
\end{cases}
$$

For the d-dimensional Brownian motion $(B_t)_{t \in \mathbb{R}_+}$, the Itô formula reads

$$f(B_t) = f(B_0) + \int_0^t \langle \nabla f(B_s), dB_s \rangle_H + \frac{1}{2}\int_0^t \Delta f(B_s)ds,$$

for $\mathcal{C}_b^2$ functions $f : \mathbb{R}^d \to \mathbb{R}$, where ∇ and Δ are respectively the gradient and Laplace operators on $\mathbb{R}^d$. Consider now two Itô processes $(X_t)_{t \in \mathbb{R}_+}$ and $(Y_t)_{t \in \mathbb{R}_+}$ of the form

$$X_t = X_0 + \int_0^t u_s dB_s^{(1)} + \int_0^t v_s ds, \qquad 0 \leqslant t \leqslant T,$$

and

$$Y_t = Y_0 + \int_0^t \xi_s dB_s^{(2)} + \int_0^t \zeta_s ds, \qquad 0 \leqslant t \leqslant T,$$

where $(u_t)_{t \in \mathbb{R}_+}$, $(v_t)_{t \in \mathbb{R}_+}$, $(\xi_t)_{t \in \mathbb{R}_+}$, $(\zeta_t)_{t \in \mathbb{R}_+}$ are $(\mathcal{F}_t)_{t \in \mathbb{R}_+}$-adapted and sufficiently integrable processes, and $\left(B_t^{(1)}\right)_{t \in \mathbb{R}_+}$, $\left(B_t^{(2)}\right)_{t \in \mathbb{R}_+}$, are two Brownian motions with correlation $\rho \in [-1, 1]$, *i.e.* their covariation is

$$dB_t^{(1)} \boldsymbol{\cdot} \ dB_t^{(2)} = \rho dt.$$

The Itô formula in two variables reads

$$f(t, X_t, Y_t) = f(0, X_0, Y_0) \tag{1.12}$$

$$+ \int_0^t \frac{\partial f}{\partial s}(s, X_s, Y_s) ds + \int_0^t \frac{\partial f}{\partial x}(s, X_s, Y_s) dX_s + \int_0^t \frac{\partial f}{\partial y}(s, X_s, Y_s) dY_s$$

$$+ \frac{1}{2} \int_0^t \frac{\partial^2 f}{\partial x^2}(s, X_s, Y_s) dX_s \boldsymbol{\cdot} \ dX_s + \frac{1}{2} \int_0^t \frac{\partial^2 f}{\partial y^2}(s, X_s, Y_s) dY_s \boldsymbol{\cdot} \ dY_s$$

$$+ \int_0^t \frac{\partial^2 f}{\partial x \partial y}(s, X_s, Y_s) dX_s \boldsymbol{\cdot} \ dY_s$$

$$= f(0, X_0, Y_0) + \int_0^t u_s \frac{\partial f}{\partial x}(s, X_s, Y_s) dB_s^{(1)} + \int_0^t \xi_s \frac{\partial f}{\partial y}(s, X_s, Y_s) dB_s^{(2)}$$

$$+ \int_0^t \frac{\partial f}{\partial s}(s, X_s, Y_s) ds + \int_0^t v_s \frac{\partial f}{\partial x}(s, X_s, Y_s) ds + \int_0^t \zeta_s \frac{\partial f}{\partial y}(s, X_s, Y_s) ds$$

$$+ \frac{1}{2} \int_0^t u_s^2 \frac{\partial^2 f}{\partial x^2}(s, X_s, Y_s) ds + \frac{1}{2} \int_0^t \xi_s^2 \frac{\partial^2 f}{\partial y^2}(s, X_s, Y_s) ds$$

$$+ \rho \int_0^t u_s \xi_s \frac{\partial^2 f}{\partial x \partial y}(s, X_s, Y_s) ds,$$

with

$$dX_t \boldsymbol{\cdot} \ dY_t = (v_t dt + u_t dB_t) \boldsymbol{\cdot} \ (\zeta_t dt + \xi_t dB_t) = u_t \xi_t dB_t^{(1)} \boldsymbol{\cdot} \ dB_t^{(2)} = \rho u_t \xi_t dt$$

according to the correlated Itô multiplication table.

$\cdot$	dt	$dB_t^{(1)}$	$dB_t^{(2)}$
dt	0	0	0
$dB_t^{(1)}$	0	dt	ρdt
$dB_t^{(2)}$	0	ρdt	dt

Table 1.2: Correlated Itô multiplication table.

We close this chapter by quoting a classical result on Stochastic Differential Equations (SDEs) see *e.g.* Protter (2004), Theorem V-7. Let the coefficients

$$\sigma : \mathbb{R}_+ \times \mathbb{R}^n \to \mathbb{R}^n \otimes \mathbb{R}^d$$

where $\mathbb{R}^n \otimes \mathbb{R}^d$ denotes the space of $d \times n$ matrices, and

$$b : \mathbb{R}_+ \times \mathbb{R}^n \to \mathbb{R}^n$$

satisfy the global Lipschitz condition

$$\|\sigma(t,x) - \sigma(t,y)\|^2 + \|b(t,x) - b(t,y)\|^2 \leqslant K^2 \|x - y\|^2,$$

$t \in \mathbb{R}_+$, $x, y \in \mathbb{R}^n$. Then there exists a unique strong solution to the stochastic differential equation

$$X_t = X_0 + \int_0^t \sigma(s, X_s) dB_s + \int_0^t b(s, X_s) ds,$$

where $(B_t)_{t \in \mathbb{R}_+}$ is a d-dimensional Brownian motion.

Exercises

Exercise 1.1. Let $c > 0$. Using the definition of the Brownian motion $(B_t)_{t \in \mathbb{R}_+}$, show that:

(1) $(B_{c+t} - B_c)_{t \in \mathbb{R}_+}$ is a Brownian motion.
(2) $(cB_{t/c^2})_{t \in \mathbb{R}_+}$ is a Brownian motion.

Exercise 1.2. Solve the stochastic differential equation

$$dS_t = \mu S_t dt + \sigma S_t dB_t, \tag{1.13}$$

defining the geometric Brownian motion $(S_t)_{t \in \mathbb{R}_+}$, where $\mu, \sigma \in \mathbb{R}$. *Hint:* Look for a solution of the form $S_t = f(t, B_t)$.

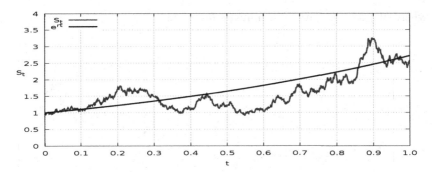

Fig. 1.3: Sample path of (1.13) with $r = 1$ and $\sigma^2 = 0.5$.

Exercise 1.3. Solve the stochastic differential equation

$$dX_t = -\alpha X_t dt + \sigma dB_t, \qquad (1.14)$$

with $\alpha > 0$ and $\sigma > 0$. *Hint*: Look for a solution of the form

$$X_t = a(t)\left(X_0 + \int_0^t b(s)dB_s\right),$$

where $a(\cdot)$ and $b(\cdot)$ are deterministic functions.

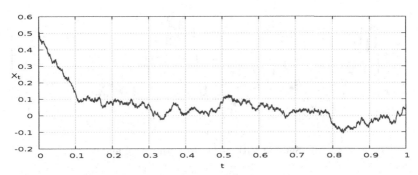

Fig. 1.4: Sample path of (1.14) with $\alpha = 10$ and $\sigma = 0.2$.

Exercise 1.4. Solve the stochastic differential equation

$$dX_t = tX_t dt + e^{t^2/2}dB_t, \qquad X_0 = x_0. \qquad (1.15)$$

Hint: Look for a solution of the form

$$X_t = a(t)\left(X_0 + \int_0^t b(s)dB_s\right),$$

where $a(\cdot)$ and $b(\cdot)$ are deterministic functions.

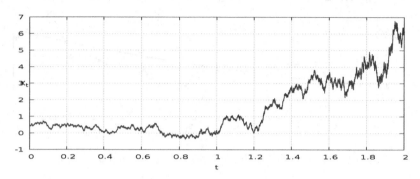

Fig. 1.5: Sample path of (1.15).

Exercise 1.5. Solve the stochastic differential equation
$$dY_t = \big(2\alpha Y_t + \sigma^2\big)dt + 2\sigma\sqrt{Y_t}dB_t, \qquad (1.16)$$
where $\alpha \in \mathbb{R}$ and $\sigma > 0$. *Hint*: Let $X_t = \sqrt{Y_t}$.

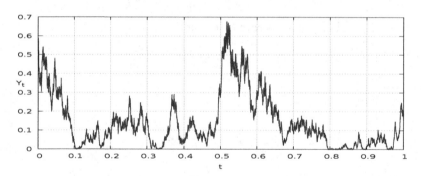

Fig. 1.6: Sample path of (1.16) with $\alpha = 5$ and $\sigma = 1$.

Exercise 1.6. Let $f \in L^2([0,T])$. Compute the conditional expectation
$$\mathbb{E}\left[e^{\int_0^T f(s)dB_s} \,\middle|\, \mathcal{F}_t\right], \qquad 0 \leqslant t \leqslant T,$$
where $(\mathcal{F}_t)_{t\in\mathbb{R}_+}$ denotes the filtration generated by $(B_t)_{t\in\mathbb{R}_+}$.

Exercise 1.7. Compute the expected value
$$\mathbb{E}\left[e^{\beta \int_0^T B_t dB_t}\right]$$
for all $\beta < 1/T$. *Hint*: expand $(B_T)^2$ using the Itô formula.

Exercise 1.8. Stochastic bridge. Given $T > 0$, let $(X_t^T)_{t \in [0,T)}$ denote the solution of the stochastic differential equation

$$dX_t^T = \sigma dB_t - \frac{X_t^T}{T - t} dt, \qquad 0 \leqslant t < T,$$

under the initial condition $X_0^T = 0$ and $\sigma > 0$.

(1) Show that

$$X_t^T = \sigma \int_0^t \frac{T - t}{T - s} dB_s, \qquad 0 \leqslant t < T. \qquad (1.17)$$

Hint: Start by computing $d(X_t^T/(T - t))$ using the Itô formula.

(2) Show that $\mathbb{E}\left[X_t^T\right] = 0$ for all $t \in [0, T]$.

(3) Show that $\operatorname{Var}\left[X_t^T\right] = \sigma^2 t(T - t)/T$ for all $t \in [0, T]$.

(4) Show that $\lim_{t \to T} \|X_t^T\|_{L^2(\Omega)} = 0$. The process $(X_t^T)_{t \in [0,T]}$ is called a *Brownian bridge*.

Chapter 2

A Review of Black-Scholes Pricing and Hedging

This chapter reviews the Black-Scholes framework for the pricing and hedging of financial derivatives. The complexity of the interest rate models makes it in general difficult to obtain closed-form expressions, and in many situations one has to rely on the Black-Scholes formula to price interest rate derivatives. In that sense, the Black-Scholes formula can be considered as a building block for the pricing of financial derivatives, and its importance is not restricted to the pricing of options on stocks.

2.1 Call and Put Options

An important concern for the buyer of a stock at time $t \in [0, T)$ is whether its price S_T can fall down at some future date T. The buyer may seek protection from a market crash by buying a contract that allows him to sell his asset at time T at a guaranteed price K fixed at an initial time t.

This contract is called a put option with strike price K and exercise date T. In case the price S_T falls down below the level K, exercising the contract will give the buyer of the option a gain equal to $K - S_T$ in comparison to others who did not subscribe the option. In turn, the issuer of the option will register a loss also equal to $K - S_T$, assuming the absence of transaction costs and other fees.

In the general case, the payoff of a (so-called European) put option will be of the form

$$\phi(S_T) = (K - S_T)^+ = \begin{cases} K - S_T & \text{if } S_T \leqslant K, \\ 0 & \text{if } S_T \geqslant K. \end{cases}$$

In order for this contract to be fair, the buyer of the option should pay a fee (similar to an insurance fee) at the signature of the contract. The computation of this fee is an important issue, which is known as option pricing.

Two possible scenarios, with S_T finishing above K or below K, are illustrated in Figure 2.1.

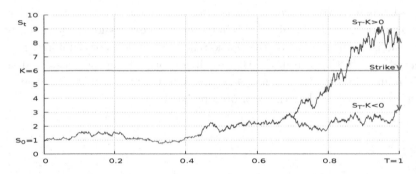

Fig. 2.1: Sample price processes modeled using geometric Brownian motion.

On the other hand, if the trader aims at buying some stock or commodity, his interest will be in prices not going up and he might want to purchase a call option, which is a contract allowing him to buy the considered asset at time T at a price not higher than a level K fixed at time $t \in [0, T)$.

Here, in the event that S_T goes above K, the buyer of the option will register a potential gain equal to $S_T - K$ in comparison to an agent who did not subscribe to the call option.

In general, a (so-called European) call option is an option with payoff function

$$\phi(S_T) = (S_T - K)^+ = \begin{cases} S_T - K & \text{if } S_T \geqslant K, \\ 0, & \text{if } S_T \leqslant K. \end{cases}$$

In preparation for the setting of interest rate models to be presented in subsequent chapters, we note at the present stage that similar contracts can be applied to interest rates.

A contract protecting a borrower at variable rate $r(t)$ by forcing his offered rate not to go above a level κ will result into an interest rate equal to $\min(r(t), \kappa)$. The corresponding contract is called an interest rate cap and potentially gives its buyer an advantage $(r(t) - \kappa)^+$, measured in terms of interest rate points. The counterpart of an interest rate cap, called a floor, offers a similar protection, this time against interest rates going down, for the benefit of lenders.

The classical Black-Scholes formula is of importance for the pricing of interest rates derivatives since some of the interest rate models that we will consider will be based on geometric Brownian motion.

2.2 Market Model and Self-Financing Portfolio

Let $(r(t))_{t \in \mathbb{R}_+}$, $(\mu(t))_{t \in \mathbb{R}_+}$ and $(\sigma(t))_{t \in \mathbb{R}_+}$ be deterministic nonnegative bounded functions. Let $(A_t)_{t \in \mathbb{R}_+}$ be a risk-free asset with price given by

$$\frac{dA_t}{A_t} = r(t)dt, \qquad A_0 = 1, \qquad t \geqslant 0,$$

i.e.

$$A_t = A_0 \, e^{\int_0^t r(s)ds}, \qquad t \geqslant 0.$$

Let $(S_t)_{t \in [0,T]}$ be the price process defined by the stochastic differential equation

$$dS_t = \mu(t)S_t dt + \sigma(t)S_t dB_t, \qquad t \geqslant 0,$$

i.e., in integral form,

$$S_t = S_0 + \int_0^t \mu(u)S_u du + \int_0^t \sigma(u)S_u dB_u, \qquad t \geqslant 0,$$

with solution

$$S_t = S_0 \exp\left(\int_0^t \sigma(s)dB_s + \int_0^t \left(\mu(s) - \frac{1}{2}\sigma^2(s) \right) ds \right),$$

$t \in \mathbb{R}_+$, see Exercise 1.1.2.

Let η_t and ζ_t be the numbers of units invested at time $t \geqslant 0$, respectively in the assets priced $(S_t)_{t \in \mathbb{R}_+}$ and $(A_t)_{t \in \mathbb{R}_+}$. The value of the portfolio V_t at time $t \geqslant 0$ is given by

$$V_t = \zeta_t A_t + \eta_t S_t, \qquad t \geqslant 0.$$

Definition 2.1. *The portfolio process $(\eta_t, \zeta_t)_{t \in \mathbb{R}_+}$ is said to be self-financing if*

$$dV_t = \zeta_t dA_t + \eta_t dS_t. \tag{2.1}$$

Note that the self-financing condition (2.1) can be written as

$$A_t d\zeta_t + S_t d\eta_t = 0, \qquad 0 \leqslant t \leqslant T, \tag{2.2}$$

see *e.g.* § 5.1 in Privault (2014).

2.3   PDE Method

In the standard Black-Scholes model, it is possible to determine a portfolio strategy for the hedging of European claims. First, we note that the self-financing condition (2.1) implies

$$
\begin{aligned}
dV_t &= \zeta_t dA_t + \eta_t dS_t \\
&= r(t)\zeta_t A_t dt + \mu(t)\eta_t S_t dt + \sigma(t)\eta_t S_t dB_t \\
&= r(t)V_t dt + (\mu(t) - r(t))\eta_t S_t dt + \sigma(t)\eta_t S_t dB_t,
\end{aligned} \tag{2.3}
$$

$t \in \mathbb{R}_+$. Assume now that the value V_t of the portfolio at time $t \geqslant 0$ is given by a function $C(t, x)$ as

$$V_t = C(t, S_t), \qquad t \geqslant 0.$$

An application of the Itô formula (1.9) leads to

$$
\begin{aligned}
dC(t, S_t) &= \frac{\partial C}{\partial t}(t, S_t)dt + \mu(t)S_t \frac{\partial C}{\partial x}(t, S_t)dt + \frac{1}{2}\sigma^2(t)S_t^2 \frac{\partial^2 C}{\partial x^2}(t, S_t)dt \\
&\quad + \sigma(t)S_t \frac{\partial C}{\partial x}(t, S_t)dB_t.
\end{aligned} \tag{2.4}
$$

Therefore, after respective identification of the terms in dB_t and dt in (2.3) and (2.4), we obtain

$$
\begin{cases}
r(t)C(t, S_t) = \left(\dfrac{\partial C}{\partial t} + r(t)S_t \dfrac{\partial C}{\partial x} + \dfrac{1}{2}\sigma^2(t)S_t^2 \dfrac{\partial^2 C}{\partial x^2} \right)(t, S_t), & (2.5a) \\[4mm]
\eta_t S_t \sigma(t)dB_t = S_t \sigma(t)\dfrac{\partial C}{\partial x}(t, S_t)dB_t, & (2.5b)
\end{cases}
$$

hence

$$\eta_t = \frac{\partial C}{\partial x}(t, S_t). \tag{2.6}$$

The process $(\eta_t)_{t \in \mathbb{R}_+}$ is called the Delta. In addition to computing the Delta, we have also derived the Black-Scholes partial differential equation (PDE) (2.5a), as stated in the next proposition.

Proposition 2.1. *The Black-Scholes PDE for the price of a European call is written as*

$$\frac{\partial C}{\partial t}(t, x) + r(t)x\frac{\partial C}{\partial x}(t, x) + \frac{1}{2}x^2\sigma^2(t)\frac{\partial^2 C}{\partial x^2}(t, x) = r(t)C(t, x),$$

under the terminal condition $C(T, x) = (x - K)^+$.

The solution of this PDE is given by the Black-Scholes formula

$$C(t, x) = \mathrm{Bl}(K, x, \tilde{\sigma}(t), \tilde{r}(t), T - t)$$
$$:= x\Phi(d_+(t, T)) - K\,\mathrm{e}^{-(T-t)\tilde{r}(t)}\Phi(d_-(t, T)), \tag{2.7}$$

where

$$\Phi(x) = \frac{1}{\sqrt{2\pi}}\int_{-\infty}^{x} \mathrm{e}^{-y^2/2}dy, \qquad x \in \mathbb{R},$$

denotes the standard normal cumulative distribution function,

$$\begin{cases} d_+(t, T) = \dfrac{\log(x/K) + (\tilde{r}(t) + \tilde{\sigma}^2(t)/2)(T - t)}{\tilde{\sigma}(t)\sqrt{T - t}}, \\[3mm] d_-(t, T) = \dfrac{\log(x/K) + (\tilde{r}(t) - \tilde{\sigma}^2(t)/2)(T - t)}{\tilde{\sigma}(t)\sqrt{T - t}}, \end{cases}$$

and

$$\tilde{\sigma}^2(t) := \frac{1}{T - t}\int_t^T \sigma^2(s)ds, \quad \tilde{r}(t) := \frac{1}{T - t}\int_t^T r(s)ds, \quad 0 \leqslant t < T. \tag{2.8}$$

We refer to Mikosch (1998), Øksendal (2003), Privault (2014) for more detailed expositions of these topics.

2.4 The Girsanov Theorem

Before proceeding to the pricing of options using the martingale approach, we need to review the Girsanov theorem. Let us come back to the informal interpretation (1.1) of the Brownian motion via its infinitesimal increments:

$$\Delta B_t = \pm\sqrt{dt},$$

with

$$\mathbb{P}(\Delta B_t = +\sqrt{dt}) = \mathbb{P}(\Delta B_t = -\sqrt{dt}) = \frac{1}{2}.$$

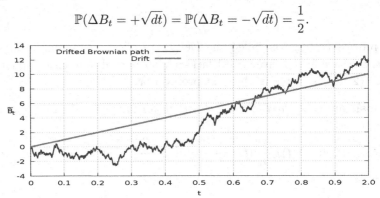

Fig. 2.2: Drifted Brownian path.

Clearly, given $\nu \in \mathbb{R}$, the drifted process $\nu t + B_t$ is no longer a standard Brownian motion because it is not centered:

$$\mathbb{E}[\nu t + B_t] = \nu t + \mathbb{E}[B_t] = \nu t \neq 0,$$

cf. Figure 2.2. This identity can be formulated in terms of infinitesimal increments as

$$\mathbb{E}[\nu dt + dB_t] = \frac{1}{2}(\nu dt + \sqrt{dt}) + \frac{1}{2}(\nu dt - \sqrt{dt}) = \nu dt \neq 0.$$

In order to make $\nu t + B_t$ a centered process (*i.e.* a standard Brownian motion, since $\nu t + B_t$ conserves all the other properties (1)-(3) in Definition 1.1, one may change the probabilities of ups and downs, which have been fixed so far equal to $1/2$.

That is, the problem is now to find two numbers $p^*, q^* \in (0,1)$ such that

$$\begin{cases} p^*(\nu dt + \sqrt{dt}) + q^*(\nu dt - \sqrt{dt}) = 0 \\[2mm] p^* + q^* = 1. \end{cases}$$

The solution to this problem is given by

$$p^* = \frac{1}{2}(1 - \nu\sqrt{dt}) \quad \text{and} \quad q^* = \frac{1}{2}(1 + \nu\sqrt{dt}).$$

Coming back to the Brownian motion considered as a discrete random walk with independent increments $\pm\sqrt{\Delta t}$, we intend to construct a new probability measure denoted $\mathbb{P}^*$, under which the drifted process $\widehat{B}_t := \nu t + B_t$

will be a standard Brownian motion. This probability measure will be defined through its Radon-Nikodym density

$$
\begin{aligned}
\frac{d\mathbb{P}^*}{d\mathbb{P}} &:= \frac{\mathbb{P}^*(\Delta B_{t_1} = \epsilon_1\sqrt{\Delta t}, \ldots, \Delta B_{t_N} = \epsilon_N\sqrt{\Delta t})}{\mathbb{P}(\Delta B_{t_1} = \epsilon_1\sqrt{\Delta t}, \ldots, \Delta B_{t_N} = \epsilon_N\sqrt{\Delta t})} \\
&= \frac{\mathbb{P}^*(\Delta B_{t_1} = \epsilon_1\sqrt{\Delta t}) \cdots \mathbb{P}^*(\Delta B_{t_N} = \epsilon_N\sqrt{\Delta t})}{\mathbb{P}(\Delta B_{t_1} = \epsilon_1\sqrt{\Delta t}) \cdots \mathbb{P}(\Delta B_{t_N} = \epsilon_N\sqrt{\Delta t})} \\
&= \frac{1}{(1/2)^N}\mathbb{P}^*(\Delta B_{t_1} = \epsilon_1\sqrt{\Delta t}) \cdots \mathbb{P}^*(\Delta B_{t_N} = \epsilon_N\sqrt{\Delta t}),
\end{aligned}
$$

$\epsilon_1, \ldots, \epsilon_N \in \{-1, 1\}$, with respect to the historical probability measure $\mathbb{P}$, obtained by taking the product of the above probabilities divided by the reference probability $1/2^N$ corresponding to the symmetric random walk.

Interpreting $N = T/\Delta t$ as an (infinitely large) number of discrete time steps and under the identification $\{0 < t < T\} \simeq \{t_1, t_2, \ldots, t_N\}$, this Radon-Nikodym density can be rewritten as

$$
\frac{d\mathbb{P}^*}{d\mathbb{P}} \simeq \frac{1}{(1/2)^N} \prod_{0<t<T} \left(\frac{1}{2} \mp \frac{1}{2}\nu\sqrt{\Delta t} \right)
$$

where 2^N becomes a normalization factor. Using the expansion

$$
\log\left(1 + \nu\sqrt{\Delta t}\right) = \nu\sqrt{\Delta t} - \frac{1}{2}\left(\nu\sqrt{\Delta t}\right)^2 + o(\Delta t),
$$

for small values of Δt, this Radon-Nikodym density can be informally shown to converge as follows as N tends to infinity, *i.e.* as the time step $\Delta t = T/N$ tends to zero:

$$
\begin{aligned}
2^N \prod_{0<t<T} \left(\frac{1}{2} \mp \frac{1}{2}\nu\sqrt{\Delta t} \right) &= \prod_{0<t<T} \left(1 \mp \nu\sqrt{\Delta t} \right) \\
&= \exp\left(\log \prod_{0<t<T} \left(1 \mp \nu\sqrt{\Delta t} \right) \right) \\
&= \exp\left(\sum_{0<t<T} \log\left(1 \mp \nu\sqrt{\Delta t} \right) \right) \\
&\simeq \exp\left(\nu \sum_{0<t<T} \mp\sqrt{\Delta t} - \frac{1}{2} \sum_{0<t<T} (\mp\nu\sqrt{\Delta t})^2 \right) \\
&= \exp\left(-\nu \sum_{0<t<T} \pm\sqrt{\Delta t} - \frac{\nu^2}{2} \sum_{0<t<T} \Delta t \right)
\end{aligned}
$$

$$= \exp\left(-\nu \sum_{0<t<T} \Delta B_t - \frac{\nu^2}{2} \sum_{0<t<T} \Delta t\right)$$

$$= \exp\left(-\nu B_T - \frac{\nu^2}{2}T\right),$$

based on the identifications

$$B_T \simeq \sum_{0<t<T} \pm\sqrt{\Delta t} \quad \text{and} \quad T \simeq \sum_{0<t<T} \Delta t.$$

The Girsanov theorem can be restated in a more rigorous way, using Radon-Nikodym derivatives and changes of probability measures. Recall that, given $\mathbb{Q}$ a probability measure on Ω, the notation

$$\frac{d\mathbb{Q}}{d\mathbb{P}} = F$$

means that the probability measure $\mathbb{Q}$ has a Radon-Nikodym density F with respect to $\mathbb{P}$, where $F : \Omega \longrightarrow \mathbb{R}$ is a nonnegative random variable such that $\mathbb{E}[F] = 1$. We also write

$$d\mathbb{Q} = Fd\mathbb{P},$$

which is equivalent to stating that

$$\mathbb{E}_{\mathbb{Q}}[G] = \int_\Omega G(\omega)d\mathbb{Q}(\omega)$$

$$= \int_\Omega G(\omega)\frac{d\mathbb{Q}}{d\mathbb{P}}(\omega)d\mathbb{P}(\omega)$$

$$= \int_\Omega G(\omega)F(\omega)d\mathbb{P}(\omega)$$

$$= \mathbb{E}[FG],$$

where G is an integrable random variable. In addition we say that $\mathbb{Q}$ is *equivalent* to $\mathbb{P}$ when $F > 0$ with $\mathbb{P}$-probability one.

On the Wiener space $\Omega = \mathcal{C}_0([0,T])$, where $\omega \in \Omega$ is a continuous function on $[0,T]$ starting at 0 in $t = 0$, Brownian motion $(B_t)_{t\in[0,T]}$ is constructed as the canonical process $B_t(\omega) = \omega(t)$, $\omega \in \Omega$, $t \in \mathbb{R}_+$. Given the probability measure $\mathbb{Q}$ defined by

$$d\mathbb{Q}(\omega) = \exp\left(-\nu B_T(\omega) - \frac{\nu^2}{2}T\right)d\mathbb{P}(\omega),$$

the Girsanov theorem can be restated by saying that the process $(\nu t + B_t)_{t\in[0,T]}$ is a standard (centered) Brownian motion under the probability measure $\mathbb{Q}$.

For example, the fact that $\nu T + B_T$ has a standard (centered) Gaussian distribution under $\mathbb{Q}$ can be recovered as follows:

$$
\begin{aligned}
\mathbb{E}_{\mathbb{Q}}[f(\nu T + B_T)] &= \int_\Omega f(\nu T + B_T) d\mathbb{Q} \\
&= \int_\Omega f(\nu T + B_T) \exp\left(-\nu B_T - \frac{\nu^2}{2}T\right) d\mathbb{P} \\
&= \int_{-\infty}^\infty f(\nu T + x) \exp\left(-\nu x - \frac{\nu^2}{2}T\right) e^{-x^2/(2T)} \frac{dx}{\sqrt{2\pi T}} \\
&= \int_{-\infty}^\infty f(y) e^{-y^2/(2T)} \frac{dy}{\sqrt{2\pi T}} \\
&= \int_\Omega f(B_T) d\mathbb{P} \\
&= \mathbb{E}[f(B_T)],
\end{aligned}
$$

i.e.

$$
\mathbb{E}_{\mathbb{Q}}[f(\nu T + B_T)] = \int_\Omega f(\nu T + B_T) d\mathbb{Q} = \int_\Omega f(B_T) d\mathbb{P} = \mathbb{E}[f(B_T)]. \quad (2.9)
$$

More precisely, the Girsanov theorem can be extended to shifts by $(\mathcal{F}_t)_{t \in \mathbb{R}_+}$-adapted processes as follows, see *e.g.* Theorem III-42, page 141 of Protter (2004).

Theorem 2.1. *Let $(\psi_t)_{t \in [0,T]}$ be an $(\mathcal{F}_t)_{t \in \mathbb{R}_+}$-adapted process satisfying the Novikov integrability condition*

$$
\mathbb{E}\left[\exp\left(\frac{1}{2}\int_0^T \psi_t^2 dt\right)\right] < \infty, \quad (2.10)
$$

and let $\mathbb{Q}$ denote the probability measure defined by

$$
\frac{d\mathbb{Q}}{d\mathbb{P}} = \exp\left(-\int_0^T \psi_t dB_t - \frac{1}{2}\int_0^T \psi_t^2 dt\right).
$$

Then

$$
\widehat{B}_t := B_t + \int_0^t \psi_s ds, \qquad 0 \leqslant t \leqslant T,
$$

is a standard Brownian motion under $\mathbb{Q}$.

The Girsanov Theorem allows us to extend (2.9) as

$$
\mathbb{E}[F] = \mathbb{E}\left[F\left(B_\cdot + \int_0^\cdot \psi_s ds\right) \exp\left(-\int_0^T \psi_t dB_t - \frac{1}{2}\int_0^T \psi_t^2 dt\right)\right],
$$

for all random variables $F \in L^1(\Omega)$.

2.5   Martingale Method

In this section we give the expression of the Black-Scholes price $C(t, S_t)$ using expectations of discounted payoffs.

Definition 2.2. *A market is said without arbitrage if there exists (at least) a probability $\mathbb{P}^*$ under which the discounted price process*

$$\widetilde{S}_t := e^{-\int_0^t r(s)ds} S_t, \qquad t \geqslant 0,$$

is a martingale under $\mathbb{P}^*$.

Such a probability $\mathbb{P}^*$ is usually called a *risk-neutral* probability measure, or a *martingale* measure. When the martingale measure is unique, the market is said to be *complete*.

We will now show that the Black-Scholes model admits a unique martingale measure, which shows that the market is without arbitrage and complete. When applied to the market price of risk

$$\psi_t := \frac{\mu - r}{\sigma},$$

with constant volatility, interest rate and drift $\sigma = \sigma(t)$, $r = r(t)$ and $\mu = \mu(t)$, the Girsanov theorem shows that

$$\widehat{B}_t := \frac{\mu - r}{\sigma} t + B_t, \qquad 0 \leqslant t \leqslant T,$$

is a standard Brownian motion under the probability measure $\mathbb{P}^*$ defined by

$$\frac{d\mathbb{P}^*}{d\mathbb{P}} = \exp\left(-\frac{\mu - r}{\sigma} B_T - \frac{(\mu - r)^2}{2\sigma^2} T\right).$$

From the Girsanov Theorem 2.1, we know that

$$\widehat{B}_t := B_t + \int_0^t \psi_s ds, \qquad 0 \leqslant t \leqslant T,$$

is a Brownian motion under $\mathbb{P}^*$. Hence, by Proposition 1.3 the discounted price process given by

$$dX_t = (\mu - r)X_t dt + \sigma X_t dB_t = \sigma X_t d\widehat{B}_t, \qquad t \geqslant 0,$$

is a martingale under $\mathbb{P}^*$, and $\mathbb{P}^*$ is therefore a risk-neutral measure. We obviously have $\mathbb{P} = \mathbb{P}^*$ when $\mu = r$.

Let also

$$\widetilde{V}_t := e^{-\int_0^t r(s)ds} V_t, \quad \text{and} \quad \widetilde{S}_t := e^{-\int_0^t r(s)ds} S_t, \quad t \geqslant 0,$$

respectively denote the discounted portfolio value process and the underlying asset price process.

Lemma 2.1. *The following statements are equivalent:*

i) the portfolio process $(\eta_t, \zeta_t)_{t \in \mathbb{R}_+}$ *is self-financing,*

ii) we have

$$\widetilde{V}_t = \widetilde{V}_0 + \int_0^t \sigma(s) \eta_u \widetilde{S}_u d\widehat{B}_u, \qquad t \geqslant 0, \tag{2.11}$$

iii) we have

$$V_t = V_0 e^{\int_0^t r(s)ds} + \int_0^t \sigma(u) \eta_u S_u e^{\int_u^t r(s)ds} d\widehat{B}_u, \qquad t \geqslant 0. \tag{2.12}$$

Proof. First, note that (2.11) is clearly equivalent to (2.12). Now, the self-financing condition (2.1) shows that

$$\begin{aligned} dV_t &= \zeta_t dA_t + \eta_t dS_t \\ &= \zeta_t A_t r(t) dt + \eta_t r(t) S_t dt + \sigma(t) \eta_t S_t d\widehat{B}_t \\ &= r(t) V_t dt + \sigma(t) \eta_t S_t d\widehat{B}_t, \qquad t \geqslant 0, \end{aligned}$$

hence

$$\begin{aligned} d\widetilde{V}_t &= d\big(e^{-\int_0^t r(s)ds} V_t \big) \\ &= -r(t) e^{-\int_0^t r(s)ds} V_t dt + e^{-\int_0^t r(s)ds} dV_t \\ &= e^{-\int_0^t r(s)ds} \sigma(t) \eta_t S_t d\widehat{B}_t, \qquad t \geqslant 0, \end{aligned}$$

i.e. (2.11) holds. Conversely, if (2.11) is satisfied we have

$$\begin{aligned} dV_t &= d\big(e^{\int_0^t r(s)ds} \widetilde{V}_t \big) \\ &= r(t) e^{\int_0^t r(s)ds} \widetilde{V}_t dt + e^{\int_0^t r(s)ds} d\widetilde{V}_t \\ &= r(t) e^{\int_0^t r(s)ds} \widetilde{V}_t dt + \sigma(t) \eta_t S_t d\widehat{B}_t \\ &= V_t r(t) dt + \sigma(t) \eta_t S_t d\widehat{B}_t \\ &= \zeta_t A_t r(t) dt + \eta_t S_t r(t) dt + \sigma(t) \eta_t S_t d\widehat{B}_t \\ &= \zeta_t dA_t + \eta_t dS_t, \end{aligned}$$

hence the portfolio process $(\eta_t, \zeta_t)_{t \in \mathbb{R}_+}$ is self-financing. $\qquad \square$

In the next proposition we compute a self-financing hedging strategy leading to an arbitrary square-integrable random variable F admitting a predictable representation of the form

$$F = \mathbb{E}^*[F] + \int_0^T \xi_t d\widehat{B}_t, \tag{2.13}$$

where $(\xi_t)_{t\in[0,t]}$ is a square-integrable $(\mathcal{F}_t)_{t\in\mathbb{R}_+}$-adapted process and $\mathbb{E}^*$ denotes the expectation under the probability measure $\mathbb{P}^*$.

Proposition 2.2. *Given* $F \in L^2(\Omega)$, *let*

$$\eta_t := \frac{e^{-\int_t^T r(s)ds}}{\sigma(t)S_t}\xi_t, \tag{2.14}$$

$$\zeta_t := \frac{e^{-\int_t^T r(s)ds}\,\mathbb{E}^*[F \mid \mathcal{F}_t] - \eta_t S_t}{A_t}, \qquad 0 \leqslant t \leqslant T. \tag{2.15}$$

Then the portfolio process $(\eta_t, \zeta_t)_{t\in[0,T]}$ *is self-financing, and letting*

$$V_t := \zeta_t A_t + \eta_t S_t, \qquad 0 \leqslant t \leqslant T, \tag{2.16}$$

we have

$$V_t = e^{-\int_t^T r(s)ds}\,\mathbb{E}^*[F \mid \mathcal{F}_t], \qquad 0 \leqslant t \leqslant T. \tag{2.17}$$

In particular, we have

$$V_T = F,$$

i.e. the portfolio process $(\eta_t, \zeta_t)_{t\in\mathbb{R}_+}$ *yields a hedging strategy leading to* F, *starting from the initial value*

$$V_0 = e^{-\int_0^T r(s)ds}\,\mathbb{E}^*[F].$$

Proof.   Applying (2.15) and (2.16) at $t = 0$, we get

$$e^{-\int_0^T r(s)ds}\,\mathbb{E}^*[F] = \zeta_0 A_0 + \eta_0 S_0 = V_0,$$

hence from (2.15) again, the definition (2.14) of η_t and (2.13), we obtain

$$\begin{aligned}
V_t &= \zeta_t A_t + \eta_t S_t \\
&= e^{-\int_t^T r(s)ds}\,\mathbb{E}^*[F \mid \mathcal{F}_t] \\
&= e^{-\int_t^T r(s)ds}\left(\mathbb{E}^*[F] + \int_0^t \xi_s d\widehat{B}_s\right) \\
&= V_0\,e^{\int_0^t r(s)ds} + e^{-\int_t^T r(s)ds}\int_0^t \xi_s d\widehat{B}_s \\
&= V_0\,e^{\int_0^t r(s)ds} + \int_0^t \eta_u \sigma(u)S_u\,e^{\int_u^t r(s)ds}d\widehat{B}_u, \qquad 0 \leqslant t \leqslant T,
\end{aligned}$$

and from Lemma 2.1 this also implies that the portfolio process $(\eta_t, \zeta_t)_{t\in[0,T]}$ is self-financing. □

The above proposition shows that there always exists a hedging strategy starting from

$$V_0 = e^{-\int_0^T r(s)ds} \, \mathbb{E}^*[F].$$

In addition, since there exists a hedging strategy leading to

$$\widetilde{V}_T = e^{-\int_0^T r(s)ds} F,$$

then by (2.11), the portfolio value process $\left(\widetilde{V}_t\right)_{t\in[0,T]}$ is necessarily a martingale, with

$$\widetilde{V}_t = \mathbb{E}^* \left[\widetilde{V}_T \mid \mathcal{F}_t\right] = e^{-\int_0^T r(s)ds} \, \mathbb{E}^*[F \mid \mathcal{F}_t], \qquad 0 \leqslant t \leqslant T,$$

and initial value

$$\widetilde{V}_0 = \mathbb{E}^* \left[\widetilde{V}_T\right] = \mathbb{E}^*[F] \, e^{-\int_0^T r(s)ds}.$$

In practice, the hedging problem can be reduced to the computation of the process $(\xi_t)_{t\in[0,T]}$ appearing in (2.13). This computation, called the Delta hedging, can be performed by application of the Itô formula and the Markov property, see *e.g.* Protter (2001). Consider the (non-homogeneous) semi-group $(P_{s,t})_{0\leqslant s\leqslant t\leqslant T}$ associated to $(S_t)_{t\in[0,T]}$, and defined by

$$\begin{aligned} P_{s,t}f(S_s) &= \mathbb{E}^*[f(S_t) \mid S_s] \\ &= \mathbb{E}^*[f(S_t) \mid \mathcal{F}_s], \quad 0 \leqslant s \leqslant t \leqslant T, \end{aligned}$$

which acts on $\mathcal{C}_b^2(\mathbb{R})$ functions, with

$$P_{s,t}P_{t,u} = P_{s,u}, \qquad 0 \leqslant s \leqslant t \leqslant u \leqslant T.$$

Note that $(P_{t,T}f(S_t))_{t\in[0,T]}$ is an $(\mathcal{F}_t)_{t\in[0,T]}$-martingale, *i.e.*:

$$\begin{aligned} \mathbb{E}^*[P_{t,T}f(S_t) \mid \mathcal{F}_s] &= \mathbb{E}^*[\mathbb{E}^*[f(S_T) \mid \mathcal{F}_t] \mid \mathcal{F}_s] \\ &= \mathbb{E}^*[f(S_T) \mid \mathcal{F}_s] \\ &= P_{s,T}f(S_s), \qquad 0 \leqslant s \leqslant t \leqslant T. \end{aligned} \tag{2.18}$$

The next Lemma 2.2 allows us to compute the process $(\xi_t)_{t\in[0,T]}$ in case the payoff F is of the form $F = \phi(S_T)$ for some function ϕ.

Lemma 2.2. *Let $\phi \in \mathcal{C}_b^2(\mathbb{R})$. The predictable representation*

$$\phi(S_T) = \mathbb{E}^*[\phi(S_T)] + \int_0^T \xi_t d\widehat{B}_t \tag{2.19}$$

is given by

$$\xi_t = \sigma(t)S_t \frac{\partial}{\partial x}(P_{t,T}\phi)(S_t), \qquad 0 \leqslant t \leqslant T. \tag{2.20}$$

Proof.   Since $P_{t,T}\phi$ is in $\mathcal{C}_b^2(\mathbb{R})$, we can apply the Itô formula (1.9) to the process

$$t \mapsto P_{t,T}\phi(S_t) = \mathbb{E}^*[\phi(S_T) \mid \mathcal{F}_t],$$

which is a martingale from (2.18), cf. also the appendix. From the fact that the finite variation term in the Itô formula vanishes when $(P_{t,T}\phi(S_t))_{t\in[0,T]}$ is a martingale, (see *e.g.* Corollary 1, page 72 of Protter (2004)), we obtain:

$$P_{t,T}\phi(S_t) = P_{0,T}\phi(S_0) + \int_0^t \sigma(s)S_s\frac{\partial}{\partial x}(P_{s,T}\phi)(S_s)d\widehat{B}_s, \quad 0 \leqslant t \leqslant T,$$
$$(2.21)$$

with $P_{0,T}\phi(S_0) = \mathbb{E}^*[\phi(S_T)]$. Letting $t = T$, we obtain (2.20) by uniqueness of the predictable representation (2.19) of $F = \phi(S_T)$.   □

Let now $(S_{t,s}^x)_{s\in[t,\infty)}$ be the price process solution of the stochastic differential equation

$$\frac{dS_{t,s}^x}{S_{t,s}^x} = r(s)ds + \sigma(s)d\widehat{B}_s, \qquad s \in [t, \infty),$$

with initial condition $S_{t,t}^x = x \in (0, \infty)$.

The value V_t of the portfolio at time $t \in [0,T]$ can be computed from (2.17) as

$$V_t = e^{-\int_t^T r(s)ds}\,\mathbb{E}^*[\phi(S_T) \mid \mathcal{F}_t] = C(t, S_t),$$

where

$$\begin{aligned}
C(t,x) &= e^{-\int_t^T r(s)ds}\,\mathbb{E}^*[\phi(S_T) \mid S_t = x] \\
&= e^{-\int_t^T r(s)ds}\,P_{t,T}\phi(x) \\
&= e^{-\int_t^T r(s)ds}\,\mathbb{E}^*[\phi(S_{t,T}^x)], \qquad 0 \leqslant t \leqslant T,
\end{aligned}$$

from Relation (11.5) in the appendix. Again, from the fact that the finite variation term vanishes in (2.21) we recover the fact that $C(t,x)$ solves the Black-Scholes PDE

$$\begin{cases}
\dfrac{\partial C}{\partial t}(t,x) + xr(t)\dfrac{\partial C}{\partial x}(t,x) + \dfrac{1}{2}x^2\sigma^2(t)\dfrac{\partial^2 C}{\partial x^2}(t,x) = r(t)C(t,x), \\[2mm]
C(T,x) = \phi(x).
\end{cases}$$

In the case of European options with payoff function $\phi(x) = (x - K)^+$ we recover Relation (2.7), *i.e.*

$$C(t, x) = \text{Bl}(K, x, \tilde{\sigma}(t), \tilde{r}(t), T - t),$$

where $\tilde{\sigma}(t)$ and $\tilde{r}(t)$ are defined in (2.8), as a consequence of (2.17) and the following lemma.

Lemma 2.3. *Let X be a centered Gaussian random variable with variance $v^2 > 0$. We have*

$$\mathbb{E}^* \left[\left(e^{m+X} - K \right)^+ \right] = e^{m+v^2/2} \Phi \left(v + \frac{m - \log K}{v} \right) - K \Phi \left(\frac{m - \log K}{v} \right).$$

Proof. We have

$$\mathbb{E}^* \left[\left(e^{m+X} - K \right)^+ \right] = \int_{-\infty}^{\infty} \left(e^{m+x} - K \right)^+ e^{-x^2/(2v^2)} \frac{dx}{\sqrt{2\pi v^2}}$$

$$= \int_{-m+\log K}^{\infty} \left(e^{m+x} - K \right) e^{-x^2/(2v^2)} \frac{dx}{\sqrt{2\pi v^2}}$$

$$= e^m \int_{-m+\log K}^{\infty} e^{x-x^2/(2v^2)} \frac{dx}{\sqrt{2\pi v^2}} - K \int_{-m+\log K}^{\infty} e^{-x^2/(2v^2)} \frac{dx}{\sqrt{2\pi v^2}}$$

$$= e^{m+v^2/2} \int_{-m+\log K}^{\infty} e^{-(v^2-x)^2/(2v^2)} \frac{dx}{\sqrt{2\pi v^2}} - K \int_{(-m+\log K)/v}^{\infty} e^{-x^2/2} \frac{dx}{\sqrt{2\pi}}$$

$$= e^{m+v^2/2} \int_{-v^2-m+\log K}^{\infty} e^{-x^2/(2v^2)} \frac{dx}{\sqrt{2\pi v^2}} - K \Phi((m - \log K)/v)$$

$$= e^{m+v^2/2} \Phi \left(v + \frac{m - \log K}{v} \right) - K \Phi \left(\frac{m - \log K}{v} \right). \qquad \square$$

Moreover, still in the case of European options, the process $(\xi)_{t\in[0,T]}$ can be computed via the next proposition.

Proposition 2.3. *Assume that $F = (S_T - K)^+$. Then we have*

$$\xi_t = \sigma(t) \, \mathbb{E}^* \left[S_{t,T}^x \mathbb{1}_{[K,\infty[}(S_{t,T}^x) \right]_{x=S_t}, \qquad 0 \leqslant t \leqslant T.$$

Proof. This result is consequence of Lemma 2.2 and of the relation $P_{t,T}f(x) = \mathbb{E}^*[f(S_{t,T}^x)]$, after approximation of the function $x \mapsto f(x) := (x - K)^+$ with $\mathcal{C}_b^2$ functions. $\qquad \square$

From the above Proposition 2.3 we recover the formula for the Delta of a European call option in the Black-Scholes model.

Proposition 2.4. *The Delta of a European call option with payoff $F = (S_T - K)^+$ is given by*

$$\eta_t = \Phi \left(\frac{\log(S_t/K) + \left(\tilde{r}(t) + \tilde{\sigma}^2(t)/2 \right)(T - t)}{\tilde{\sigma}(t)\sqrt{T - t}} \right), \qquad 0 \leqslant t \leqslant T.$$

Proof. By (2.14) and Proposition 2.3 we have, taking $x = S_t$,

$$\eta_t = \frac{1}{\sigma(t)S_t} \, e^{-(T-t)\tilde{r}(t)} \xi_t$$

$$= e^{-(T-t)\tilde{r}(t)} \, \mathbb{E}^* \left[\frac{S_{t,T}^x}{x} \mathbb{1}_{[K,\infty[}(S_{t,T}^x) \right]$$

$$= e^{-(T-t)\tilde{r}(t)} \, \mathbb{E}^* \left[e^{\tilde{\sigma}(t)\widehat{B}_{T-t}+(T-t)\tilde{r}(t)-(T-t)\tilde{\sigma}^2(t)/2} \right.$$

$$\left. \times \mathbb{1}_{[K,\infty[}\left(x \, e^{\tilde{\sigma}(t)\widehat{B}_{T-t}+(T-t)\tilde{r}(t)-(T-t)\tilde{\sigma}^2(t)/2}\right) \right]$$

$$= \frac{e^{-(T-t)\tilde{r}(t)}}{\sqrt{2\pi(T-t)}}$$

$$\times \int_{(T-t)\tilde{\sigma}(t)/2-(T-t)\tilde{r}(t)/\tilde{\sigma}(t)+(\log(K/x))/\tilde{\sigma}(t)}^{\infty} e^{y\tilde{\sigma}(t)+(T-t)\tilde{r}(t)-(T-t)\tilde{\sigma}^2(t)/2} \, e^{-y^2/(2(T-t))} dy$$

$$= \frac{1}{\sqrt{2\pi(T-t)}} \int_{-d_-(t,T)/\sqrt{T-t}}^{\infty} e^{-(y-(T-t)\tilde{\sigma}(t))^2 \, (2(T-t))} dy$$

$$= \frac{1}{\sqrt{2\pi}} \int_{-d_+(t,T)}^{\infty} e^{-y^2/2} dy$$

$$= \frac{1}{\sqrt{2\pi}} \int_{-\infty}^{d_+(t,T)} e^{-y^2/2} dy$$

$$= \Phi(d_+(t,T)).$$

$\square$

The result of Proposition 2.4 can also be recovered by (2.6) and a direct differentiation of the Black-Scholes function (2.7), as follows:

$$\frac{\partial C}{\partial x}(x,t) = \frac{\partial}{\partial x}\left(x\Phi\left(\frac{\log(x/K) + \left(\tilde{r}(t) + \tilde{\sigma}^2(t)/2\right)(T-t)}{\tilde{\sigma}(t)\sqrt{T-t}} \right) \right)$$

$$-K e^{-(T-t)\tilde{r}(t)} \frac{\partial}{\partial x} \Phi\left(\frac{\log(x/K) + \left(\tilde{r}(t) - \tilde{\sigma}^2(t)/2\right)(T-t)}{\tilde{\sigma}(t)\sqrt{T-t}} \right)$$

$$= \Phi\left(\frac{\log(x/K) + \left(\tilde{r}(t) - \tilde{\sigma}^2(t)/2\right)(T-t)}{\tilde{\sigma}(t)\sqrt{T-t}} \right)$$

$$+x\frac{\partial}{\partial x} \Phi\left(\frac{\log(x/K) + \left(\tilde{r}(t) + \tilde{\sigma}^2(t)/2\right)(T-t)}{\tilde{\sigma}(t)\sqrt{T-t}} \right)$$

$$-K e^{-(T-t)\tilde{r}(t)} \frac{\partial}{\partial x} \Phi\left(\frac{\log(x/K) + \left(\tilde{r}(t) - \tilde{\sigma}^2(t)/2\right)(T-t)}{\tilde{\sigma}(t)\sqrt{T-t}} \right)$$

$$= \Phi\left(\frac{\log(x/K) + \left(\tilde{r}(t) - \tilde{\sigma}^2(t)/2\right)(T-t)}{\tilde{\sigma}(t)\sqrt{T-t}} \right)$$

$$+ \frac{1}{\sqrt{2\pi}\tilde{\sigma}(t)\sqrt{T-t}} \exp\left(-\frac{1}{2}\left(\frac{\log(x/K) + \big(\tilde{r}(t) + \tilde{\sigma}^2(t)/2\big)(T-t)}{\tilde{\sigma}(t)\sqrt{T-t}}\right)^2\right)$$

$$- \frac{K\,\mathrm{e}^{-(T-t)\tilde{r}(t)}}{x\sqrt{2\pi}\tilde{\sigma}(t)\sqrt{T-t}} \exp\left(-\frac{1}{2}\left(\frac{\log(x/K) + \big(\tilde{r}(t) - \tilde{\sigma}^2(t)/2\big)(T-t)}{\tilde{\sigma}(t)\sqrt{T-t}}\right)^2\right)$$

$$= \Phi\left(\frac{\log(x/K) + \big(\tilde{r}(t) - \tilde{\sigma}^2(t)/2\big)(T-t)}{\tilde{\sigma}(t)\sqrt{T-t}}\right)$$

$$+ \frac{1}{\sqrt{2\pi}\tilde{\sigma}(t)\sqrt{T-t}} \exp\left(-\frac{1}{2}\left(\frac{\log(x/K) + \big(\tilde{r}(t) + \tilde{\sigma}^2(t)/2\big)(T-t)}{\tilde{\sigma}(t)\sqrt{T-t}}\right)^2\right)$$

$$- \frac{K}{x\sqrt{2\pi}\tilde{\sigma}(t)\sqrt{T-t}}$$

$$\times \exp\left(-\frac{1}{2}\left(\frac{\log(x/K) + \big(\tilde{r}(t) + \tilde{\sigma}^2(t)/2\big)(T-t)}{\tilde{\sigma}(t)\sqrt{T-t}}\right)^2 + \log\frac{x}{K}\right)$$

$$= \Phi\left(\frac{\log(x/K) + \big(\tilde{r}(t) - \tilde{\sigma}^2(t)/2\big)(T-t)}{\tilde{\sigma}(t)\sqrt{T-t}}\right), \qquad 0 \leqslant t < T. \qquad (2.22)$$

Exercises

Exercise 2.1.

(1) Solve the stochastic differential equation
$$dS_t = \alpha S_t dt + \sigma dB_t$$
in terms of $\alpha, \sigma > 0$, and the initial condition S_0.
(2) For which values α_M of α is the discounted price process $\tilde{S}_t = \mathrm{e}^{-rt}S_t$, $t \in [0,T]$, a martingale under $\mathbb{P}$?
(3) Compute the arbitrage price $C(t, S_t) = \mathrm{e}^{-(T-t)r}\,\mathbb{E}[\exp(S_T) \mid \mathcal{F}_t]$ at time $t \in [0,T]$ of the contingent claim of $\exp(S_T)$, with $\alpha = \alpha_M$.
(4) Explicitly compute the strategy $(\zeta_t, \eta_t)_{t\in[0,T]}$ that hedges the contingent claim $\exp(S_T)$.

Exercise 2.2. Consider the price process $(S_t)_{t\in[0,T]}$ given by
$$\frac{dS_t}{S_t} = \alpha dt + \sigma dB_t$$

and a risk-free asset with value $A_t = A_0 e^{rt}$ at time $t \in [0,T]$, with $r > 0$. Let $(\zeta_t, \eta_t)_{t\in[0,T]}$ a self-financing portfolio process with value

$$V_t = \eta_t A_t + \zeta_t S_t, \qquad 0 \leqslant t \leqslant T.$$

(1) Using the Girsanov Theorem 2.1, construct a probability $\mathbb{P}^*$ under which the process $\widetilde{S}_t := S_t/A_t$, $t \in [0,T]$ is an $(\mathcal{F}_t)_{t\in[0,T]}$-martingale.

(2) Given $T_0 \in [0,T]$, compute the arbitrage price

$$C(t, S_t) = e^{-(T-t)r}\, \mathbb{E}^* \left[\frac{S_T}{S_{T_0}} \,\Big|\, \mathcal{F}_t \right],$$

at time $t \in [0,T]$, for the claim of payoff S_T/S_{T_0}.
Hint: Consider separately the cases $t \in [0, T_0]$ and $t \in (T_0, T]$.

(3) Compute the portfolio strategy $(\zeta_t, \eta_t)_{t\in[0,T]}$ hedging the claim S_T/S_{T_0}. Check that the portfolio strategy $(\eta_t, \zeta_t)_{t\in[0,T]}$ is self-financing.

Exercise 2.3. (Exercise 2.2 continued).

(1) Compute the arbitrage price

$$C(t, S_t) = e^{-(T-t)r}\, \mathbb{E}^* \left[|S_T|^2 \,\big|\, \mathcal{F}_t \right],$$

at time $t \in [0,T]$, of the contingent claim of payoff $|S_T|^2$.

(2) Compute the portfolio strategy $(\zeta_t, \eta_t)_{t\in[0,T]}$ hedging the claim $|S_T|^2$.

Exercise 2.4. Let $\alpha \in \mathbb{R}$, and consider the solution $\left(X_t^{(\alpha)}\right)_{t\in[0,T]}$ of the stochastic differential equation

$$dX_t^{(\alpha)} = \alpha X_t^{(\alpha)} dt + dB_t, \qquad 0 \leqslant t \leqslant T.$$

(1) Using the Girsanov Theorem 2.1, construct a probability measure $\mathbb{Q}$ under which the process $\left(X_t^{(\alpha)}\right)_{t\in[0,T]}$ becomes a standard Brownian motion.

(2) Compute the expected value

$$\mathbb{E} \left[\exp \left((\beta - \alpha) \int_0^T X_t^{(\alpha)} dX_t^{(\alpha)} + \frac{\alpha^2}{2} \int_0^T \left(X_t^{(\alpha)} \right)^2 dt \right) \right]$$

for all $\beta < 1/T$.

(3) Compute the expected value

$$\mathbb{E} \left[\exp \left(\frac{\alpha^2}{2} \int_0^T \left(X_t^{(\alpha)} \right)^2 dt \right) \right], \qquad \alpha < \frac{1}{T}.$$

Chapter 3

Short-Term Interest Rate Models

This chapter is a short introduction to some common short-term interest rate models (Vasicek, CIR, CEV, affine models), for which the mean reversion property plays an important role. We also cover the calibration of the Vasicek model to market data using time series. The Vasicek model will be often used in examples in the subsequent chapters, as it allows for explicit calculations.

3.1 Mean-Reverting Models

Money market accounts with price $(A_t)_{t \in \mathbb{R}_+}$ can be defined from a short-term interest rate process $(r_t)_{t \in \mathbb{R}_+}$ as

$$\frac{A_{t+dt} - A_t}{A_t} = r_t dt, \qquad \frac{dA_t}{A_t} = r_t dt, \qquad \frac{dA_t}{dt} = r_t A_t, \quad t \geqslant 0,$$

with

$$A_t = A_0 \, \mathrm{e}^{\int_0^t r_s ds}, \qquad t \geqslant 0.$$

As short-term interest rates behave differently from stock prices, they require the development of specific models to account for properties such as positivity, boundedness, and return to equilibrium.

Vasicek model

The first model to capture the mean reversion property of interest rates, a property not possessed by geometric Brownian motion, is the Vasicek Vašíček (1977) model, which is based on the Ornstein-Uhlenbeck process. Here, the short-term interest rate process $(r_t)_{t \in \mathbb{R}_+}$ solves the equation

$$dr_t = (a - br_t)dt + \sigma dB_t, \tag{3.1}$$

where $a, \sigma \in \mathbb{R}$, $b > 0$, and $(B_t)_{t \in \mathbb{R}_+}$ is a standard Brownian motion, with solution

$$r_t = r_0 e^{-bt} + \frac{a}{b}(1 - e^{-bt}) + \sigma \int_0^t e^{-(t-s)b} dB_s, \qquad t \geqslant 0, \qquad (3.2)$$

see Exercise 3.1. The probability distribution of r_t is Gaussian at all times $t > 0$, with mean

$$\mathbb{E}[r_t] = r_0 e^{-bt} + \frac{a}{b}(1 - e^{-bt}),$$

and variance

$$\begin{aligned}
\mathrm{Var}[r_t] &= \mathrm{Var}\left[\sigma \int_0^t e^{-(t-s)b} dB_s\right] \\
&= \sigma^2 \int_0^t \left(e^{-(t-s)b}\right)^2 ds \\
&= \sigma^2 \int_0^t e^{-2bs} ds \\
&= \frac{\sigma^2}{2b}(1 - e^{-2bt}), \qquad t \geqslant 0,
\end{aligned}$$

i.e.

$$r_t \simeq \mathcal{N}\left(r_0 e^{-bt} + \frac{a}{b}(1 - e^{-bt}), \frac{\sigma^2}{2b}(1 - e^{-2bt})\right), \qquad t > 0.$$

In particular, the probability density function $f_t(x)$ of r_t at time $t > 0$ is given by

$$f_t(x) = \frac{\sqrt{b/\pi}}{\sigma\sqrt{1 - e^{-2bt}}} \exp\left(-\frac{\left(x - r_0 e^{-bt} - a\left(1 - e^{-bt}\right)/b\right)^2}{\sigma^2\left(1 - e^{-2bt}\right)/b}\right), \qquad x \in \mathbb{R}.$$

When $b > 0$ and in the long run, *i.e.* as time t becomes large, we have

$$\lim_{t \to \infty} \mathbb{E}[r_t] = \frac{a}{b} \qquad \text{and} \qquad \lim_{t \to \infty} \mathrm{Var}[r_t] = \frac{\sigma^2}{2b}, \qquad (3.3)$$

and this distribution converges to the Gaussian $\mathcal{N}(a/b, \sigma^2/(2b))$ distribution, which is also the *invariant* (or stationary) distribution of $(r_t)_{t \in \mathbb{R}_+}$, and the process tends to revert to its long term mean $a/b = \lim_{t \to \infty} \mathbb{E}[r_t]$ which makes the average drift vanish:

$$\lim_{t \to \infty} \mathbb{E}[a - br_t] = a - b \lim_{t \to \infty} \mathbb{E}[r_t] = 0.$$

Figure 3.1 presents a random simulation of $t \mapsto r_t$ in the Vasicek model with $r_0 = 3\%$, and illustrates the mean-reverting property of this process with respect to $a/b = 2.5\%$.

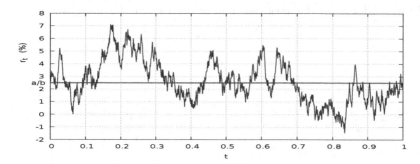

Fig. 3.1: Graph of Vasicek short rates $t \mapsto r_t$ with $a = 2.5\%$, $b = 1$, and $\sigma = 0.1$.

As can be checked from the simulation of Figure 3.1 the value of r_t in the Vasicek model may become negative due to its Gaussian distribution. Although real interest rates can sometimes fall below zero,[1] this can be regarded as a potential drawback of the Vasicek model. The next R code provides a numerical solution of the stochastic differential equation (3.1) using the Euler method, see Figure 3.1.

```
N=10000;t<-0:(N-1);d <-1.0/N;nsim<-2; a=0.025;b=1;sigma=0.1;sd=sqrt(sigma^2/2/b)
X <- matrix(rnorm(n = nsim * N, sd = sqrt(dt)), nsim, N)
R <- matrix(0,nsim,N);R[,1]=0.03
for (i in 1:nsim){for (j in 2:N){R[i,j]=R[i,j-1]+(a-b*R[i,j-1])*dt+sigma*X[i,j]}}
plot(t,R[1,],xlab = "time",ylab = "",type = "l",ylim = c(R[1,1]-0.2,R[1,1]+0.2),
    col = 0,axes=FALSE)
axis(2, pos=0);for (i in 1:nsim){lines(t, R[i, ], xlab = "time", type = "l", col =
    i+3)}
abline(h=a/b,col="blue",lwd=3);abline(h=0)
```

Example - TNX yield

We consider the yield of the 10 Year Treasury Note on the Chicago Board Options Exchange (CBOE), for later use in the calibration of the Vasicek model. Treasury notes usually have a maturity between one and 10 years, whereas treasury bonds have maturities beyond 10 years).

```
library(quantmod)
getSymbols("^TNX",from="2012-01-01",to="2016-01-01",src="yahoo")
rate=Ad(`TNX`)
rate<-rate[!is.na(rate)]
chartSeries(rate,up.col="blue",theme="white")
n = sum(!is.na(rate))
```

---

[1] Eurozone interest rates turned negative in 2014.

The next Figure 3.2 displays the yield of the 10 Year Treasury Note.

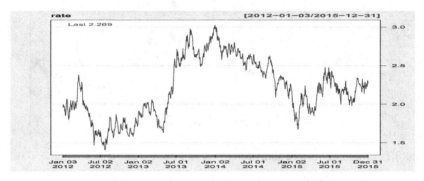

Fig. 3.2: CBOE 10 Year Treasury Note (TNX) yield.

Cox-Ingersoll-Ross (CIR) model

The Cox-Ingersoll-Ross (CIR) Cox *et al.* (1985) model brings a solution to the positivity problem encountered with the Vasicek model, by the use the nonlinear stochastic differential equation

$$dr_t = \beta(\alpha - r_t)dt + \sigma\sqrt{r_t}dB_t, \tag{3.4}$$

with $\alpha > 0$, $\beta > 0$, $\sigma > 0$.

The probability distribution of r_t at time $t > 0$ admits the noncentral Chi square probability density function $f_t(x)$ given for $x > 0$ by

$$f_t(x) = 2\beta \frac{\exp\left(-\frac{2\beta(x + r_0\,e^{-\beta t})}{\sigma^2(1 - e^{-\beta t})}\right)}{\sigma^2(1 - e^{-\beta t})} \left(\frac{x}{r_0\,e^{-\beta t}}\right)^{\alpha\beta/\sigma^2 - 1/2} I_{2\alpha\beta/\sigma^2 - 1}\left(\frac{4\beta\sqrt{r_0 x\,e^{-\beta t}}}{\sigma^2(1 - e^{-\beta t})}\right), \tag{3.5}$$

where

$$I_\lambda(z) := \left(\frac{z}{2}\right)^\lambda \sum_{k \geqslant 0} \frac{(z^2/4)^k}{k!\Gamma(\lambda + k + 1)}, \qquad z \in \mathbb{R},$$

is the modified Bessel function of the first kind, see Lemma 9 in Feller (1951) and Corollary 24 in Albanese and Lawi (2005). Note that $f_t(x)$ is not defined at $x = 0$ if $\alpha\beta/\sigma^2 - 1/2 < 0$, *i.e.* $\sigma^2 > 2\alpha\beta$, in which case the probability distribution of r_t admits a point mass at $x = 0$. On the other hand, r_t remains almost surely strictly positive under the Feller condition $2\alpha\beta \geqslant \sigma^2$, *cf.* the study of the associated probability density function in Lemma 4 of Feller (1951) for $\alpha, \beta \in \mathbb{R}$.

Figure 3.3 presents a random simulation of $t \mapsto r_t$ in the CIR model in the case $\sigma^2 > 2\alpha\beta$, in which the process is mean reverting with respect to $\alpha = 2.5\%$ and has a nonzero probability of hitting 0.

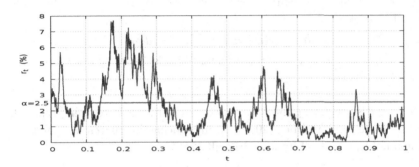

Fig. 3.3: Graph of the CIR short rate $t \mapsto r_t$ with $\alpha = 2.5\%$, $\beta = 1$, and $\sigma = 1.3$.

The next R code provides a numerical solution of the stochastic differential equation (3.4) using the Euler method, see Figure 3.3.

```
N=10000; t <- 0:(N-1); dt <- 1.0/N; nsim <- 2
a=0.025; b=1; sigma=0.1; sd=sqrt(sigma^2/2/b)
X <- matrix(rnorm(n = nsim * N, sd = sqrt(dt)), nsim, N)
R <- matrix(0,nsim,N);R[,1]=0.03
for (i in 1:nsim){for (j in 2:N){R[i,j]=max(0,R[i,j-1]+(a-b*R[i,j-1])*dt+sigma*
    sqrt(R[i,j-1])*X[i,j])}}
plot(t,R[1,],xlab="time",ylab="",type="l",ylim=c(0,R[1,1]+sd/5),col = 0,axes=FALSE
    )
axis(2, pos=0)
for (i in 1:nsim){lines(t, R[i, ], xlab = "time", type = "l", col = i+8)}
abline(h=a/b,col="blue",lwd=3);abline(h=0)
```

In large time $t \to \infty$, using the asymptotics

$$I_\lambda(z) \simeq_{z \to 0} \frac{1}{\Gamma(\lambda + 1)} \left(\frac{z}{2}\right)^\lambda,$$

the probability density function (3.5) becomes the Gamma density function

$$f(x) = \lim_{t \to \infty} f_t(x) = \frac{1}{\Gamma(2\alpha\beta/\sigma^2)} \left(\frac{2\beta}{\sigma^2}\right)^{2\alpha\beta/\sigma^2} x^{-1+2\alpha\beta/\sigma^2} e^{-2\beta x/\sigma^2}, \quad x > 0,$$

with shape parameter $2\alpha\beta/\sigma^2$ and scale parameter $\sigma^2/(2\beta)$, which is also the *invariant distribution* of r_t.

The family of classical mean-reverting models also includes the Courtadon (1982) model

$$dr_t = \beta(\alpha - r_t)dt + \sigma r_t dB_t$$

where α, β, σ are nonnegative coefficients, see Exercise 3.2, and the exponential Vasicek model

$$dr_t = r_t(\eta - a \log r_t)dt + \sigma r_t dB_t,$$

where a, η, σ are nonnegative coefficients, which is discussed in Exercises 3.3 and 3.4 in this chapter.

Other models preserving the positivity of interest rates have been proposed, cf. eg. James and Webber (2001), using stochastic differential equations on manifolds, see also Lim and Privault (2016).

3.2 Constant Elasticity of Variance (CEV) Models

Constant Elasticity of Variance models are designed to take into account non-constant volatilities that can vary as a power of the underlying asset. The Marsh and Rosenfeld (1983) model

$$dr_t = \left(\beta r_t^{\gamma-1} + \alpha r_t\right)dt + \sigma r_t^{\gamma/2}dB_t, \tag{3.6}$$

where $\alpha \in \mathbb{R}$ and $\beta, \sigma > 0$ are constants and $\gamma > 0$ is the variance (or diffusion) elasticity coefficient, covers most of the CEV models. Denoting by $v(r) := \sigma r^{\gamma/2}$ the diffusion coefficient in (3.6), constant elasticity refers to the constant ratio

$$\frac{dv(r)/v(r)}{dr/r} = \frac{r}{v(r)}\frac{dv(r)}{dr} = \frac{d\log v(r)}{d\log r} = \frac{d\log r^{\gamma/2}}{d\log r} = \frac{\gamma}{2}$$

between the relative change $dv(r)/v(r)$ in the variance $v(r)$ and the relative change dr/r in r.

For $\gamma = 1$, (3.6) yields the Cox *et al.* (1985) (CIR) equation

$$dr_t = (\beta + \alpha r_t)dt + \sigma\sqrt{r_t}dB_t.$$

For $\beta = 0$ we get the standard CEV model

$$dr_t = \alpha r_t dt + \sigma r_t^{\gamma/2}dB_t,$$

and for $\gamma = 2$ and $\beta = 0$ this yields the Dothan (1978) model

$$dr_t = \alpha r_t dt + \sigma r_t dB_t,$$

which is a version of geometric Brownian motion used for short-term interest rate modeling.

3.3 Time-Dependent Affine Models

Most of the models discussed in the above sections admit time-dependent extensions. This includes the class of affine models of the form

$$dr_t = (\eta(t) + \lambda(t)r_t)dt + \sqrt{\delta(t) + \gamma(t)r_t}dB_t. \tag{3.7}$$

Such models are called affine because the associated bonds can be priced using an *affine* PDE of the type (4.14) below, as will be seen after Proposition 4.2.

The family of affine models also includes:

(1) the Ho and Lee (1986) model

$$dr_t = \theta(t)dt + \sigma dB_t, \tag{3.8}$$

where $\theta(t)$ is a deterministic function of time, as an extension of the Merton model $dr_t = \theta dt + \sigma dB_t$,

(2) the Hull-White model Hull and White (1990),   cf. Section 3.3,

$$dr_t = (\theta(t) - \alpha(t)r_t)dt + \sigma(t)dB_t$$

which is itself a time-dependent extension of the Vasicek model, see Section 7.4.

Moreover, such time-dependent models can be used to fit an initial curve of instantaneous forward rates, see *e.g.* Problem 6.3-(6) below under absence of arbitrage.

3.4 Calibration of the Vasicek Model

The Vasicek equation (3.1), *i.e.*

$$dr_t = (a - br_t)dt + \sigma dB_t,$$

can be discretized according to a discrete-time sequence $(t_k)_{k=0,1,\ldots,n} = (t_0, t_1, \ldots, t_n)$ of $n+1$ time instants as

$$r_{t_{k+1}} - r_{t_k} = (a - br_{t_k})\Delta t + \sigma Z_k, \qquad k \geqslant 0,$$

where $\Delta t := t_{k+1} - t_k$ and $(Z_k)_{k\geqslant 0}$ is a Gaussian white noise sequence with variance Δt, *i.e.* a sequence of independent, centered and identically distributed $\mathcal{N}(0, \Delta t)$ Gaussian random variables, which yields

$$r_{t_{k+1}} = r_{t_k} + (a - br_{t_k})\Delta t + \sigma Z_k = a\Delta t + (1 - b\Delta t)r_{t_k} + \sigma Z_k, \quad k \geqslant 0.$$

Based on a set $\left(\tilde{r}_{t_k}\right)_{k=0,1,\ldots,n}$ of *market data,* we consider the quadratic residual

$$\sum_{k=0}^{n-1} \left(\tilde{r}_{t_{k+1}} - a\Delta t - (1 - b\Delta t)\tilde{r}_{t_k}\right)^2 \tag{3.9}$$

which represents the quadratic distance between the observed data sequence $\left(\tilde{r}_{t_k}\right)_{k=1,2,\ldots,n}$ and its predictions $\left(a\Delta t + (1 - b\Delta t)\tilde{r}_{t_k}\right)_{k=0,1,\ldots,n-1}.$

In order to minimize the residual (3.9) over a and b we use Ordinary Least Square (OLS) regression, and equate the following derivatives to zero. Namely, we have

$$\frac{\partial}{\partial a} \sum_{l=0}^{n-1} \left(\tilde{r}_{t_{l+1}} - a\Delta t - (1 - b\Delta t)\tilde{r}_{t_l}\right)^2$$

$$= -2\Delta t\left(-an\Delta t + \sum_{l=0}^{n-1} \left(\tilde{r}_{t_{l+1}} - (1 - b\Delta t)\tilde{r}_{t_l}\right)\right)$$

$$= 0,$$

hence

$$a\Delta t = \frac{1}{n} \sum_{l=0}^{n-1} \left(\tilde{r}_{t_{l+1}} - (1 - b\Delta t)\tilde{r}_{t_l}\right),$$

and

$$\frac{1}{2}\frac{\partial}{\partial b} \sum_{k=0}^{n-1} \left(\tilde{r}_{t_{k+1}} - a\Delta t - (1 - b\Delta t)\tilde{r}_{t_k}\right)^2$$

$$= \Delta t \sum_{k=0}^{n-1} \tilde{r}_{t_k}\left(-a\Delta t + \tilde{r}_{t_{k+1}} - (1 - b\Delta t)\tilde{r}_{t_k}\right)$$

$$= \Delta t \sum_{k=0}^{n-1} \tilde{r}_{t_k}\left(\tilde{r}_{t_{k+1}} - (1 - b\Delta t)\tilde{r}_{t_k} - \frac{1}{n}\sum_{l=0}^{n-1}\left(\tilde{r}_{t_{l+1}} - (1 - b\Delta t)\tilde{r}_{t_l}\right)\right)$$

$$= \Delta t \sum_{k=0}^{n-1} \tilde{r}_{t_k}\tilde{r}_{t_{k+1}} - \frac{\Delta t}{n}\sum_{k,l=0}^{n-1}\tilde{r}_{t_k}\tilde{r}_{t_{l+1}}$$

$$-\Delta t(1 - b\Delta t)\left(\sum_{k=0}^{n-1}\left(\tilde{r}_{t_k}\right)^2 - \frac{1}{n}\sum_{k,l=0}^{n-1}\tilde{r}_{t_k}\tilde{r}_{t_l}\right)$$

$$= 0.$$

This leads to estimators for the parameters a and b, respectively as the empirical mean and covariance of $\left(\tilde{r}_{t_k}\right)_{k=0,1,\ldots,n}$, *i.e.*

$$
\left\{
\begin{aligned}
&\hat{a}\Delta t = \frac{1}{n}\sum_{k=0}^{n-1}\tilde{r}_{t_{k+1}} - \frac{1}{n}\left(1-\hat{b}\Delta t\right)\sum_{k=0}^{n-1}\tilde{r}_{t_k} \\[2em]
&\text{and} \\[1em]
&1-\hat{b}\Delta t = \frac{\displaystyle\sum_{k=0}^{n-1}\tilde{r}_{t_k}\tilde{r}_{t_{k+1}} - \frac{1}{n}\sum_{k=0}^{n-1}\tilde{r}_{t_k}\sum_{l=0}^{n-1}\tilde{r}_{t_{l+1}}}{\displaystyle\sum_{k=0}^{n-1}\tilde{r}_{t_k}\tilde{r}_{t_k} - \frac{1}{n}\sum_{k=0}^{n-1}\tilde{r}_{t_k}\sum_{l=0}^{n-1}\tilde{r}_{t_l}} \\[2em]
&\qquad\qquad = \frac{\displaystyle\sum_{k=0}^{n-1}\left(\tilde{r}_{t_k}-\frac{1}{n}\sum_{l=0}^{n-1}\tilde{r}_{t_l}\right)\left(\tilde{r}_{t_{k+1}}-\frac{1}{n}\sum_{l=0}^{n-1}\tilde{r}_{t_{l+1}}\right)}{\displaystyle\sum_{k=0}^{n-1}\left(\tilde{r}_{t_k}-\frac{1}{n}\sum_{k=0}^{n-1}\tilde{r}_{t_k}\right)^2}.
\end{aligned}
\right.
$$

This also yields

$$
\begin{aligned}
\sigma^2\Delta t &= \operatorname{Var}[\sigma Z_k] \\
&\simeq \mathbb{E}\left[\left(\tilde{r}_{t_{k+1}} - (1-b\Delta t)\tilde{r}_{t_k} - a\Delta t\right)^2\right], \qquad k\geqslant 0,
\end{aligned}
$$

hence σ can be estimated as

$$
\hat{\sigma}^2\Delta t = \frac{1}{n}\sum_{k=0}^{n-1}\left(\tilde{r}_{t_{k+1}} - \tilde{r}_{t_k}\left(1-\hat{b}\Delta t\right) - \hat{a}\Delta t\right)^2,
$$

as follows from minimizing the residual

$$
\eta\mapsto\sum_{k=0}^{n-1}\left(\left(\tilde{r}_{t_{k+1}} - \tilde{r}_{t_k}\left(1-\hat{b}\Delta t\right) - \hat{a}\Delta t\right)^2 - \eta\Delta t\right)^2
$$

as a function of $\eta > 0$, see also Exercise 3.6.

3.5 Time Series Modeling

Defining $\hat{r}_{t_n} := r_{t_n} - a/b$, $n\in\mathbb{N}$, we have

$$
\begin{aligned}
\hat{r}_{t_{n+1}} &= r_{t_{n+1}} - \frac{a}{b} \\
&= r_{t_n} - \frac{a}{b} + (a - br_{t_n})\Delta t + \sigma Z_n
\end{aligned}
$$

$$= r_{t_n} - \frac{a}{b} - b\left(r_{t_n} - \frac{a}{b}\right)\Delta t + \sigma Z_n$$
$$= \hat{r}_{t_n} - b\hat{r}_{t_n}\Delta t + \sigma Z_n$$
$$= (1 - b\Delta t)\hat{r}_{t_n} + \sigma Z_n, \qquad n \geqslant 0.$$

In other words, the sequence $\left(\hat{r}_{t_n}\right)_{n\geqslant 0}$ is modeled according to an autoregressive AR(1) time series $(X_n)_{n\geqslant 0}$ with parameter $\alpha = 1 - b\Delta t$, in which the current state X_n of the system is expressed as the linear combination

$$X_n := \alpha X_{n-1} + \sigma Z_{n-1}, \qquad n \geqslant 1, \tag{3.10}$$

where $(Z_n)_{n\in\mathbb{Z}}$ is another Gaussian white noise sequence with variance Δt. This equation can be solved recursively as the causal series

$$X_n = \sigma Z_n + \alpha(\sigma Z_{n-1} + \alpha X_{n-2})$$
$$= \cdots$$
$$= \sigma \sum_{k\geqslant 0} \alpha^k Z_{n-k},$$

which converges when $|\alpha| < 1$, *i.e.* $|1 - b\Delta t| < 1$, in which case the time series $(X_n)_{n\geqslant 0}$ is weakly stationary, with

$$\mathbb{E}[X_n] = \sigma \sum_{k\geqslant 0} \alpha^k \mathbb{E}[Z_{n-k}] = \sigma \mathbb{E}[Z_0] \sum_{k\geqslant 0} \alpha^k$$
$$= \frac{\sigma}{1 - \alpha} \mathbb{E}[Z_0]$$
$$= 0, \qquad n \geqslant 0.$$

The variance of X_n is given by

$$\mathrm{Var}[X_n] = \sigma^2 \mathrm{Var}\left[\sum_{k\geqslant 0} \alpha^k Z_{n-k}\right]$$
$$= \sigma^2 \Delta t \sum_{k\geqslant 0} \alpha^{2k} = \sigma^2 \Delta t \sum_{k\geqslant 0}(1 - b\Delta t)^{2k}$$
$$= \frac{\sigma^2 \Delta t}{1 - (1 - b\Delta t)^2} = \frac{\sigma^2 \Delta t}{2b\Delta t - b^2(\Delta t)^2}$$
$$\simeq \frac{\sigma^2}{2b},$$

which coincides with the expected variance (3.3) of the Vasicek process in the stationary regime.

Example - TNX yield calibration

The next R code is estimating the parameters of the Vasicek model using the 10 Year Treasury Note yield data of Figure 3.2.

```
ratek=as.vector(rate);ratekplus1 <- c(ratek[-1],0)
b<-(sum(ratek*ratekplus1) - sum(ratek)*sum(ratekplus1)/n)/(sum(ratek*ratek) - sum(
    ratek)*sum(ratek)/n)
a<-sum(ratekplus1)/n-b*sum(ratek)/n;sigma <- sqrt(sum((ratekplus1-b*ratek-a)^2)/n)
```

Parameter estimation can also be implemented using the linear regression command

$$\text{lm}(c(\text{diff}(\text{ratek})) \sim \text{ratek}[1:\text{length}(\text{ratek})-1])$$

in R, which estimates the values of $a\Delta t$ and $-b\Delta t$ in the regression

$$r_{t_{k+1}} - r_{t_k} = (a - br_{t_k})\Delta t + \sigma Z_k, \qquad k \geqslant 0,$$

Coefficients:

(Intercept)   ratek[1:length(ratek) - 1]

0.017110   -0.007648

```
x11();for (i in 1:100) {ar.sim<-arima.sim(model=list(ar=c(b)),n.start=100,n)
y=a/b+sigma*ar.sim;y=y+ratek[1]-y[1]
time <- as.POSIXct(time(rate), format = "%Y-%m-%d")
yield <- xts(x = y, order.by = time)
chartSeries(yield,up.col="blue",theme="white",yrange=c(0,max(ratek)))
Sys.sleep(0.5)}
```

The above R code is generating Vasicek random samples according to the AR(1) time series (3.10), see Figure 3.4.

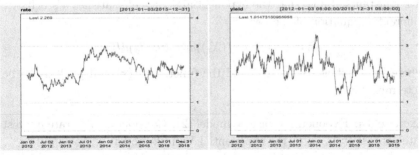

(a) CBOE TNX market yield.   (b) Calibrated Vasicek sample path.

Fig. 3.4: Calibrated Vasicek simulation *vs* market data.

Exercises

Exercise 3.1. Show that the solution of the equation

$$dr_t = (a - br_t)dt + \sigma dB_t, \tag{3.11}$$

where $a, \sigma \in \mathbb{R}$, $b > 0$, is given by

$$r_t = r_0 e^{-bt} + \frac{a}{b}(1 - e^{-bt}) + \sigma \int_0^t e^{-(t-s)b} dB_s, \quad t \geq 0.$$

Exercise 3.2. Consider the Courtadon (1982) model

$$dr_t = \beta(\alpha - r_t)dt + \sigma r_t dB_t, \tag{3.12}$$

where α, β, σ are nonnegative coefficients, which is a particular case of the Chan-Karolyi-Longstaff-Sanders (CKLS) model (Chan *et al.* (1992)) with $\gamma = 1$. Show that the solution of (3.12) is given by

$$r_t = \alpha\beta \int_0^t \frac{S_t}{S_u} du + r_0 S_t, \quad t \geq 0, \tag{3.13}$$

where $(S_t)_{t \in \mathbb{R}_+}$ is the geometric Brownian motion solution of $dS_t = -\beta S_t dt + \sigma S_t dB_t$ with $S_0 = 1$.

Exercise 3.3. Exponential Vasicek model (1). Consider a Vasicek process $(r_t)_{t \in \mathbb{R}_+}$ solution of the stochastic differential equation

$$dr_t = (a - br_t)dt + \sigma dB_t, \quad t \geq 0,$$

where $(B_t)_{t \in \mathbb{R}_+}$ is a standard Brownian motion and $\sigma, a, b > 0$ are positive constants. Show that the exponential $X_t := e^{r_t}$ satisfies a stochastic differential equation of the form

$$dX_t = X_t(\tilde{a} - \tilde{b}f(X_t))dt + \sigma g(X_t)dB_t,$$

where the coefficients $\tilde{a}$ and $\tilde{b}$ and the functions $f(x)$ and $g(x)$ are to be determined.

Exercise 3.4. Exponential Vasicek model (2). Consider a short rate interest rate process $(r_t)_{t \in \mathbb{R}_+}$ in the exponential Vasicek model:

$$dr_t = r_t(\eta - a \log r_t)dt + \sigma r_t dB_t, \tag{3.14}$$

where η, a, σ are positive parameters.

(1) Find the solution $(Y_t)_{t \in \mathbb{R}_+}$ of the stochastic differential equation
$$dY_t = (\theta - aY_t)dt + \sigma dB_t$$
as a function of the initial condition y_0, where θ, a, σ are positive parameters. *Hint*: Let $Z_t = Y_t - \theta/a$, $t \in \mathbb{R}_+$.

(2) Let $X_t = e^{Y_t}$, $t \in \mathbb{R}_+$. Determine the stochastic differential equation satisfied by $(X_t)_{t \in \mathbb{R}_+}$.

(3) Find the solution $(r_t)_{t \in \mathbb{R}_+}$ of (3.14) in terms of the initial condition r_0.

(4) Compute the conditional mean $\mathbb{E}[r_t \mid \mathcal{F}_u]$ of r_t, $0 \leqslant u \leqslant t$, where $(\mathcal{F}_u)_{u \in \mathbb{R}_+}$ denotes the filtration generated by the Brownian motion $(B_t)_{t \in \mathbb{R}_+}$.

(5) Compute the conditional variance $\mathrm{Var}[r_t \mid \mathcal{F}_u] := \mathbb{E}\left[r_t^2 \mid \mathcal{F}_u\right] - (\mathbb{E}[r_t \mid \mathcal{F}_u])^2$ of r_t, $0 \leqslant u \leqslant t$.

(6) Compute the asymptotic mean and variance $\lim_{t \to \infty} \mathbb{E}[r_t]$ and $\lim_{t \to \infty} \mathrm{Var}[r_t]$.

Exercise 3.5. Cox-Ingersoll-Ross (CIR) model. Consider the equation
$$dr_t = (\alpha - \beta r_t)dt + \sigma \sqrt{r_t} dB_t \qquad (3.15)$$
which models the variations of the short rate process r_t, where α, β, σ and r_0 are positive parameters and $(B_t)_{t \in \mathbb{R}_+}$ is a standard Brownian motion.

(1) Write down the equation (3.15) in integral form.

(2) Let $u(t) = \mathbb{E}[r_t \mid \mathcal{F}_s]$, $0 \leqslant s \leqslant t$. Show, using the integral form of (3.15), that $u(t)$ satisfies the differential equation
$$u'(t) = \alpha - \beta u(t), \qquad 0 \leqslant s \leqslant t.$$

(3) By an application of Itô's formula to r_t^2, show that
$$dr_t^2 = r_t(2\alpha + \sigma^2 - 2\beta r_t)dt + 2\sigma r_t^{3/2} dB_t. \qquad (3.16)$$

(4) Using the integral form of (3.16), find a differential equation satisfied by $v(t) = E[r_t^2 \mid \mathcal{F}_s]$, $0 \leqslant s \leqslant t$, and compute $E[r_t^2 \mid \mathcal{F}_s]$, $0 \leqslant s \leqslant t$. You may assume that $a = 0$ to simplify the computation.
Hint: The function $f(t) = ce^{-\beta t}/\beta$ solves the differential equation $f'(t) + 2\beta f(t) = ce^{-\beta t}$ for all $c \in \mathbb{R}$.

(5) Let
$$X_t = e^{-\beta t/2}\left(x_0 + \frac{\sigma}{2}\int_0^t e^{\beta s/2} dB_s\right), \qquad t \geqslant 0.$$
Show that X_t satisfies the equation
$$dX_t = \frac{\sigma}{2}dB_t - \frac{\beta}{2}X_t dt.$$

(6) Let $R_t = X_t^2$ and

$$W_t = \int_0^t \text{sign}(X_s)dB_s, \qquad t \geqslant 0, \tag{3.17}$$

where $\text{sign}(x) = \mathbb{1}_{\{x \geqslant 0\}} - \mathbb{1}_{\{x < 0\}}$, $x \in \mathbb{R}$. Show that

$$dR_t = \left(\frac{\sigma^2}{4} - \beta R_t\right)dt + \sigma\sqrt{R_t}dW_t.$$

Exercise 3.6.   Consider the Chan-Karolyi-Longstaff-Sanders (CKLS) interest rate model (Chan *et al.* (1992)) parametrized as

$$dr_t = (a - br_t)dt + \sigma r_t^\gamma dB_t,$$

and time-discretized as

$$\begin{aligned}
r_{t_{k+1}} &= r_{t_k} + (a - br_{t_k})\Delta t + \sigma r_{t_k}^\gamma Z_k \\
&= a\Delta t + (1 - b\Delta t)r_{t_k} + \sigma r_{t_k}^\gamma Z_k, \qquad k \geqslant 0,
\end{aligned}$$

where $\Delta t := t_{k+1} - t_k$ and $(Z_k)_{k \geqslant 0}$ is an *i.i.d.* sequence of $\mathcal{N}(0, \Delta t)$ random variables. Assuming that $a, b, \gamma > 0$ are known, find an unbiased estimator $\hat{\sigma}^2$ for the variance coefficient σ^2, based on a market data set $\left(\tilde{r}_{t_k}\right)_{k=0,1,\dots,n}$.

Exercise 3.7.   Consider the Vasicek process $(r_t)_{t \in \mathbb{R}_+}$ solution of the equation

$$dr_t = (a - br_t)dt + \sigma dB_t. \tag{3.18}$$

(1) Consider the discretization

$$r_{t_{k+1}} := r_{t_k} + (a - br_{t_k})\Delta t \pm \sigma\sqrt{\Delta t}, \qquad k = 0, 1, 2, \dots$$

of the equation (3.18) with $p(r_{t_0}) = p(r_{t_1}) = 1/2$ and

$$\mathbb{E}[\Delta r_{t_1}] = (a - br_{t_0})\Delta t + \sigma p(r_{t_0})\sqrt{\Delta t} - \sigma q(r_{t_0})\sqrt{\Delta t} = (a - br_{t_0})\Delta t$$

and

$$\mathbb{E}[\Delta r_{t_2}] = (a - br_{t_1})\Delta t + \sigma p(r_{t_0})\sqrt{\Delta t} - \sigma q(r_{t_0})\sqrt{\Delta t} = (a - br_{t_1})\Delta t.$$

Does this discretization lead to a (recombining) binomial tree?

(2) Using the Girsanov Theorem, find a probability measure $\mathbb{Q}$ under which the process

$$\frac{r_t}{\sigma} = \frac{r_0}{\sigma} + \int_0^t \frac{a - br_s}{\sigma}ds + B_t, \qquad t \geqslant 0,$$

is a standard Brownian motion.

Hint: By the Girsanov Theorem, the process $X_t = X_0 + \int_0^t u_s ds + B_t$ is a martingale under the probability measure $\mathbb{Q}$ with Radon-Nikodym density

$$\frac{d\mathbb{Q}}{d\mathbb{P}} = \exp\left(-\int_0^T u_t dB_t - \frac{1}{2}\int_0^T (u_t)^2 dt \right)$$

with respect to $\mathbb{P}$.

(3) Using the approximation

$$\exp\left(\frac{1}{\sigma^2}\int_0^T (a - br_t)dr_t - \frac{1}{2\sigma^2}\int_0^T (a - br_t)^2 dt \right)$$

$$\simeq 2^{T/\Delta t} \prod_{0 < t < T} \left(\frac{1}{2} \pm \frac{a - br_t}{2\sigma}\sqrt{\Delta t} \right),$$

show that the Vasicek process can be discretized along the *binomial* tree

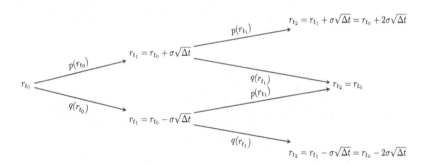

with

$$\mathbb{E}[\Delta r_{t_1} \mid r_{t_0}] = (a - br_{t_0})\Delta t \quad \text{and} \quad \mathbb{E}[\Delta r_{t_2} \mid r_{t_1}] = (a - br_{t_1})\Delta t,$$

where $\Delta r_{t_1} := r_{t_1} - r_{t_0}$, $\Delta r_{t_2} := r_{t_2} - r_{t_1}$, and the probabilities $p(r_{t_0})$, $q(r_{t_0})$, $p(r_{t_1})$, $q(r_{t_1})$ will be computed explicitly.

The use of binomial (or recombining) trees can make the implementation of the Monte Carlo method easier as their size grows linearly instead of exponentially.

Chapter 4

Pricing of Zero-Coupon and Coupon Bonds

In this chapter, we describe the basics of bond pricing in the absence of arbitrage opportunities. Explicit calculations are carried out in the Vasicek model, using both the probabilistic and PDE approaches. Numerical simulations are also presented, together with market examples. The definition of zero-coupon bonds will be used in Chapter 5 for the construction of forward rate processes.

4.1 Definition and Basic Properties

A zero-coupon bond is a contract priced $P(t,T)$ at time $t < T$ to deliver $P(T,T) = \$1$ at time T. In addition to its value at maturity, a bond may yield a periodic *coupon* payment at regular time intervals until the maturity date. The computation of the arbitrage price $P_0(t,T)$ of a zero-coupon bond based on an underlying short-term interest rate process $(r_t)_{t \in \mathbb{R}_+}$ is a basic and important issue in interest rate modeling.

Constant short rate

In case the short-term interest rate is a constant $r_t = r$, $t \in \mathbb{R}_+$, a standard arbitrage argument shows that the price $P(t,T)$ of the bond is given by

$$P(t,T) = e^{-(T-t)r}, \qquad 0 \leqslant t \leqslant T.$$

Indeed, if $P(t,T) > e^{-(T-t)r}$ we could issue a bond at the price $P(t,T)$ and invest this amount at the compounded risk free rate r, which would yield $P(t,T) e^{(T-t)r} > 1$ at time T.

On the other hand, if $P(t,T) < e^{-(T-t)r}$ we could borrow $P(t,T)$ at the rate r to purchase a bond priced $P(t,T)$. At maturity time T we would receive \$1 and refund only $P(t,T)\,e^{(T-t)r} < 1$.

The price $P(t,T) = e^{-(T-t)r}$ of the bond is the value of $P(t,T)$ that makes the potential profit $P(t,T)\,e^{(T-t)r} - 1$ vanish for both traders.

Time-dependent deterministic short rates

Similarly to the above, when the short-term interest rate process $(r_t)_{t\in\mathbb{R}_+}$ is a deterministic function of time, a similar argument shows that

$$P(t,T) = \exp\left(-\int_t^T r_s ds \right), \qquad 0 \leqslant t \leqslant T. \tag{4.1}$$

Stochastic short rates

In case $(r_t)_{t\in\mathbb{R}_+}$ is an $(\mathcal{F}_t)_{t\in\mathbb{R}_+}$-adapted random process, the formula (4.1) is no longer valid as it relies on future information, and we replace it with the averaged discounted payoff

$$P(t,T) = \mathbb{E}^*\left[e^{-\int_t^T r_s ds} \,\middle|\, \mathcal{F}_t \right], \qquad 0 \leqslant t \leqslant T, \tag{4.2}$$

under a risk-neutral measure $\mathbb{P}^*$. It is natural to write $P(t,T)$ as a conditional expectation under a martingale measure, as the use of conditional expectation helps to "filter out" the future information past time t contained in $\displaystyle\int_t^T r_s ds$. The expression (4.2) makes sense as the "best possible estimate" of the future quantity $e^{-\int_t^T r_s ds}$ in mean square sense, given the information known up to time t.

Coupon bonds

Pricing bonds with nonzero coupon is not difficult since in general the amount and periodicity of coupons are deterministic.[1] In the case of a succession of coupon payments $c_1, c_2, \ldots, c_n$ at times $T_1, T_2, \ldots, T_n \in (t,T]$, another application of the above absence of arbitrage argument shows that the price $P_c(t,T)$ of the coupon bond with discounted (deterministic) coupon payments is given by the linear combination of zero-coupon bond prices

$$P_c(t,T) := \mathbb{E}^*\left[\sum_{k=1}^n c_k\, e^{-\int_t^{T_k} r_s ds} \,\middle|\, \mathcal{F}_t \right] + \mathbb{E}^*\left[e^{-\int_t^T r_s ds} \,\middle|\, \mathcal{F}_t \right]$$

---

[1] However, coupon default cannot be excluded.

$$= \sum_{k=1}^{n} c_k \, \mathbb{E}^* \left[e^{-\int_t^{T_k} r_s ds} \,\bigg|\, \mathcal{F}_t \right] + P_0(t, T)$$

$$= P_0(t, T) + \sum_{k=1}^{n} c_k P_0(t, T_k), \qquad 0 \leqslant t \leqslant T_1, \tag{4.3}$$

which represents the present value at time t of future $\$c_1, \$c_2, \ldots, \$c_n$ receipts respectively at times $T_1, T_2, \ldots, T_n$, in addition to a terminal $\$1$ principal payment.

In the case of a constant coupon rate $c > 0$ paid at regular time intervals $\tau = T_{k+1} - T_k$, $k = 0, 1, \ldots, n$, with $T_0 = t$ and a constant deterministic short rate r, we find

$$P_c(t, T) = e^{-rn\tau} + c \sum_{k=1}^{n} e^{-(T_k - t)r}$$

$$= e^{-rn\tau} + c \sum_{k=1}^{n} e^{-kr\tau}$$

$$= e^{-rn\tau} + c \frac{e^{-r\tau} - e^{-r(n+1)\tau}}{1 - e^{-r\tau}}.$$

In terms of the discrete-time interest rate $\tilde{r} := e^{r\tau} - 1$, we have

$$P_c(t, T) = \frac{1}{(1 + \tilde{r})^n} + \frac{c}{\tilde{r}} \left(1 - \frac{1}{(1 + \tilde{r})^n} \right).$$

In the case of a continuous-time coupon rate $c > 0$, the above discrete-time calculation (4.3) can be reinterpreted as follows:

$$P_c(t, T) = P_0(t, T) + c \int_t^T P_0(t, u) du \tag{4.4}$$

$$= P_0(t, T) + c \int_t^T e^{-(u-t)r} du$$

$$= e^{-(T-t)r} + c \int_0^{T-t} e^{-ru} du$$

$$= e^{-(T-t)r} + \frac{c}{r} \left(1 - e^{-(T-t)r} \right),$$

$$= \frac{c}{r} + \frac{r - c}{r} e^{-(T-t)r}, \qquad 0 \leqslant t \leqslant T, \tag{4.5}$$

where the coupon bond price $P_c(t, T)$ solves the ordinary differential equation

$$dP_c(t, T) = (r - c) e^{-(T-t)r} dt = -c \, dt + r P_c(t, T) dt, \qquad 0 \leqslant t \leqslant T,$$

see also Figure 4.5-(b) below.

In the sequel, we will mostly consider zero-coupon bonds priced as $P(t,T) = P_0(t,T)$, $0 \leqslant t \leqslant T$, in the setting of stochastic short rates.

Absence of arbitrage and the Markov property

The bond price $P(t,T)$ is defined as the conditional expectation

$$P(t,T) = \mathbb{E}_{\mathbb{Q}} \left[e^{-\int_t^T r_s ds} \,\middle|\, \mathcal{F}_t \right], \tag{4.6}$$

which makes sense as the "best possible estimate" of the future quantity $e^{-\int_t^T r_s ds}$ given the information known up to time t, under a martingale (or risk-neutral) measure $\mathbb{Q}$. The use of conditional expectation is natural in this framework since it helps in "filtering out" future information beyond time t contained in $\displaystyle\int_t^T r_s ds$.

Assume from now on that the underlying short rate process $(r_t)_{t \in \mathbb{R}_+}$ is solution to the stochastic differential equation

$$dr_t = \mu(t, r_t)dt + \sigma(t, r_t)dB_t \tag{4.7}$$

where $(B_t)_{t \in \mathbb{R}_+}$ is a standard Brownian motion under $\mathbb{P}$. For example, in the Vasicek model (3.1), we have

$$\mu(t, x) = a - bx \quad \text{and} \quad \sigma(t, x) = \sigma.$$

Assume now that the probability measure $\mathbb{Q}$ is equivalent to $\mathbb{P}$, with the Radon-Nikodym density

$$\frac{d\mathbb{Q}}{d\mathbb{P}} = \exp\left(-\int_0^\infty K_s dB_s - \frac{1}{2} \int_0^\infty |K_s|^2 ds \right),$$

where $(K_s)_{s \in \mathbb{R}_+}$ is an $(\mathcal{F}_t)_{t \in \mathbb{R}_+}$-adapted process satisfying the Novikov integrability condition (2.10). By the Girsanov Theorem 2.1, it is known that

$$\widehat{B}_t := B_t + \int_0^t K_s ds$$

is a standard Brownian motion under $\mathbb{Q}$, thus (4.7) can be rewritten as

$$dr_t = \tilde{\mu}(t, r_t)dt + \sigma(t, r_t)d\widehat{B}_t$$

where

$$\tilde{\mu}(t, r_t) := \mu(t, r_t) - \sigma(t, r_t)K_t.$$

The process K_t, which is called the "market price of risk" (MPoR), has to be specified, usually via statistical estimation based on market data.

In the sequel we will assume for simplicity that $K_t = 0$; in other terms we assume that $\mathbb{P} = \mathbb{P}^*$ is the martingale measure used for pricing.

Proposition 4.1. *The discounted bond price process*

$$t \mapsto \widetilde{P}(t, T) := e^{-\int_0^t r_s ds} P(t, T)$$

is a martingale under $\mathbb{P}^*$.

Proof. By (4.6), we have

$$
\begin{aligned}
e^{-\int_0^t r_s ds} P(t, T) &= e^{-\int_0^t r_s ds} \, \mathbb{E}^* \left[e^{-\int_t^T r_s ds} \,\middle|\, \mathcal{F}_t \right] \\
&= \mathbb{E}^* \left[e^{-\int_0^t r_s ds} \, e^{-\int_t^T r_s ds} \,\middle|\, \mathcal{F}_t \right] \\
&= \mathbb{E}^* \left[e^{-\int_0^T r_s ds} \,\middle|\, \mathcal{F}_t \right], \qquad 0 \leqslant t \leqslant T,
\end{aligned}
$$

and this suffices in order to conclude, since by the "tower property" (11.7) of conditional expectations, any process $(X_t)_{t \in \mathbb{R}_+}$ of the form $t \mapsto X_t := \mathbb{E}^*[F \mid \mathcal{F}_t]$, $F \in L^1(\Omega)$, is a martingale, see (11.7). $\qquad \square$

4.2 Bond Pricing PDE

The *Markov* property, see the appendix, states that the future after time t of a Markov process $(X_s)_{s \in \mathbb{R}_+}$ depends only on its present state t and not on the whole history of the process up to time t. It can be stated as follows using conditional expectations:

$$\mathbb{E}[f(X_{t_1}, \ldots, X_{t_n}) \mid \mathcal{F}_t] = \mathbb{E}[f(X_{t_1}, \ldots, X_{t_n}) \mid X_t]$$

for all times $t_1, \ldots, t_n$ greater than t and all sufficiently integrable function f on $\mathbb{R}^n$, see the appendix for details.

We will make use of the following fundamental property of stochastic differential equations of the form (4.7), see *e.g.* Theorem V-32 in Protter (2004).

Property 4.1. *All solutions of stochastic differential equations such as* (4.7) *have the* Markov *property.*

We assume from now on that the underlying short rate process solves a stochastic differential equation of the form (4.7), *i.e.*

$$dr_t = \mu(t, r_t)dt + \sigma(t, r_t)dB_t, \tag{4.8}$$

where $(B_t)_{t \in \mathbb{R}_+}$ is a standard Brownian motion under $\mathbb{P}^*$. Note that specifying the dynamics of $(r_t)_{t \in \mathbb{R}_+}$ under the historical probability measure $\mathbb{P}$ will also lead to a notion of market price of risk (MPoR) for the modeling of short rates.

Since all solutions of stochastic differential equations such as (4.8) have the Markov property by Property 4.1, the arbitrage price $P(t, T)$ can be rewritten as a function $F(t, r_t)$ of r_t, *i.e.*

$$\begin{aligned}
P(t, T) &= \mathbb{E}^* \left[e^{-\int_t^T r_s ds} \,\middle|\, \mathcal{F}_t \right] \\
&= \mathbb{E}^* \left[e^{-\int_t^T r_s ds} \,\middle|\, r_t \right] \\
&= F(t, r_t),
\end{aligned} \tag{4.9}$$

and depends on (t, r_t) only, instead of depending on the whole information available in $\mathcal{F}_t$ up to time t. In other words, the pricing problem can now be formulated as a search for a function $F(t, x)$ such that

$$P(t, T) = F(t, r_t), \qquad 0 \leqslant t \leqslant T.$$

In the next proposition, we derive a Partial Differential Equation (PDE) for the function $F(t, x)$, based on an absence of arbitrage argument.

Proposition 4.2. *(Bond pricing PDE). Consider a short rate process* $(r_t)_{t \in \mathbb{R}_+}$ *modeled by a diffusion equation of the form*

$$dr_t = \mu(t, r_t)dt + \sigma(t, r_t)dB_t.$$

The bond pricing PDE for $P(t, T) = F(t, r_t)$ *as in (4.9) is written as*

$$xF(t, x) = \frac{\partial F}{\partial t}(t, x) + \mu(t, x)\frac{\partial F}{\partial x}(t, x) + \frac{1}{2}\sigma^2(t, x)\frac{\partial^2 F}{\partial x^2}(t, x), \tag{4.10}$$

$t \in \mathbb{R}_+$, $x \in \mathbb{R}$, *subject to the terminal condition*

$$F(T, x) = 1, \qquad x \in \mathbb{R}. \tag{4.11}$$

Proof. By Itô's formula, we have

$$d\left(e^{-\int_0^t r_s ds} P(t,T)\right) = -r_t e^{-\int_0^t r_s ds} P(t,T)dt + e^{-\int_0^t r_s ds} dP(t,T)$$

$$= -r_t e^{-\int_0^t r_s ds} F(t,r_t)dt + e^{-\int_0^t r_s ds} dF(t,r_t)$$

$$= -r_t e^{-\int_0^t r_s ds} F(t,r_t)dt + e^{-\int_0^t r_s ds} \frac{\partial F}{\partial x}(t,r_t)dr_t$$

$$+ \frac{1}{2} e^{-\int_0^t r_s ds} \frac{\partial^2 F}{\partial x^2}(t,r_t)(dr_t)^2 + e^{-\int_0^t r_s ds} \frac{\partial F}{\partial t}(t,r_t)dt$$

$$= -r_t e^{-\int_0^t r_s ds} F(t,r_t)dt + e^{-\int_0^t r_s ds} \frac{\partial F}{\partial x}(t,r_t)(\mu(t,r_t)dt + \sigma(t,r_t)dB_t)$$

$$+ e^{-\int_0^t r_s ds} \left(\frac{1}{2}\sigma^2(t,r_t)\frac{\partial^2 F}{\partial x^2}(t,r_t) + \frac{\partial F}{\partial t}(t,r_t) \right) dt$$

$$= e^{-\int_0^t r_s ds} \sigma(t,r_t)\frac{\partial F}{\partial x}(t,r_t)dB_t$$

$$+ e^{-\int_0^t r_s ds} \left(-r_t F(t,r_t) + \mu(t,r_t)\frac{\partial F}{\partial x}(t,r_t) \right.$$

$$\left. + \frac{1}{2}\sigma^2(t,r_t)\frac{\partial^2 F}{\partial x^2}(t,r_t) + \frac{\partial F}{\partial t}(t,r_t) \right) dt.$$

$$(4.12)$$

Given that $t \mapsto e^{-\int_0^t r_s ds} P(t,T)$ is a martingale by Proposition 4.1, the above expression (4.12) should only contain terms in dB_t (cf. Corollary II-1, page 72 of Protter (2004)), and all terms in dt should vanish inside (4.12). This leads to the identities

$$\begin{cases} r_t F(t,r_t) = \mu(t,r_t)\dfrac{\partial F}{\partial x}(t,r_t) + \dfrac{1}{2}\sigma^2(t,r_t)\dfrac{\partial^2 F}{\partial x^2}(t,r_t) + \dfrac{\partial F}{\partial t}(t,r_t) \\[2em] d\left(e^{-\int_0^t r_s ds} P(t,T)\right) = e^{-\int_0^t r_s ds} \sigma(t,r_t)\dfrac{\partial F}{\partial x}(t,r_t)dB_t, \end{cases} \quad (4.13a)$$

which recover (4.10). Condition (4.11) is due to the fact that $P(T,T) = \$1$. $\square$

Condition (4.11) is due to the fact that $P(T,T) = \$1$. On the other hand,

$$\left(e^{-\int_0^t r_s ds} P(t,T) \right)_{t \in [0,T]} \qquad \text{and} \qquad (P(t,T))_{t \in [0,T]}$$

respectively satisfy the stochastic differential equations

$$d\left(e^{-\int_0^t r_s ds} P(t,T)\right) = e^{-\int_0^t r_s ds} \sigma(t,r_t)\frac{\partial F}{\partial x}(t,r_t)dB_t$$

and

$$dP(t,T) = P(t,T)r_t dt + \sigma(t,r_t)\frac{\partial F}{\partial x}(t,r_t)dB_t,$$

i.e.

$$\frac{dP(t,T)}{P(t,T)} = r_t dt + \frac{\sigma(t,r_t)}{P(t,T)}\frac{\partial F}{\partial x}(t,r_t)dB_t$$

$$= r_t dt + \sigma(t,r_t)\frac{\partial}{\partial x}\log F(t,r_t)dB_t.$$

In the case of an interest rate process modeled by (3.7) we have

$$\mu(t,x) = \eta(t) + \lambda(t)x \qquad \text{and} \qquad \sigma(t,x) = \sqrt{\delta(t) + \gamma(t)x},$$

hence (4.10) yields the (time-dependent) *affine* PDE

$$xF(t,x) = \frac{\partial F}{\partial t}(t,x) + (\eta(t) + \lambda(t)x)\frac{\partial F}{\partial x}(t,x) + \frac{1}{2}(\delta(t) + \gamma(t)x)\frac{\partial^2 F}{\partial x^2}(t,x),$$

$$(4.14)$$

$t \in \mathbb{R}_+$, $x \in \mathbb{R}$. By (4.13a), the proof of the above Proposition 4.2 also shows that

$$\frac{dP(t,T)}{P(t,T)} = \frac{1}{P(t,T)}d\left(e^{\int_0^t r_s ds}e^{-\int_0^t r_s ds}P(t,T)\right)$$

$$= \frac{1}{P(t,T)}\left(r_t P(t,T)dt + e^{\int_0^t r_s ds}d\left(e^{-\int_0^t r_s ds}P(t,T)\right)\right)$$

$$= r_t dt + \frac{1}{P(t,T)}e^{\int_0^t r_s ds}d\left(e^{-\int_0^t r_s ds}P(t,T)\right)$$

$$= r_t dt + \frac{1}{F(t,r_t)}\frac{\partial F}{\partial x}(t,r_t)\sigma(t,r_t)dB_t$$

$$= r_t dt + \sigma(t,r_t)\frac{\partial}{\partial x}\log F(t,r_t)dB_t. \qquad (4.15)$$

4.3   Probabilistic Solution

We solve the bond pricing PDE (4.10), written with $\mu(t,x) = a - bx$ and $\sigma(t,x) = \sigma$ in the Vašíček (1977) model

$$dr_t = (a - br_t)dt + \sigma dB_t \qquad (4.16)$$

as

$$\begin{cases} xF(t,x) = \dfrac{\partial F}{\partial t}(t,x) + (a - bx)\dfrac{\partial F}{\partial x}(t,x) + \dfrac{\sigma^2}{2}\dfrac{\partial^2 F}{\partial x^2}(t,x), \\[4mm] F(T,x) = 1, \end{cases} \qquad (4.17)$$

by a direct computation of the conditional expectation

$$F(t,r_t) = P(t,T) = \mathbb{E}^*\left[e^{-\int_t^T r_s ds}\,\Big|\,\mathcal{F}_t\right].$$

See also Exercise 4.4 for a bond pricing formula in the Cox *et al.* (1985) (CIR) model.

Proposition 4.3. *The zero-coupon bond price in the Vasicek model* (4.16) *can be expressed as*

$$P(t,T) = e^{A(T-t)+r_t C(T-t)}, \qquad 0 \leqslant t \leqslant T, \qquad (4.18)$$

where $A(x)$ *and* $C(x)$ *are functions of time to maturity given by*

$$C(x) := -\frac{1}{b}\left(1 - e^{-bx}\right), \qquad (4.19)$$

and

$$
\begin{aligned}
A(x) &:= \frac{4ab - 3\sigma^2}{4b^3} + \frac{\sigma^2 - 2ab}{2b^2}x + \frac{\sigma^2 - ab}{b^3}e^{-bx} - \frac{\sigma^2}{4b^3}e^{-2bx} \qquad (4.20) \\
&= -\left(\frac{a}{b} - \frac{\sigma^2}{2b^2}\right)(x + C(x)) - \frac{\sigma^2}{4b}C^2(x), \qquad x \geqslant 0.
\end{aligned}
$$

Proof. Recall that in the Vasicek model, the short rate process $(r_t)_{t \in \mathbb{R}_+}$ solution of (4.16) has the expression

$$r_t = g(t) + \int_0^t h(t,s)dB_s = r_0 e^{-bt} + \frac{a}{b}\left(1 - e^{-bt}\right) + \sigma \int_0^t e^{-(t-s)b}dB_s,$$

where g and h are the deterministic functions

$$g(t) := r_0 e^{-bt} + \frac{a}{b}\left(1 - e^{-bt}\right), \qquad t \geqslant 0,$$

and

$$h(t,s) := \sigma e^{-(t-s)b}, \qquad 0 \leqslant s \leqslant t.$$

Using the fact that Wiener integrals are Gaussian random variables and the Gaussian moment generating function, and exchanging the order of integration between ds and du over $[t,T]$ according to the following picture,

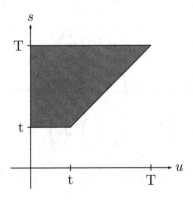

we have

$$P(t,T) = \mathbb{E}^* \left[e^{-\int_t^T r_s ds} \,\middle|\, \mathcal{F}_t \right]$$

$$= \mathbb{E}^* \left[e^{-\int_t^T (g(s) + \int_0^s h(s,u)dB_u)ds} \,\middle|\, \mathcal{F}_t \right]$$

$$= e^{-\int_t^T g(s)ds} \, \mathbb{E}^* \left[e^{-\int_t^T \int_0^s h(s,u)dB_u ds} \,\middle|\, \mathcal{F}_t \right]$$

$$= e^{-\int_t^T g(s)ds} \, \mathbb{E}^* \left[e^{-\int_0^T \int_{\mathrm{Max}(u,t)}^T h(s,u)dsdB_u} \,\middle|\, \mathcal{F}_t \right]$$

$$= \exp\left(-\int_t^T g(s)ds - \int_0^t \int_{\mathrm{Max}(u,t)}^T h(s,u)dsdB_u \right)$$

$$\times \, \mathbb{E}^* \left[e^{-\int_t^T \int_{\mathrm{Max}(u,t)}^T h(s,u)dsdB_u} \,\middle|\, \mathcal{F}_t \right]$$

$$= \exp\left(-\int_t^T g(s)ds - \int_0^t \int_t^T h(s,u)dsdB_u \right)$$

$$\times \, \mathbb{E}^* \left[e^{-\int_t^T \int_u^T h(s,u)dsdB_u} \,\middle|\, \mathcal{F}_t \right]$$

$$= \exp\left(-\int_t^T g(s)ds - \int_0^t \int_t^T h(s,u)dsdB_u \right) \mathbb{E}^* \left[e^{-\int_t^T \int_u^T h(s,u)dsdB_u} \right]$$

$$= \exp\left(-\int_t^T g(s)ds - \int_0^t \int_t^T h(s,u)dsdB_u + \frac{1}{2} \int_t^T \left(\int_u^T h(s,u)ds \right)^2 du \right)$$

$$= \exp\left(-\int_t^T \left(r_0 e^{-bs} + \frac{a}{b}(1 - e^{-bs}) \right)ds - \sigma \int_0^t \int_t^T e^{-(s-u)b}dsdB_u \right)$$

$$\times \exp\left(\frac{\sigma^2}{2} \int_t^T \left(\int_u^T e^{-(s-u)b}ds \right)^2 du \right)$$

$$= \exp\left(-\int_t^T \left(r_0 e^{-bs} + \frac{a}{b}(1 - e^{-bs}) \right)ds \right)$$

$$\times \exp\left(-\frac{\sigma}{b}(1 - e^{-(T-t)b}) \int_0^t e^{-(t-u)b}dB_u \right)$$

$$\times \exp\left(\frac{\sigma^2}{2} \int_t^T e^{2bu} \left(\frac{e^{-bu} - e^{-bT}}{b} \right)^2 du \right)$$

$$= \exp\left(-\frac{r_t}{b}(1 - e^{-(T-t)b}) + \frac{1}{b}(1 - e^{-(T-t)b}) \left(r_0 e^{-bt} + \frac{a}{b}(1 - e^{-bt}) \right) \right)$$

$$\times \exp\left(-\int_t^T \left(r_0 e^{-bs} + \frac{a}{b}\left(1 - e^{-bs}\right)\right) ds\right)$$

$$\times \exp\left(\frac{\sigma^2}{2} \int_t^T e^{2bu} \left(\frac{e^{-bu} - e^{-bT}}{b}\right)^2 du\right)$$

$$= \exp\left(-\frac{r_t}{b}\left(1 - e^{-(T-t)b}\right) + \frac{a}{b^2}\left(1 - e^{-(T-t)b}\right) - (T-t)\frac{a}{b}\right)$$

$$\times \exp\left(\frac{\sigma^2}{2} \int_t^T e^{2bu} \left(\frac{e^{-bu} - e^{-bT}}{b}\right)^2 du\right)$$

$$= e^{A(T-t)+r_t C(T-t)}, \tag{4.21}$$

where $A(x)$ and $C(x)$ are the functions given by (4.19) and (4.20). $\square$

See Problem 4.8 for another way to calculate $P(t,T)$ in the Vasicek model.

In the Vasicek model

$$dr_t = (a - br_t)dt + \sigma dB_t,$$

the bond price takes the form

$$F(t, r_t) = P(t,T) = e^{A(T-t)+r_t C(T-t)},$$

where $A(\cdot)$ and $C(\cdot)$ are functions of time, cf. (4.21), and (4.15) yields

$$\frac{dP(t,T)}{P(t,T)} = r_t dt - \frac{\sigma}{b}\left(1 - e^{-(T-t)b}\right)dB_t, \tag{4.22}$$

since $F(t, x) = e^{A(T-t)+xC(T-t)}$. Note that more generally, all affine short rate models as defined in Relation (3.7), including the Vasicek model, will yield a bond pricing formula of the form

$$P(t,T) = e^{A(T-t)+r_t C(T-t)}, \qquad 0 \leqslant t \leqslant T.$$

Analytical solution of the Vasicek PDE

In order to solve the PDE (4.17) analytically, we may start by looking for a solution of the form

$$F(t, x) = e^{A(T-t)+xC(T-t)}, \tag{4.23}$$

where $A(\cdot)$ and $C(\cdot)$ are functions to be determined under the conditions $A(0) = 0$ and $C(0) = 0$. Substituting (4.23) into the PDE (4.10) with the Vasicek coefficients $\mu(t, x) = (a - bx)$ and $\sigma(t, x) = \sigma$ shows that

$$x e^{A(T-t)+xC(T-t)} = -(A'(T-t) + xC'(T-t))e^{A(T-t)+xC(T-t)}$$

$$+ (a - bx)C(T-t)e^{A(T-t)+xC(T-t)} + \frac{\sigma^2}{2}C^2(T-t)e^{A(T-t)+xC(T-t)},$$

i.e.

$$x = -A'(T-t) - xC'(T-t) + (a - bx)C(T-t) + \frac{\sigma^2}{2}C^2(T-t).$$

By identification of terms for $x = 0$ and $x \neq 0$, this yields the system of Riccati and linear differential equations

$$\begin{cases} A'(s) = aC(s) + \dfrac{\sigma^2}{2}C^2(s) \\ C'(s) = -1 - bC(s), \end{cases}$$

which can be solved to recover the above value of $P(t,T) = F(t, r_t)$ via

$$C(s) = -\frac{1}{b}\left(1 - e^{-bs}\right)$$

and

$$\begin{aligned} A(t) &= A(0) + \int_0^t A'(s)ds \\ &= \int_0^t \left(aC(s) + \frac{\sigma^2}{2}C^2(s)\right)ds \\ &= \int_0^t \left(\frac{a}{b}(1 - e^{-bs}) + \frac{\sigma^2}{2b^2}(1 - e^{-bs})^2\right)ds \\ &= \frac{a}{b}\int_0^t (1 - e^{-bs})ds + \frac{\sigma^2}{2b^2}\int_0^t (1 - e^{-bs})^2 ds \\ &= \frac{4ab - 3\sigma^2}{4b^3} + \frac{\sigma^2 - 2ab}{2b^2}t + \frac{\sigma^2 - ab}{b^3}e^{-bt} - \frac{\sigma^2}{4b^3}e^{-2bt}, \qquad t \geqslant 0. \end{aligned}$$

As a verification, we easily check that $C(x)$ and $A(x)$ given in (4.19) and (4.20) above do satisfy

$$bC(x) + 1 = -e^{-bx} = -C'(x),$$

and

$$\begin{aligned} aC(x) + \frac{\sigma^2 C^2(x)}{2} &= -\frac{a}{b}(1 - e^{-bx}) + \frac{\sigma^2}{2b^2}(1 - e^{-bx})^2 \\ &= \frac{\sigma^2 - 2ab}{2b^2} - \frac{\sigma^2 - ab}{b^2}e^{-bx} + \frac{\sigma^2}{2b^2}e^{-2bx} \\ &= A'(x), \qquad x \geqslant 0. \end{aligned}$$

See also Exercise 4.4 for a bond pricing formula in the CIR model. Finally, we consider the graphs of the functions A and C in Figure 4.1, with $\sigma = 1\%$, $r_0 = 5\%$, $b = 0.5$, $a = 5\%$, and maturity $T = 10$.

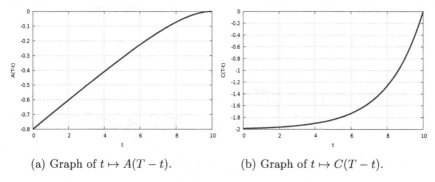

(a) Graph of $t \mapsto A(T - t)$.   (b) Graph of $t \mapsto C(T - t)$.

Fig. 4.1: Graph of the functions $t \mapsto A(T - t)$ and $t \mapsto C(T - t)$.

The solution of the pricing PDE, which can be useful for calibration purposes, is plotted in Figure 4.2.

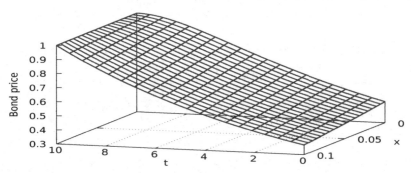

Fig. 4.2: Graph of the PDE solution $(x, t) \mapsto e^{A(T-t)+xC(T-t)}$ with $T = 10$.

4.4 Numerical Simulations

Vasicek bond price simulations

In this section we consider again the Vasicek model, in which the short rate process $(r_t)_{t \in \mathbb{R}_+}$ is solution to (3.1). The next Figure 4.3 compares the Monte Carlo and analytical estimation of Vasicek bond prices respectively obtained from (4.6) and (4.18).

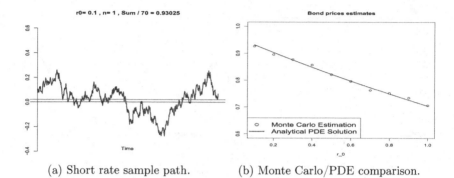

(a) Short rate sample path.   (b) Monte Carlo/PDE comparison.

Fig. 4.3: Comparison of Monte Carlo and analytical PDE solutions.

Given the Brownian path represented in Figure 4.4-(a), Figure 4.4-(b) presents the corresponding random simulation of $t \mapsto r_t$ in the Vasicek model with $r_0 = a/b = 5\%$, *i.e.* the reverting property of the process is with respect to its initial value $r_0 = 5\%$.

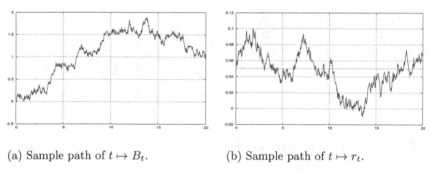

(a) Sample path of $t \mapsto B_t$.   (b) Sample path of $t \mapsto r_t$.

Fig. 4.4: Sample paths of Brownian motion and the Vasicek process.

Note that the interest rate in Figure 4.4-(b) becomes negative for a short period of time, which may nevertheless happen in practice.

Figure 4.5-(a) presents a random simulation of the zero-coupon bond price (4.18) in the Vasicek model with $\sigma = 10\%$, $r_0 = 2.96\%$, $b = 0.5$, and $a = 0.025$. The graph of the corresponding deterministic zero-coupon bond price with $r = r_0 = 2.96\%$ is also shown in Figure 4.5-(a).

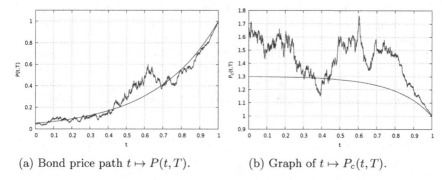

(a) Bond price path $t \mapsto P(t, T)$.   (b) Graph of $t \mapsto P_c(t, T)$.

Fig. 4.5: Zero coupon and 5% coupon bond price graphs.

Figure 4.5-(b) presents a random simulation of the coupon bond price (4.4) in the Vasicek model with $\sigma = 2\%$, $r_0 = 3.5\%$, $b = 0.5$, $a = 0.025$, and coupon rate $c = 5\%$. The graph of the corresponding deterministic coupon bond price (4.5) with $r = r_0 = 3.5\%$ is also shown in Figure 4.5-(b), see Figure 4.6 for a comparison with market data.

4.5 Bond Prices and Yield Data

The following zero coupon public bond price data was downloaded from EMMA at the Municipal Securities Rulemaking Board.

Dated Date: 06/12/1996 (June 12, 1996)
Maturity Date: 09/01/2016 (September 1st, 2016)
Interest Rate: 0.0 %
Principal Amount at Issuance: \$26,056,000
Initial Offering Price: 19.465

```
1  library(quantmod)
2  bondprice <- read.table("bond_data_R.txt",col.names = c("Date","HighPrice","
       LowPrice","HighYield","LowYield","Count","Amount"))
   head(bondprice)
4  time <- as.POSIXct(bondprice$Date, format = "%Y-%m-%d")
   price <- xts(x = bondprice$HighPrice, order.by = time)
6  yield <- xts(x = bondprice$HighYield, order.by = time)
   chartSeries(price,up.col="blue",theme="white")
8  chartSeries(yield,up.col="blue",theme="white")
```

	Date	HighPrice	LowPrice	HighYield	LowYield	Count	Amount
1	2016-01-13	99.082	98.982	1.666	1.501	2	20000
2	2015-12-29	99.183	99.183	1.250	1.250	1	10000
3	2015-12-21	97.952	97.952	3.014	3.014	1	10000
4	2015-12-17	99.141	98.550	2.123	1.251	5	610000
5	2015-12-07	98.770	98.770	1.714	1.714	2	10000
6	2015-12-04	98.363	98.118	2.628	2.280	2	10000

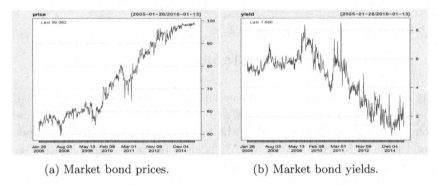

(a) Market bond prices.   (b) Market bond yields.

Fig. 4.6: Graphs of market bond prices and yields.

Figure 4.6-(b) plots the bond yields $y(t,T)$ estimated from the market bond prices $P(t,T)$ in Figure 4.6-(a) as

$$y(t,T) = -\frac{\log P(t,T)}{T-t}, \quad \text{or} \quad P(t,T) = e^{-(T-t)y(t,T)}, \quad 0 \leqslant t \leqslant T.$$

Exercises

Exercise 4.1. Let $(B_t)_{t\in\mathbb{R}_+}$ denote a standard Brownian motion started at 0 under the risk-neutral probability measure $\mathbb{P}^*$. We consider a short-term interest rate process $(r_t)_{t\in\mathbb{R}_+}$ in a Ho-Lee model with constant deterministic volatility, defined by

$$dr_t = \theta dt + \sigma dB_t,$$

where $a \in \mathbb{R}$ and $\sigma > 0$. Let $P(t,T)$ will denote the arbitrage price of a zero-coupon bond in this model:

$$P(t,T) = \mathbb{E}^*\left[e^{-\int_t^T r_s ds} \,\middle|\, \mathcal{F}_t\right], \qquad 0 \leqslant t \leqslant T. \tag{4.24}$$

(1) State the bond pricing PDE satisfied by the function $F(t,x)$ defined via

$$F(t,x) := \mathbb{E}^* \left[e^{- \int_t^T r_s ds} \,\middle|\, r_t = x \right], \qquad 0 \leqslant t \leqslant T.$$

(2) Compute the arbitrage price $F(t, r_t) = P(t,T)$ from its expression (4.24) as a conditional expectation.

Hint: One may use the *integration by parts* relation

$$\int_t^T B_s ds = T B_T - t B_t - \int_t^T s dB_s$$

$$= (T-t)B_t + (B_T - B_t)T - \int_t^T s dB_s$$

$$= (T-t)B_t + \int_t^T (T-s) dB_s,$$

and the Gaussian moment generating function $\mathbb{E}\left[e^{\lambda X} \right] = e^{\lambda^2 \eta^2 / 2}$ for $X \simeq \mathcal{N}(0, \eta^2)$.

(3) Check that the function $F(t,x)$ computed in Question (2) does satisfy the PDE derived in Question (1).

Exercise 4.2. Consider the Marsh and Rosenfeld (1983) interest rate model

$$dr_t = (\beta r_t^{\gamma-1} + \alpha r_t)dt + \sigma r_t^{\gamma/2} dB_t$$

where $\alpha \in \mathbb{R}$ and $\beta, \sigma, \gamma > 0$.

(1) Letting $R_t := r_t^{2-\gamma}$, $t \in \mathbb{R}_+$, find the stochastic differential equation satisfied by $(R_t)_{t \in \mathbb{R}_+}$.

(2) Knowing that the discounted bond price process is a martingale, derive the bond pricing PDE satisfied by the function $F(t,x)$ such that

$$F(t, r_t) = P(t,T) = \mathbb{E}^* \left[e^{- \int_t^T r_s ds} \,\middle|\, \mathcal{F}_t \right] = \mathbb{E}^* \left[e^{- \int_t^T r_s ds} \,\middle|\, r_t \right].$$

Exercise 4.3. (Exercise 3.5 continued). Write down the bond pricing PDE in the CIR model for the function $F(t,x)$ defined by

$$F(t,x) := \mathbb{E} \left[e^{- \int_t^T r_s ds} \,\middle|\, r_t = x \right],$$

and show that in case $\alpha = 0$ the corresponding bond price $P(t,T)$ equals

$$P(t,T) = e^{-B(T-t)r_t}, \qquad 0 \leqslant t \leqslant T,$$

where

$$B(s) = \frac{2(e^{\gamma s} - 1)}{2\gamma + (\beta + \gamma)(e^{\gamma s} - 1)}, \qquad s \geqslant 0,$$

with $\gamma := \sqrt{\beta^2 + 2\sigma^2}$.

Exercise 4.4. Consider the CIR process $(r_t)_{t \in \mathbb{R}_+}$ solution of

$$dr_t = -ar_t dt + \sigma\sqrt{r_t} dB_t,$$

where $a, \sigma > 0$ are constants $(B_t)_{t \in \mathbb{R}_+}$ is a standard Brownian motion started at 0.

(1) Write down the bond pricing PDE for the function $F(t, x)$ given by

$$F(t, x) := \mathbb{E}^*\left[e^{-\int_t^T r_s ds} \mid r_t = x \right], \qquad 0 \leqslant t \leqslant T.$$

   Hint: Use Itô calculus and the fact that the discounted bond price is a martingale.

(2) Show that the PDE of Question (1) admits a solution of the form $F(t, x) = e^{A(T-t)+xC(T-t)}$ where the functions $A(s)$ and $C(s)$ satisfy ordinary differential equations to be also written down together with the values of $A(0)$ and $C(0)$.

Exercise 4.5. Consider a zero-coupon bond with prices $P(1, 2) = 91.74\%$ and $P(0, 2) = 83.40\%$ at times $t = 0$ and $t = 1$.

(1) Compute the corresponding yields $y_{0,1}$, $y_{0,2}$ and $y_{1,2}$ at times $t = 0$ and $t = 1$.

(2) Assume that \$0.1 coupons are paid at times $t = 1$ and $t = 2$. Price the corresponding coupon bond at times $t = 0$ and $t = 1$ using the yields y_0 and y_1.

Exercise 4.6.   Black-Derman-Toy model. Consider a two-step interest rate model in which the short-term interest rate r_0 on $[0, 1]$ can turn into two possible values $r_1^u = r_0 e^{\mu \Delta t + \sigma \sqrt{\Delta t}}$ and $r_1^d = r_0 e^{\mu \Delta t - \sigma \sqrt{\Delta t}}$ on the time interval $[1, 2]$ with equal probabilities $1/2$ at time $\Delta t = 1$ year and volatility $\sigma = 22\%$ per year, and two zero coupon bonds with prices $P(0, 1)$ and $P(0, 2)$ at time $t = 0$.

(1) Write down the value of $P(1,2)$ using r_1^u and r_1^d.
(2) Write down the value of $P(0,2)$ using r_1^u, r_1^d and r_0.
(3) Estimate the value of r_0 from the market price $P(0,1) = 91.74$.
(4) Estimate the values of r_1^u and r_1^d from the market price $P(0,2) = 83.40$.

Exercise 4.7. Bond duration. Compute the duration

$$D_c(0,n) := -\frac{1+r}{P_c(0,n)}\frac{\partial}{\partial r}P_c(0,n)$$

of a discrete-time coupon bond priced as

$$P_c(0,n) = \frac{1}{(1+r)^n} + c\sum_{k=1}^{n}\frac{1}{(1+r)^k}$$

$$= \frac{1}{(1+r)^n} + \frac{c}{r}\left(1 - \frac{1}{(1+r)^n}\right),$$

where $r > 0$, and $c \geqslant 0$ denotes the coupon rate. What happens when n becomes large?

Problem 4.8. Consider the Vasicek stochastic differential equation

$$\begin{cases} dX_t = -bX_tdt + \sigma dB_t, & t > 0, \\[2mm] X_0 = 0, \end{cases} \tag{4.25}$$

where b and σ are positive parameters and $(B_t)_{t\in\mathbb{R}_+}$ is a standard Brownian motion under $\mathbb{P}^*$, generating the filtration $(\mathcal{F}_t)_{t\in\mathbb{R}_+}$. Let the short-term interest rate process $(r_t)_{t\in\mathbb{R}_+}$ be given by

$$r_t = r + X_t, \qquad t \geqslant 0,$$

where $r > 0$ is a given constant. Recall that from the Markov property, the arbitrage price

$$P(t,T) = \mathbb{E}^*\left[e^{-\int_t^T r_s ds}\,\middle|\,\mathcal{F}_t\right], \qquad 0 \leqslant t \leqslant T,$$

of a zero-coupon bond is a function $F(t,X_t) = P(t,T)$ of t and X_t.

(1) Using Itô's calculus, derive the PDE satisfied by the function $(t,x)\mapsto F(t,x)$.
(2) Solve the stochastic differential equation (4.25).
(3) Show that

$$\int_0^t X_s ds = \frac{\sigma}{b}\int_0^t \left(1 - e^{-(t-s)b}\right)dB_s, \qquad t > 0.$$

(4) Show that for all $0 \leqslant t \leqslant T$,

$$\int_t^T X_s ds = -\frac{\sigma}{b} \left(\int_0^t \left(e^{-(T-s)b} - e^{-(t-s)b} \right) dB_s + \int_t^T \left(e^{-(T-s)b} - 1 \right) dB_s \right).$$

(5) Show that

$$\mathbb{E}^* \left[\int_t^T X_s ds \,\middle|\, \mathcal{F}_t \right] = -\frac{\sigma}{b} \int_0^t \left(e^{-(T-s)b} - e^{-(t-s)b} \right) dB_s.$$

(6) Show that

$$\mathbb{E}^* \left[\int_t^T X_s ds \,\middle|\, \mathcal{F}_t \right] = \frac{X_t}{b} \left(1 - e^{-(T-t)b} \right).$$

(7) Show that

$$\mathrm{Var} \left[\int_t^T X_s ds \,\middle|\, \mathcal{F}_t \right] = \frac{\sigma^2}{b^2} \int_t^T \left(e^{-(T-s)b} - 1 \right)^2 ds.$$

(8) What is the distribution of $\int_t^T X_s ds$ given $\mathcal{F}_t$?

(9) Compute the arbitrage price $P(t,T)$ from its expression

$$P(t,T) = \mathbb{E}^* \left[e^{-\int_t^T r_s ds} \,\middle|\, \mathcal{F}_t \right], \qquad 0 \leqslant t \leqslant T,$$

as a conditional expectation, and show that

$$P(t,T) = e^{A(t,T)-(T-t)r+X_t C(t,T)},$$

where $C(t,T) = \left(e^{-(T-t)b} - 1 \right)/b$ and

$$A(t,T) = \frac{\sigma^2}{2b^2} \int_t^T \left(e^{-(T-s)b} - 1 \right)^2 ds.$$

(10) Check explicitly that the function $F(t,x) = e^{A(t,T)+(T-t)r+xC(t,T)}$ computed in Question (9) does solve the PDE derived in Question (1).

Chapter 5

Forward Rates and Swap Rates

Forward rates are interest rates used in Forward Rate Agreements (FRA) for financial transactions, such as loans, that can take place at a future date. In this chapter, we define the forward and instantaneous forward rates from absence of arbitrage arguments, based on the construction of zero-coupon bonds presented in Chapter 4. This is followed by the definition of LIBOR rates using linear compounding, the construction of swap rates, and examples of market data.

5.1 Forward Rates

Financial institutions often require the possibility to agree at a present time t for a loan to be delivered over a future period of time $[T, S]$ at a rate $r(t, T, S)$, $t \leqslant T \leqslant S$. This type of forward interest rate contract (or Forward Rate Agreement, FRA) gives to its holder the possibility to lock an interest rate denoted by $f(t, T, S)$ at present time t for a loan to be delivered over a future period of time $[T, S]$, with $t \leqslant T \leqslant S$.

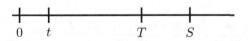

The rate $f(t, T, S)$ is called a forward interest rate. When $T = t$, the *spot* forward rate $f(t, t, T)$ coincides with the *yield*, see Relation (5.3) below. Figure 5.1 presents a typical yield curve on the LIBOR (London Interbank Offered Rate) market with $t = 07$ May 2003.

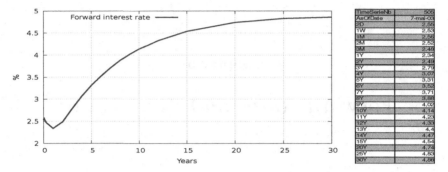

Fig. 5.1: Graph of market forward rates $S \mapsto f(t, t, S)$.

Long maturities usually correspond to higher rates as they carry an increased risk. The dip observed with short maturities can correspond to a lower motivation to lend/invest in the short-term. *Maturity transformation, i.e.,* the ability to transform short-term borrowing (debt with short maturities, such as deposits) into long term lending (credits with long maturities, such as loans), is among the roles of banks. Profitability is then dependent on the difference between long rates and short rates.

Forward rates from bond prices

Let us determine the arbitrage or "fair" value of the forward interest rate $f(t, T, S)$ by implementing the Forward Rate Agreement using the instruments available in the market, which are bonds priced at $P(t, T)$ for various maturity dates $T > t$.

The loan can be realized using the available instruments (here, bonds) on the market, by proceeding in two steps:

1) At time t, borrow the amount $P(t, S)$ by issuing (or short selling) one bond with maturity S, which means refunding \$1 at time S.

2) Since the money is only needed at time T, the rational investor will invest the amount $P(t, S)$ over the period $[t, T]$ by buying a (possibly fractional) quantity $P(t, S)/P(t, T)$ of a bond with maturity T priced $P(t, T)$ at time t. This will yield the amount

$$\$1 \times \frac{P(t, S)}{P(t, T)}$$

at time $T > 0$.

As a consequence, the investor will actually receive $P(t,S)/P(t,T)$ at time T, to refund \$1 at time S.

The corresponding forward rate $f(t,T,S)$ is then given by the relation

$$\frac{P(t,S)}{P(t,T)} e^{(S-T)f(t,T,S)} = \$1, \qquad 0 \leqslant t \leqslant T \leqslant S, \tag{5.1}$$

where we used exponential compounding, which leads to the following definition (5.2).

Definition 5.1. *The forward rate $f(t,T,S)$ at time t for a loan on $[T,S]$ is given by*

$$f(t,T,S) = \frac{\log P(t,T) - \log P(t,S)}{S-T}. \tag{5.2}$$

The *spot* forward rate $f(t,t,T)$ coincides with the *yield* $y(t,T)$ with

$$f(t,t,T) = y(t,T) = -\frac{\log P(t,T)}{T-t}, \quad \text{or} \quad P(t,T) = e^{-(T-t)f(t,t,T)},$$
$$\tag{5.3}$$
$$0 \leqslant t \leqslant T.$$

5.2 Instantaneous Forward Rates

Proposition 5.1. *The instantaneous forward rate $f(t,T) = f(t,T,T)$ is defined by taking the limit of $f(t,T,S)$ as $S \searrow T$, and satisfies*

$$f(t,T) := \lim_{S \searrow T} f(t,T,S) = -\frac{1}{P(t,T)} \frac{\partial}{\partial T} P(t,T). \tag{5.4}$$

Proof. We have

$$f(t,T) := \lim_{S \searrow T} f(t,T,S)$$

$$= -\lim_{S \searrow T} \frac{\log P(t,S) - \log P(t,T)}{S-T}$$

$$= -\lim_{\varepsilon \searrow 0} \frac{\log P(t,T+\varepsilon) - \log P(t,T)}{\varepsilon}$$

$$= -\frac{\partial}{\partial T} \log P(t,T)$$

$$= -\frac{1}{P(t,T)} \frac{\partial}{\partial T} P(t,T).$$

□

The above equation (5.4) can be viewed as a differential equation to be solved for $\log P(t,T)$ under the initial condition $P(T,T) = 1$, which yields the following proposition.

Proposition 5.2. *The bond price $P(t,T)$ can be recovered from the instantaneous forward rate $f(t,s)$, as*

$$P(t,T) = \exp\left(-\int_t^T f(t,s)ds \right), \qquad 0 \leqslant t \leqslant T. \tag{5.5}$$

Proof. We check that

$$
\begin{aligned}
\log P(t,T) &= \log P(t,T) - \log P(t,t) \\
&= \int_t^T \frac{\partial}{\partial s} \log P(t,s)ds \\
&= -\int_t^T f(t,s)ds, \qquad 0 \leqslant t \leqslant T.
\end{aligned}
$$

□

Proposition 5.2 also shows that

$$
\begin{aligned}
f(t,t,t) &= f(t,t) \\
&= \frac{\partial}{\partial T} \int_t^T f(t,s)ds_{|T=t} = -\frac{\partial}{\partial T} \log P(t,T)_{|T=t} \\
&= -\frac{1}{P(t,T)}_{|T=t} \frac{\partial}{\partial T} P(t,T)_{|T=t} = -\frac{1}{P(T,T)} \frac{\partial}{\partial T} \mathbb{E}^* \left[e^{-\int_t^T r_s ds} \,\middle|\, \mathcal{F}_t \right]_{|T=t} \\
&= \mathbb{E}^* \left[r_T e^{-\int_t^T r_s ds} \,\middle|\, \mathcal{F}_t \right]_{|T=t} = \mathbb{E}^*[r_t \mid \mathcal{F}_t] \\
&= r_t,
\end{aligned}
$$

i.e. the short rate r_t can be recovered from the instantaneous forward rate as

$$r_t = f(t,t) = \lim_{T \searrow t} f(t,T).$$

As a consequence of (5.1) and (5.5) the forward rate $f(t,T,S)$ can be recovered from (5.2) and the instantaneous forward rate $f(t,s)$, as:

$$f(t,T,S) = \frac{\log P(t,T) - \log P(t,S)}{S - T}$$

$$= -\frac{1}{S-T}\left(\int_t^T f(t,s)ds - \int_t^S f(t,s)ds\right)$$

$$= \frac{1}{S-T}\int_T^S f(t,s)ds, \qquad 0 \leqslant t \leqslant T < S.$$

Similarly, as a consequence of (5.3) and (5.5) we have the next proposition.

Proposition 5.3. *The* spot *forward rate, or* yield, $f(t,t,T)$ *can be written in terms of bond prices as*

$$f(t,t,T) = -\frac{\log P(t,T)}{T-t} = \frac{1}{T-t}\int_t^T f(t,s)ds, \qquad 0 \leqslant t < T.$$
$$\tag{5.6}$$

Differentiation with respect to T of the above relation shows that the yield $f(t,t,T)$ and the instantaneous forward rate $f(t,s)$ are linked by the relation

$$\frac{\partial f}{\partial T}(t,t,T) = -\frac{1}{(T-t)^2}\int_t^T f(t,s)ds + \frac{f(t,T)}{T-t}, \qquad 0 \leqslant t < T,$$

from which it follows that

$$f(t,T) = \frac{1}{T-t}\int_t^T f(t,s)ds + (T-t)\frac{\partial f}{\partial T}(t,t,T)$$

$$= f(t,t,T) + (T-t)\frac{\partial f}{\partial T}(t,t,T), \qquad 0 \leqslant t < T.$$

Vasicek forward rates

In this section we consider the Vasicek model, in which the short rate process is the solution (3.2) of (3.1) as illustrated in Figure 4.4b.

In the Vasicek model, the forward rate is given by

$$f(t,T,S) = -\frac{\log P(t,S) - \log P(t,T)}{S-T}$$

$$= -\frac{r_t(C(S-t) - C(T-t)) + A(S-t) - A(T-t))}{S-T}$$

$$= -\frac{\sigma^2 - 2ab}{2b^2} - \left(\frac{r_t}{b} + \frac{\sigma^2 - ab}{b^3}\right)\frac{e^{-(S-t)b} - e^{-(T-t)b}}{S-T}$$

$$+\sigma^2 \frac{e^{-2(S-t)b} - e^{-2(T-t)b}}{4(S-T)b^3},$$

and the spot forward rate, or yield, satisfies

$$f(t,t,T) = -\frac{\log P(t,T)}{T-t}$$

$$= -\frac{r_t C(T-t) + A(T-t)}{T-t}$$

$$= -\frac{\sigma^2 - 2ab}{2b^2} + \left(\frac{r_t}{b} + \frac{\sigma^2 - ab}{b^3}\right)\frac{1 - e^{-(T-t)b}}{T-t} - \sigma^2 \frac{1 - e^{-2(T-t)b}}{4(T-t)b^3},$$

with the mean

$$\mathbb{E}^*[f(t,t,T)] = -\frac{\sigma^2 - 2ab}{2b^2}$$

$$+\frac{1}{T-t}\left(\left(\frac{\mathbb{E}^*[r_t]}{b} + \frac{\sigma^2 - ab}{b^3}\right)\left(1 - e^{-(T-t)b}\right) - \frac{\sigma^2}{4b^3}\left(1 - e^{-2(T-t)b}\right)\right)$$

$$= -\frac{\sigma^2 - 2ab}{2b^2} + \left(\frac{r_0}{b}e^{-bt} + \frac{a}{b^2}(1 - e^{-bt}) + \frac{\sigma^2 - ab}{b^3}\right)\frac{1 - e^{-(T-t)b}}{T-t}$$

$$-\sigma^2 \frac{1 - e^{-2(T-t)b}}{4(T-t)b^3}, \qquad 0 \leqslant t \leqslant T.$$

In the Vasicek model, the forward rate $t \mapsto f(t,T,S)$ can be represented as in Figure 5.2, with $a = 0.06$, $b = 0.1$, $\sigma = 10\%$ and $r_0 = \%1$:

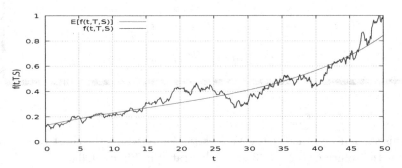

Fig. 5.2: Sample forward rate process $t \mapsto f(t,T,S)$.

We note that the Vasicek forward rate curve $t \mapsto f(t,T,S)$ appears flat for small values of t, *i.e.* longer rates are more stable, while shorter rates show higher volatility or risk. This modeling issue will be reconsidered in the framework of multifactor models in Chapter 6. Similar features can be observed in Figure 5.3 for the instantaneous forward rate given by

$$f(t,T) := -\frac{\partial}{\partial T}\log P(t,T) \tag{5.7}$$

$$= r_t \, e^{-(T-t)b} + \frac{a}{b} \left(1 - e^{-(T-t)b}\right) - \frac{\sigma^2}{2b^2} \left(1 - e^{-(T-t)b}\right)^2,$$

from which the relation $\lim_{T \searrow t} f(t, T) = r_t$ can be easily recovered. We can also evaluate the mean

$$\mathbb{E}^*[f(t,T)] = \mathbb{E}^*[r_t] \, e^{-(T-t)b} + \frac{a}{b}\left(1 - e^{-(T-t)b}\right) - \frac{\sigma^2}{2b^2}\left(1 - e^{-(T-t)b}\right)^2$$

$$= r_0 \, e^{-bT} + \frac{a}{b}\left(1 - e^{-bT}\right) - \frac{\sigma^2}{2b^2}\left(1 - e^{-(T-t)b}\right)^2.$$

The instantaneous forward rate $t \mapsto f(t, T)$ can be represented as in Figure 5.3, with $a = 0.06$, $b = 0.1$, $\sigma = 10\%$ and $r_0 = \%1$.

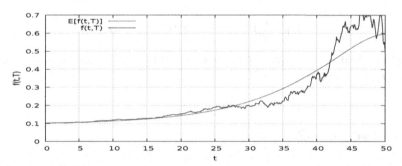

Fig. 5.3: Sample instantaneous forward rate process $t \mapsto f(t, T)$.

Similarly, the instantaneous forward rate $T \mapsto f(t, T)$ can be represented as in Figure 5.4, with here $t = 0$ and $b/a = 5/3 > r_0$.

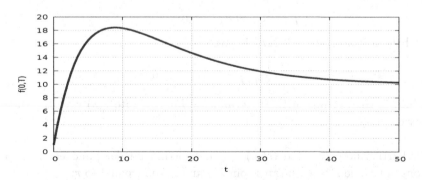

Fig. 5.4: Instantaneous forward rate curve $T \mapsto f(0, T)$.

Short rates

The underlying short-term interest rate process $(r_t)_{t \in \mathbb{R}_+}$ can be recovered from its relation to the bond price

$$P(t,T) = \mathbb{E}^* \left[e^{-\int_t^T r_s ds} \,\middle|\, r_t \right].$$

Indeed, we have

$$\begin{aligned}
\frac{\partial}{\partial T} P(t,T) &= \frac{\partial}{\partial T} \mathbb{E}^* \left[e^{-\int_t^T r_s ds} \,\middle|\, r_t \right] \\
&= \mathbb{E}^* \left[\frac{\partial}{\partial T} e^{-\int_t^T r_s ds} \,\middle|\, r_t \right] \\
&= - \mathbb{E}^* \left[r_T e^{-\int_t^T r_s ds} \,\middle|\, r_t \right],
\end{aligned}$$

hence

$$\lim_{T \searrow t} \frac{\partial}{\partial T} P(t,T) = - \mathbb{E}^* [r_t | r_t] = -r_t,$$

and the limit $\lim_{T \searrow t} f(t,T)$ of instantaneous forward rates equals the short rate r_t, *i.e.* we have

$$\lim_{T \searrow t} f(t,T) = - \lim_{T \searrow t} \frac{1}{P(t,T)} \frac{\partial}{\partial T} P(t,T) = r_t,$$

since $\lim_{T \searrow t} P(t,T) = 1$.

When the short rate $(r(s))_{s \in \mathbb{R}_+}$ is a deterministic function of time, we have

$$P(t,T) = \exp\left(-\int_t^T f(t,s) ds \right) = \exp\left(-\int_t^T r(s) ds \right), \quad 0 \leqslant t \leqslant T,$$

hence the instantaneous forward rate $f(t,T)$ is also deterministic and independent of t:

$$f(t,T) = r(T), \quad 0 \leqslant t \leqslant T,$$

and the forward rate $f(t,T,S)$ is given by

$$f(t,T,S) = \frac{1}{S-T} \int_T^S r(s) ds, \quad 0 \leqslant t \leqslant T < S,$$

which is the average of the deterministic interest rate $r(s)$ over the time period $[T,S]$.

Furthermore, in case $(r(s))_{s \in \mathbb{R}_+}$ is time-independent and equal to a constant value $r > 0$, all rates coincide and become equal to r:

$$r(s) = f(t,s) = f(t,T,S) = r, \quad 0 \leqslant t \leqslant T \leqslant s \leqslant S.$$

5.3 LIBOR Rates

Recall that the forward rate $f(t, T, S)$, $0 \leqslant t \leqslant T < S$, is defined using exponential compounding, from the relation

$$f(t, T, S) = -\frac{\log P(t, S) - \log P(t, T)}{S - T}. \tag{5.8}$$

In order to compute swaption prices one prefers to use forward rates as defined on the London InterBank Offered Rates (LIBOR) market instead of the standard forward rates given by (5.8). Other types of LIBOR rates include EURIBOR (European Interbank Offered Rates), HIBOR (Hong Kong Interbank Offered Rates), SHIBOR (Shanghai Interbank Offered Rates), SIBOR (Singapore Interbank Offered Rates), TIBOR (Tokyo Interbank Offered Rates), etc. LIBOR rates are to be replaced by alternatives, including the Secured Overnight Financing Rate (SOFR), by the end of year 2021. In the sequel, the term LIBOR refers to the linear or simple compounding used in the next definition.

Just as in Section 5.1, a forward rate agreement at time t on the LIBOR market also gives its holder an interest rate $L(t, T, S)$ over the future time period $[T, S]$. However, instead of using exponential compounding of rates, the forward LIBOR rate $L(t, T, S)$ for a loan on $[T, S]$ is defined using linear compounding, *i.e.* by replacing (5.8) with the relation

$$1 + (S - T)L(t, T, S) = \frac{P(t, T)}{P(t, S)}, \qquad 0 \leqslant t \leqslant T.$$

Equivalently, we have

$$P(t, T) - P(t, S) - P(t, S)(S - T)L(t, T, S) = 0,$$

$0 \leqslant t \leqslant T < S$, which yields the following definition.

Definition 5.2. *The forward LIBOR rate $L(t, T, S)$ at time t for a loan on $[T, S]$ is given by*

$$L(t, T, S) = \frac{1}{S - T}\left(\frac{P(t, T)}{P(t, S)} - 1\right), \qquad 0 \leqslant t \leqslant T < S. \tag{5.9}$$

Given a sequence $\{T_1 < T_2 < \cdots < T_n\}$ of maturity dates arranged according to a *tenor structure*, we note that if $1 \leqslant i < j \leqslant n$, we have

$$P(t, T_j) = P(t, T_i) \prod_{k=i}^{j-1} \frac{1}{1 + (T_{k+1} - T_k)L(t, T_k, T_{k+1})}, \qquad 0 \leqslant t \leqslant T_i,$$

and if $1 \leqslant j \leqslant i \leqslant n$,

$$P(t,T_j) = P(t,T_i) \prod_{k=j}^{i-1} (1 + (T_{k+1} - T_k)L(t,T_k,T_{k+1})), \quad 0 \leqslant t \leqslant T_j.$$

Relation (5.9) above yields the same formula for the (LIBOR) instantaneous forward rate as in (5.4), *i.e.*

$$\begin{aligned}
L(t,T) :&= \lim_{S \searrow T} L(t,T,S) \\
&= \lim_{S \searrow T} \frac{P(t,T) - P(t,S)}{(S-T)P(t,S)} \\
&= \lim_{\varepsilon \searrow 0} \frac{P(t,T) - P(t,T+\varepsilon)}{\varepsilon P(t,T+\varepsilon)} \\
&= \frac{1}{P(t,T)} \lim_{\varepsilon \searrow 0} \frac{P(t,T) - P(t,T+\varepsilon)}{\varepsilon} \\
&= -\frac{1}{P(t,T)} \frac{\partial P(t,T)}{\partial T} \\
&= -\frac{\partial}{\partial T} \log P(t,T) \\
&= f(t,T).
\end{aligned}$$

The instantaneous forward rate $f(t,T)$ can also be recovered from the forward instantaneous LIBOR rate

$$L(t,T) = \lim_{S \searrow T} L(t,T,S),$$

as

$$\begin{aligned}
f(t,T) &= -\frac{\partial}{\partial T} \log P(t,T) \\
&= -\frac{1}{P(t,T)} \frac{\partial P}{\partial T}(t,T) \\
&= -\frac{1}{P(t,T)} \lim_{S \searrow T} \frac{P(t,S) - P(t,T)}{S-T} \\
&= -\lim_{S \searrow T} \frac{P(t,S) - P(t,T)}{(S-T)P(t,S)} \\
&= \lim_{S \searrow T} L(t,T,S) \\
&= L(t,T,T) \\
&= L(t,T).
\end{aligned}$$

The short-term interest rate thus satisfies

$$r_t = L(t,t), \qquad t \geqslant 0,$$

thus the LIBOR and standard instantaneous forward rates coincide.

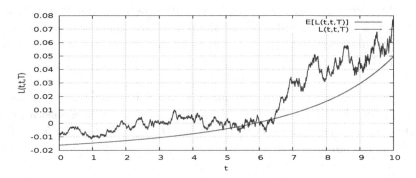

Fig. 5.5: Graph of LIBOR spot rates $t \mapsto L(t, t, T)$.

Figure 5.5 presents a simulation of the simply compounded spot rates

$$L(t, t, T) = \frac{1}{T - t}\left(\frac{1}{P(t, T)} - 1\right) = \frac{1}{T - t}\left(e^{-A(T-t) - r_t C(T-t)} - 1\right)$$

computed in the Vasicek model with $a = 0.025$, $b = 0.5$ and $\sigma = 20\%$, with $\lim_{t \to T} L(t, t, T) = r_T$.

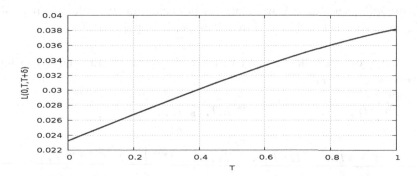

Fig. 5.6: Graph of the forward curve $T \mapsto L(0, T, T + \delta)$.

The forward curve $T \mapsto L(0, T, T + \delta)$ is plotted in Figure 5.6 for $t = 0$, also using bond prices computed in the Vasicek model.

5.4   Swap Rates

The first interest rate swap occurred in 1981 between the World Bank, which was interested in borrowing Deutsche marks and Swiss francs, and IBM, who already held large amounts of those currencies but needed to borrow U.S. dollars.

The vanilla interest rate swap makes it possible to exchange a sequence of variable LIBOR rates $L(T, T_k, T_{k+1})$, $k = 1, 2, \ldots, n - 1$, against a fixed rate κ over a succession of time intervals $[T_1, T_2), [T_2, T_3), \ldots, [T_{n-1}, T_n]$ defining a *tenor structure*.

Making the agreement fair results into an exchange of cash flows

$$\underbrace{(T_{k+1} - T_k)L(T, T_k, T_{k+1})}_{\text{floating leg}} - \underbrace{(T_{k+1} - T_k)\kappa}_{\text{fixed leg}}, \tag{5.10}$$

at the dates $T_2, T_3, \ldots, T_n$ between the two parties, generating a cumulative discounted cash flow

$$\sum_{k=1}^{n-1} e^{-\int_T^{T_{k+1}} r_s ds}(T_{k+1} - T_k)(L(T, T_k, T_{k+1}) - \kappa),$$

at time $T = T_0$, in which we used simple (or linear) interest rate compounding. This corresponds to a *payer swap* in which the swap holder receives the *floating leg* and pays the *fixed leg* κ, whereas the holder of a *seller swap* receives the *fixed leg* κ and pays the *floating leg*. In practice, the cash flow (5.10) will be multiplied by a notional principal amount, which is part of the swap contract and quoted in dollars per base point (bp).

The above cash flow is used to make the contract fair, and it can be priced *at time T* as

$$\mathbb{E}^* \left[\sum_{k=1}^{n-1} (T_{k+1} - T_k) e^{-\int_T^{T_{k+1}} r_s ds}(L(T, T_k, T_{k+1}) - \kappa) \,\Big|\, \mathcal{F}_T \right]$$

$$= \sum_{k=1}^{n-1} (T_{k+1} - T_k)(L(T, T_k, T_{k+1}) - \kappa) \mathbb{E}^* \left[e^{-\int_T^{T_{k+1}} r_s ds} \,\Big|\, \mathcal{F}_T \right]$$

$$= \sum_{k=1}^{n-1} (T_{k+1} - T_k)P(T, T_{k+1})\big(L(T, T_k, T_{k+1}) - \kappa\big). \tag{5.11}$$

The swap rate $S(T, T_1, T_n)$ is by definition the value of the rate κ that makes the contract fair by letting the above cash flow $\mathcal{C}(T)$ vanish.

Definition 5.3. *The LIBOR swap rate $S(T, T_1, T_n)$ is the value of the break-even rate κ that makes the contract fair by making the cash flow* (5.11) *vanish, i.e.*

$$\sum_{k=1}^{n-1} (T_{k+1} - T_k) P(T, T_{k+1}) \big(L(T, T_k, T_{k+1}) - \kappa \big) = 0. \tag{5.12}$$

The next Proposition 5.4 makes use of the annuity numeraire

$$P(T, T_1, T_n) := \mathbb{E}^* \left[\sum_{k=1}^{n-1} (T_{k+1} - T_k) e^{-\int_T^{T_{k+1}} r_s ds} \,\middle|\, \mathcal{F}_T \right] \tag{5.13}$$

$$= \sum_{k=1}^{n-1} (T_{k+1} - T_k) \, \mathbb{E}^* \left[e^{-\int_T^{T_{k+1}} r_s ds} \,\middle|\, \mathcal{F}_T \right]$$

$$= \sum_{k=1}^{n-1} (T_{k+1} - T_k) P(T, T_{k+1}), \qquad 0 \leqslant T \leqslant T_2,$$

which represents the present value at time T of future \$1 receipts at times $T_2, \ldots, T_n$, weighted by the lengths $T_{k+1} - T_k$ of the time intervals $(T_k, T_{k+1}]$, $k = 1, 2, \ldots, n-1$.

The time intervals $(T_{k+1} - T_k)_{k=1,2,\ldots,n-1}$ in the definition (5.13) of the annuity numeraire can be replaced by coupon payments $(c_{k+1})_{k=1,2,\ldots,n-1}$ occurring at times $(T_{k+1})_{k=1,2,\ldots,n-1}$, in which case the annuity numeraire becomes

$$P(T, T_1, T_n) := \mathbb{E}^* \left[\sum_{k=1}^{n-1} c_{k+1} e^{-\int_T^{T_{k+1}} r_s ds} \,\middle|\, \mathcal{F}_T \right]$$

$$= \sum_{k=1}^{n-1} c_{k+1} \, \mathbb{E}^* \left[e^{-\int_T^{T_{k+1}} r_s ds} \,\middle|\, \mathcal{F}_T \right]$$

$$= \sum_{k=1}^{n-1} c_{k+1} P(T, T_{k+1}), \qquad 0 \leqslant T \leqslant T_1, \tag{5.14}$$

which represents the value at time T of the future coupon payments discounted according to the bond prices $(P(T, T_{k+1}))_{k=1,2,\ldots,n-1}$. This expression can also be used to define *amortizing swaps*, for which the value of the notional decreases over time, or *accreting swaps*, for which the value of the notional increases over time.

The *swap rate* $S(t, T_1, T_n)$ is defined by solving Relation (5.12) for the forward rate $S(t, T_k, T_{k+1})$, *i.e.*

$$\sum_{k=1}^{n-1} (T_{k+1} - T_k) P(T, T_{k+1}) \big(L(T, T_k, T_{k+1}) - S(T, T_1, T_n) \big) = 0. \quad (5.15)$$

Proposition 5.4. *The LIBOR swap rate $S(T, T_1, T_n)$ is given by*

$$S(T, T_1, T_n) = \frac{1}{P(T, T_1, T_n)} \sum_{k=1}^{n-1} (T_{k+1} - T_k) P(T, T_{k+1}) L(T, T_k, T_{k+1}).$$

$$(5.16)$$

Proof.   By definition, $S(T, T_1, T_n)$ is the (fixed) break-even rate over $[T_1, T_n]$ that will be agreed in exchange for the family of forward rates $L(T, T_k, T_{k+1})$, $k = 1, 2, \ldots, n-1$, and it solves (5.15), *i.e.* we have

$$\sum_{k=1}^{n-1} (T_{k+1} - T_k) P(T, T_{k+1}) L(T, T_k, T_{k+1}) - P(T, T_1, T_n) S(T, T_1, T_n)$$

$$= \sum_{k=1}^{n-1} (T_{k+1} - T_k) P(T, T_{k+1}) L(T, T_k, T_{k+1})$$

$$- S(T, T_1, T_n) \sum_{k=1}^{n-1} (T_{k+1} - T_k) P(T, T_{k+1})$$

$$= \sum_{k=1}^{n-1} (T_{k+1} - T_k) P(T, T_{k+1}) L(T, T_k, T_{k+1}) - S(T, T_1, T_n) P(T, T_1, T_n)$$

$$= 0,$$

which shows (5.16) by solving the above equation for $S(T, T_1, T_n)$. $\square$

When $j = i+1$, the swap rate $S(t, T_i, T_{i+1})$ coincides with the forward rate $L(t, T_i, T_{i+1})$, as from (5.14) we have

$$S(t, T_i, T_{i+1}) = \frac{P(t, T_i) - P(t, T_{i+1})}{P(t, T_i, T_{i+1})} \quad (5.17)$$

$$= \frac{P(t, T_i) - P(t, T_{i+1})}{(T_{i+1} - T_i) P(t, T_{i+1})}$$

$$= L(t, T_i, T_{i+1}), \qquad 0 \leqslant t \leqslant T_i,$$

i.e. $S(t, T_i, T_{i+1}) = L(t, T_i, T_{i+1})$, $i = 1, 2, \ldots, n-1$, and in this case the LIBOR rate and the LIBOR swap rate coincide.

Using the Definition 5.9 of LIBOR rates, we obtain the next corollary.

Corollary 5.1. *The LIBOR swap rate $S(t, T_1, T_n)$ is given by*

$$S(t, T_1, T_n) = \frac{P(t, T_1) - P(t, T_n)}{P(t, T_1, T_n)}, \qquad 0 \leqslant t \leqslant T_1. \qquad (5.18)$$

Proof. By (5.16), (5.9) and a telescoping summation argument we have

$$S(t, T_1, T_n) = \frac{1}{P(t, T_1, T_n)} \sum_{k=1}^{n-1} (T_{k+1} - T_k) P(t, T_{k+1}) L(t, T_k, T_{k+1})$$

$$= \frac{1}{P(t, T_1, T_n)} \sum_{k=1}^{n-1} P(t, T_{k+1}) \left(\frac{P(t, T_k)}{P(t, T_{k+1})} - 1 \right)$$

$$= \frac{1}{P(t, T_1, T_n)} \sum_{k=1}^{n-1} (P(t, T_k) - P(t, T_{k+1}))$$

$$= \frac{P(t, T_1) - P(t, T_n)}{P(t, T_1, T_n)}.$$

$\square$

By (5.18), the bond prices $P(t, T_1)$ can be recovered from the values of the forward swap rates $S(t, T_1, T_n)$.

Clearly, a simple expression for the swap rate such as that of Corollary 5.1 cannot be obtained using the standard (*i.e.* non-LIBOR) rates defined in (5.8). Similarly, it will not be available for amortizing or accreting swaps because the telescoping summation argument does not apply to the expression (5.14) of the annuity numeraire.

The forward swap rate $S(t, T_i, T_j)$ also satisfies

$$P(t, T_i) - P(t, T_j) - S(t, T_i, T_j) \sum_{k=i}^{j-1} (T_{k+1} - T_k) P(t, T_{k+1}) = 0,$$

$0 \leqslant t \leqslant T_i$, $1 \leqslant i < j \leqslant n$, and the bond prices $P(t, T_i)$ can be recovered from the forward swap rates $S(t, T_i, T_j)$ using the relations

$$P(t, T_{i+1}) = \frac{P(t, T_i)}{1 + (T_{i+1} - T_i) S(t, T_i, T_{i+1})},$$

and

$$P(t, T_j) = \frac{P(t, T_i) - S(t, T_i, T_j) \sum_{k=i}^{j-2} (T_k - T_{k-1}) P(t, T_{k+1})}{1 + (T_j - T_{j-1}) S(t, T_i, T_j)},$$

$0 \leqslant t \leqslant T_i,\, 1 \leqslant i < j \leqslant n.$

5.5 Yield Curve Data

We refer to Chapter III-12 of Charpentier (2014) on the R package "Yield-Curve" Guirreri (2015) for the following R code, and for further details on yield curve and interest rate modeling using R.

```
install.packages("YieldCurve");require(YieldCurve)
data(FedYieldCurve);first(FedYieldCurve,'3 month');last(FedYieldCurve,'3 month')
mat.Fed=c(0.25,0.5,1,2,3,5,7,10);n=50
plot(mat.Fed, FedYieldCurve[n,], type="o",xlab="Maturities structure in years",
     ylab="Interest rates values")
title(main=paste("Federal Reserve yield curve observed at",time(FedYieldCurve[n],
     sep=" ")));grid()
```

The next Figure 5.7 is plotted using an R code adapted from the quant-mod.com website.

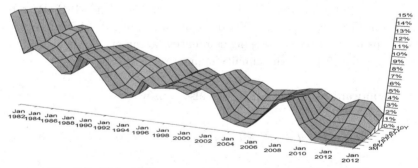

Fig. 5.7: Federal Reserve yield curves from 1982 to 2012.

European Central Bank (ECB) data can be similarly obtained by the next R code.

```
1  data(ECBYieldCurve);first(ECBYieldCurve,'3 month');last(ECBYieldCurve,'3 month')
   mat.ECB<-c(3/12,0.5,1,2,3,4,5,6,7,8,9,10,11,12,13,14,15,16,17,18,19,20,21,22,23,
       24,25,26, 27,28,29,30)
3  for (n in 200:400) {
   plot(mat.ECB, ECBYieldCurve[n,], type="o",xlab="Maturities structure in years",
       ylab="Interest rates values",ylim=c(3.1,5.1),col="blue")
5  title(main=paste("European Central Bank yield curve observed at",time(
       ECBYieldCurve[n], sep=" ")));grid();Sys.sleep(0.5)}
```

The next Figure 5.8 presents the output of the above R code.

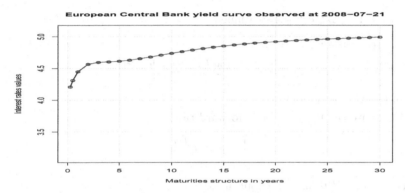

Fig. 5.8: European Central Bank yield curves.

Yield curve inversion

The next Figure 5.9 illustrates Federal Reserve (FED) yield curve inversions occurring in February and August 2019.

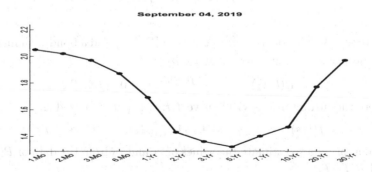

Fig. 5.9: August 2019 FED yield curve inversion.

Increasing yield curves are typical of economic expansion phases. Decreasing yield curves can occur when central banks attempt to limit inflation by tightening interest rates, such as in the case of an economic recession. In this case, uncertainty triggers increased investment in long bonds whose rates tend to drop as a consequence, while reluctance to lend in the short term can lead to higher short rates.

Exercises

Exercise 5.1. (Exercise 4.1 continued).

(1) Compute the forward rate

$$f(t, T, S) = -\frac{\log P(t, S) - \log P(t, T)}{S - T}$$

in the Ho-Lee model (3.8).

(2) Compute the instantaneous forward rate

$$f(t, T) = \lim_{S \searrow T} f(t, T, S).$$

Exercise 5.2. (Problem 4.8 continued).

(1) Compute the forward rate

$$f(t, T, S) = -\frac{\log P(t, S) - \log P(t, T)}{S - T}$$

in the Vasicek model (4.25).

(2) Compute the instantaneous forward rate

$$f(t, T) = \lim_{S \searrow T} f(t, T, S).$$

Exercise 5.3. Consider a tenor structure $\{T_1, T_2\}$ and a bond with maturity T_2 and price given at time $t \in [0, T_2]$ by

$$P(t, T_2) = e^{-\int_t^{T_2} f(t,s)ds}, \qquad 0 \leqslant t \leqslant T_2,$$

where the instantaneous yield curve $f(t, s)$ is parametrized as

$$f(t, s) = r_1 \mathbb{1}_{[0, T_1]}(s) + r_2 \mathbb{1}_{[T_1, T_2]}(s), \qquad t \leqslant s \leqslant T_2.$$

Find a formula to estimate the values of r_1 and r_2 from the data of $P(0, T_2)$ and $P(T_1, T_2)$.

Exercise 5.4. (Exercise 4.1 continued).

(1) Compute the forward rate $f(t, T, S)$ in this model.

In the next questions we take $a = 0$.

(2) Compute the instantaneous forward rate $f(t, T)$ in this model.
(3) Derive the stochastic equation satisfied by the instantaneous forward rate $f(t, T)$.

Exercise 5.5. Bridge model. (Exercise 1.8 continued). Assume that the price $P(t, T)$ of a zero coupon bond is modeled as

$$P(t, T) = e^{-\mu(T-t)+X_t^T}, \qquad 0 \leqslant t \leqslant T,$$

where $\mu > 0$.

(1) Show that the terminal condition $P(T, T) = 1$ is satisfied.
(2) Compute the forward rate

$$f(t, T, S) = -\frac{1}{S-T}(\log P(t, S) - \log P(t, T)).$$

(3) Compute the instantaneous forward rate

$$f(t, T) = -\lim_{S \searrow T} \frac{1}{S-T}(\log P(t, S) - \log P(t, T)).$$

(4) Show that the limit $\lim_{T \searrow t} f(t, T)$ does not exist in $L^2(\Omega)$.
(5) Show that $P(t, T)$ satisfies the stochastic differential equation

$$\frac{dP(t, T)}{P(t, T)} = \sigma dB_t + \frac{\sigma^2}{2} dt - \frac{\log P(t, T)}{T-t} dt, \qquad 0 \leqslant t < T.$$

(6) Rewrite the equation of Question (5) as

$$\frac{dP(t, T)}{P(t, T)} = \sigma dB_t + r_t^T dt, \qquad 0 \leqslant t \leqslant T,$$

where $\left(r_t^T\right)_{t \in [0,T]}$ is a process to be determined.

(7) Show that we have the bond price expression

$$P(t, T) = \mathbb{E}^* \left[e^{-\int_t^T r_s^T ds} \,\middle|\, \mathcal{F}_t \right], \qquad 0 \leqslant t \leqslant T.$$

Exercise 5.6. Consider a short rate process $(r_t)_{t \in \mathbb{R}_+}$ of the form $r_t = h(t) + X_t$, where $h(t)$ is a deterministic function and $(X_t)_{\mathbb{R}_+}$ is a Vasicek process started at $X_0 = 0$.

(1) Compute the price $P(0,T)$ at time $t = 0$ of a bond with maturity T, using $h(t)$ and the function $A(T)$ defined in (4.20) for the pricing of Vasicek bonds.
(2) Show how the function $h(t)$ can be estimated from the market data of the initial instantaneous forward rate curve $f(0,t)$.

Exercise 5.7.

(1) Given two LIBOR spot rates $L(t,t,T)$ and $L(t,t,S)$, compute the corresponding LIBOR forward rate $L(t,T,S)$.
(2) Assuming that $L(t,t,T) = 2\%$, $L(t,t,S) = 2.5\%$ and $t = 0$, $T = 1$, $S = 2T = 2$, would you buy a LIBOR forward contract over $[T,2T]$ with rate $L(0,T,2T)$ if $L(T,T,2T)$ remained at $L(T,T,2T) = L(0,0,T) = 2\%$?

Exercise 5.8.   Consider a yield curve $(f(t,t,T))_{0\leqslant t\leqslant T}$ and a bond paying coupons $c_1, c_2, \ldots, c_n$ at times $T_1, T_2, \ldots, T_n$ until maturity T_n, and priced as

$$P(t,T_n) = \sum_{k=1}^{n} c_k e^{-(T_k-t)f(t,t,T_k)}, \qquad 0 \leqslant t \leqslant T_1,$$

where c_n is inclusive of the last coupon payment and the nominal \$1 value of the bond. Let $\tilde{f}(t,t,T_n)$ denote the *compounded yield to maturity* defined by equating

$$P(t,T_n) = \sum_{k=1}^{n} c_k e^{-(T_k-t)\tilde{f}(t,t,T_n)}, \qquad 0 \leqslant t \leqslant T_1, \qquad (5.19)$$

i.e. $\tilde{f}(t,t,T_n)$ solves the equation

$$F\big(t, \tilde{f}(t,t,T_n)\big) = P(t,T_n), \qquad 0 \leqslant t \leqslant T_1,$$

with

$$F(t,x) := \sum_{k=1}^{n} c_k e^{-(T_k-t)x}, \qquad 0 \leqslant t \leqslant T_1.$$

The *bond duration* $D(t,T_n)$ is the relative sensitivity of $P(t,T_n)$ with respect to $\tilde{f}(t,t,T_n)$ defined as

$$D(t,T_n) := -\frac{1}{P(t,T_n)} \frac{\partial F}{\partial x}\big(t, \tilde{f}(t,t,T_n)\big), \qquad 0 \leqslant t \leqslant T_1.$$

The *bond convexity* $C(t,T_n)$ is defined as

$$C(t,T_n) := \frac{1}{P(t,T_n)} \frac{\partial^2 F}{\partial x^2}\big(t, \tilde{f}(t,t,T_n)\big), \qquad 0 \leqslant t \leqslant T_1.$$

(1) Compute the bond duration in case $n = 1$.

(2) Show that the *bond duration* $D(t, T_n)$ can be interpreted as an average of times to maturity weighted by the respective discounted bond payoffs.

(3) Show that the *bond convexity* $C(t, T_n)$ satisfies

$$C(t, T_n) = (D(t, T_n))^2 + (S(t, T_n))^2,$$

where

$$(S(t, T_n))^2 := \sum_{k=1}^{n} w_k (T_k - t - D(t, T_n))^2$$

$S(t, T_n)$ measures the dispersion of the duration of the bond payoffs around the portfolio duration $D(t, T_n)$.

(4) Consider now the zero-coupon yield defined as

$$f_\alpha(t, t, T_n) := -\frac{1}{\alpha(T_n - t)} \log P(t, t + \alpha(T_n - t)),$$

where $\alpha \in (0, 1)$. Compute the bond duration associated to the yield $f_\alpha(t, t, T_n)$ in affine bond pricing models of the form

$$P(t, T) = e^{A(T-t) + r_t B(T-t)}, \qquad 0 \leqslant t \leqslant T.$$

(5) Wu (2000) Compute the bond duration associated to the yield $f_\alpha(t, t, T_n)$ in the Vasicek model in which $B(T - t) := \left(1 - e^{-(T-t)b}\right)/b$, $t \in [0, T]$.

Chapter 6

Curve Fitting and a Two-Factor Model

This chapter is concerned with the fitting of forward interest curves to market data using various types of parametrizations. The short rate models considered in the previous chapters pose several restriction to the fitting of forward curves, and they induce undesirable correlations in the price of zero-coupon bonds with different maturities. To address this problem, we consider two-factor models which use correlated sources of randomness, and allow for a wider choice of parameters for the fitting of forward curves.

6.1 Parametrization of Forward Rates

The forward rate curve represented in Figure 5.4 has some similarities with the market data curve of Figure 5.1 (for instance, it increases on a certain interval), however it clearly misses some features often observed in forward curves, such as the hump at the left hand side of the graph. For this reason, other parametrizations of forward rates have been introduced.

In the sequel we will frequently use the Musiela convention, *i.e.*, we will write

$$g(x) = f(t, t + x) = f(t, T),$$

under the substitution $x = T - t$, $x \geqslant 0$.

Nelson-Siegel parametrization

In the Nelson and Siegel (1987) parametrization the instantaneous forward rate curves are parametrized by 4 coefficients z_1, z_2, z_3, z_4, as

$$g(x) = z_1 + (z_2 + z_3 x) e^{-x z_4}, \qquad x \geqslant 0.$$

An example of a graph obtained by the Nelson-Siegel parametrization is given in Figure 6.1, for $z_1 = 1$, $z_2 = -10$, $z_3 = 100$, $z_4 = 10$.

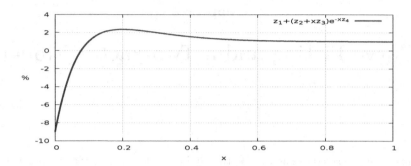

Fig. 6.1: Graph of $x \mapsto g(x)$ in the Nelson-Siegel model.

Svensson parametrization

The Svensson (1994) parametrization has the advantage to reproduce two humps instead of one, the location and height of which can be chosen *via* 6 parameters z_1, z_2, z_3, z_4, z_5, z_6 as

$$x \mapsto g(x) = z_1 + (z_2 + z_3 x)e^{-xz_4} + z_5 x e^{-xz_6}, \qquad x \geqslant 0. \tag{6.1}$$

A typical graph of a Svensson parametrization is given in Figure 6.2, for $z_1 = 6.6$, $z_2 = -5$, $z_3 = -100$, $z_4 = 10$, $z_5 = -1/2$, $z_6 = 1$.

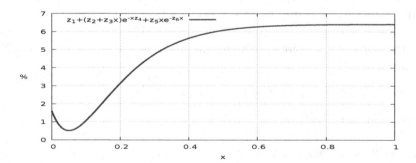

Fig. 6.2: Graph of $x \mapsto g(x)$ in the Svensson model.

Consider the market data rates displayed in Figure 6.3.

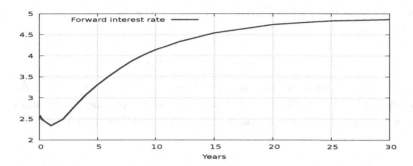

Fig. 6.3: Market data of LIBOR forward rates $T \mapsto f(t, T, T + \delta)$.

Svensson curves of the form (6.1) appear suitable for modeling of market data, see Figure 6.4 for a fit of the data of Figure 6.3 using a Svensson curve.

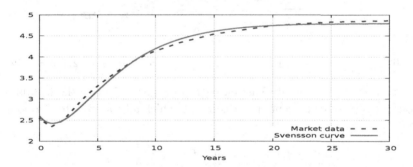

Fig. 6.4: Fit of a Svensson curve to market data.

Vasicek parametrization

In the Vasicek model, the instantaneous forward rate process is given from (5.7) as

$$f(t, T) = \frac{a}{b} - \frac{\sigma^2}{2b^2} + \left(r_t - \frac{a}{b} + \frac{\sigma^2}{b^2} \right) e^{-bx} - \frac{\sigma^2}{2b^2} e^{-2bx}, \qquad (6.2)$$

in the Musiela notation (7.1) with $x = T - t$, and we have

$$\frac{\partial f}{\partial T}(t, T) = \left(a - br_t - \frac{\sigma^2}{b} \left(1 - e^{-(T-t)b} \right) \right) e^{-(T-t)b}.$$

We check that the derivative $\partial f/\partial T$ vanishes when $a - br_t + a - \sigma^2(1 - e^{-bx})/b = 0$, *i.e.*

$$e^{-bx} = 1 + \frac{b}{\sigma^2}(br_t - a),$$

which admits at most one solution $x = T - t$, provided that $a > br_t$. As a consequence, the possible forward curves in the Vasicek model are limited to one change of "regime" per curve, as illustrated in Figure 6.5 for various values of r_t, and in Figure 6.6.

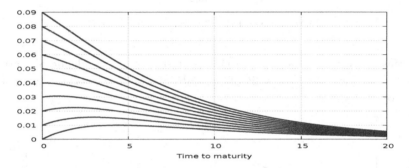

Fig. 6.5: Graphs of forward rates with $b = 0.16$, $a/b = 0.04$, $r_0 = 2\%$, $\sigma = 4.5\%$.

For example, the short rate Vasicek paths of Figure 7.5 drive the instantaneous forward rates presented in Figure 6.6 which all converge to the "long rate"

$$\lim_{x \to \infty} f(t, t + x) = \lim_{T \to \infty} f(t, T) = \frac{a}{b} - \frac{\sigma^2}{2b^2}$$

as x goes to infinity. The next Figure 6.6 is also using the parameters $b = 0.16$, $a/b = 0.04$, $r_0 = 2\%$, and $\sigma = 4.5\%$.

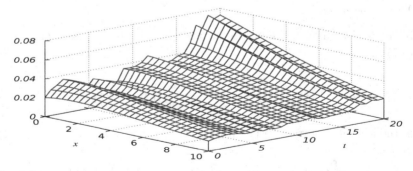

Fig. 6.6: Forward instantaneous curve $(t, x) \mapsto f(t, t + x)$ in the Vasicek model.

Fitting the Nelson-Siegel and Svensson models to Vasicek curves

From Relation (6.2) we know that the instantaneous forward curves in the Vasicek model "live" in the space of functions generated by

$$x \mapsto z_1 + z_2\, e^{z_3 x} + z_4\, e^{z_5 x}$$

with

$$
\begin{cases}
z_1 = \dfrac{a}{b} - \dfrac{\sigma^2}{2b^2}, \\[2mm]
z_2 = r_t - \dfrac{a}{b} + \dfrac{\sigma^2}{b^2}, \\[2mm]
z_3 = -b, \\[2mm]
z_4 = -\dfrac{\sigma^2}{2b}, \\[2mm]
z_5 = -2b.
\end{cases}
$$

Unfortunately, this function space is contained neither in the Nelson-Siegel space, nor in the Svensson space. In addition, it can be checked that typical realizations of Vasicek yield curves do not appear realistic enough to correctly model the behavior of actual market forward curves, by comparing Figure 6.3 to Figure 6.5.

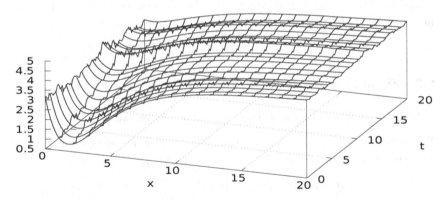

Fig. 6.7: Stochastic process of forward curves.

In order to solve this modeling problem, one may attempt to construct an instantaneous forward rate process taking values in the Svensson space, for example:

$$
\begin{aligned}
x \mapsto f(t,T) &= f(t, t+x) \\
&= z_1(t) + (z_2(t) + z_3(t)x)e^{-x z_4(t)} + z_5(t)x e^{-x z_6(t)},
\end{aligned}
$$

$x = T - t \geqslant 0$, where $z_i(t)$, $i = 1, 2, \ldots, 6$, are suitably chosen stochastic processes and $x = T - t$, see for example Figure 6.7. In this case, the short rate may be defined as

$$r_t = f(t, t + 0) = z_1(t) + z_2(t), \qquad t \geqslant 0.$$

This type of modeling raises the question of whether it is consistent with the absence of arbitrage arguments that led to the relation

$$P(t, T) = e^{-\int_t^T f(t,s)ds} \tag{6.3}$$

see Proposition 5.2, and to the bond pricing formula

$$P(t, T) = \mathbb{E}^* \left[e^{-\int_t^T r_s ds} \,\middle|\, \mathcal{F}_t \right], \tag{6.4}$$

see (4.6), which imply the relation

$$e^{-\int_t^T f(t,s)ds} = \mathbb{E}^* \left[e^{-\int_t^T f(s,s)ds} \,\middle|\, \mathcal{F}_t \right].$$

However, an instantaneous forward rate process taking values in the Svensson space is not consistent with absence of arbitrage. More precisely, it can be proved that the HJM forward instantaneous interest rate curves, see Chapter 7, cannot live in the Nelson-Siegel or Svensson spaces, see §3.5 of Björk (2004). This is consistent with the observation that the instantaneous forward curves produced by the Vasicek model are contained neither in the Nelson-Siegel space, nor in the Svensson space.

6.2   Fitting Curve Models to Market Data

Fitting the Nelson-Siegel and Svensson models to market data

Recall that in the Nelson-Siegel parametrization the instantaneous forward rate curves are parametrized by four coefficients z_1, z_2, z_3, z_4, as

$$f(t, t + x) = z_1 + (z_2 + z_3 x) e^{-x z_4}, \qquad x \geqslant 0.$$

Taking $x = T - t$, the yield $f(t, t, T)$ is given as

$$\begin{aligned}
f(t, t, T) &= \frac{1}{T - t} \int_t^T f(t, s) ds \\
&= \frac{1}{x} \int_0^x f(t, t + y) dy \\
&= z_1 + \frac{z_2}{x} \int_0^x e^{-y z_4} dy + \frac{z_3}{x} \int_0^x y e^{-y z_4} dy
\end{aligned}$$

$$= z_1 + z_2 \frac{1 - e^{-xz_4}}{xz_4} + z_3 \frac{1 - e^{-xz_4} + x e^{-xz_4}}{xz_4}.$$

In other words, the yield $f(t, t, T)$ can be reparametrized as

$$f(t, t + x) = z_1 + (z_2 + z_3 x) e^{-xz_4} = \beta_0 + \beta_1 e^{-x/\lambda} + \frac{\beta_2}{\lambda} x e^{-x/\lambda}, \quad x \geqslant 0,$$

see Charpentier (2014), with $\beta_0 = z_1$, $\beta_1 = z_2$, $\beta_2 = z_3/z_4$, $\lambda = 1/z_4$.

```
1   require(YieldCurve);data(ECBYieldCurve)
    mat.ECB<-c(3/12,0.5,1,2,3,4,5,6,7,8,9,10,11,12,13,14,15,16,17,18,19,20,21,22,23,
        24,25,26,27,28,29,30)
3   first(ECBYieldCurve, '1 month');
    Nelson.Siegel(first(ECBYieldCurve, '1 month'), mat.ECB)
```

```
1   for (n in seq(from=70, to=290, by=10)) {
2   ECB.NS <- Nelson.Siegel(ECBYieldCurve[n,], mat.ECB)
    ECB.S <- Svensson(ECBYieldCurve[n,], mat.ECB)
4   ECB.NS.yield.curve <- NSrates(ECB.NS, mat.ECB)
    ECB.S.yield.curve <- Srates(ECB.S, mat.ECB,"Spot")
6   plot(mat.ECB, as.numeric(ECBYieldCurve[n,]), type="o", lty=1, col=1,ylab="
        Interest rates", xlab="Maturity in years", ylim=c(3.2,4.8))
    lines(mat.ECB, as.numeric(ECB.NS.yield.curve), type="l", lty=3,col=2,lwd=2)
8   lines(mat.ECB, as.numeric(ECB.S.yield.curve), type="l", lty=2,col=6,lwd=2)
    title(main=paste("ECB yield curve observed at",time(ECBYieldCurve[n])))
10  legend('bottomright', legend=c("ECB data","Nelson-Siegel","Svensson"),col=c
        (1,2,6), lty=1, bg='gray90')
    grid();Sys.sleep(0.5)}
```

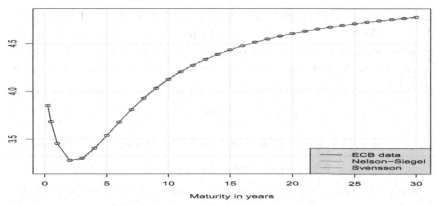

Fig. 6.8: ECB data *vs* fitted yield curve.

Curve estimation

A simple way to estimate the forward curve based on market data of bond prices $(P(t, T_k))_{k=1,2,\ldots,n}$ with maturities $T_1, T_2, \ldots, T_n$ is to assume that the instantaneous forward rate is a step function:

$$f(t, x) = g(x) = \sum_{k=1}^{n} \alpha_k \mathbb{1}_{(T_{k-1}, T_k]}(x),$$

with $T_0 = 0$. In this case we have the relation

$$\frac{P(t, T_k)}{P(t, T_{k-1})} = \mathrm{e}^{- \int_{T_{k-1}}^{T_k} g(x)dx} = \mathrm{e}^{-\alpha_k (T_k - T_{k-1})},$$

hence

$$\alpha_k = -\frac{1}{T_k - T_{k-1}} \log \frac{P(t, T_k)}{P(t, T_{k-1})}, \qquad k = 1, 2, \ldots, n.$$

A more realistic estimation can be made by requiring some additional smoothness on $g(x)$, for example requiring that it be twice differentiable. In this case the estimate $g(x)$ will be the outcome of the minimization problem

$$\min_g \left(\lambda \int_0^{T_n} |g''(x)|^2 dx + \sum_{k=1}^{n} \beta_k \left| \frac{P(t, T_k)}{P(t, T_{k-1})} - \mathrm{e}^{- \int_{T_{k-1}}^{T_k} g(x)dx} \right|^2 \right)$$

where $\beta_1, \ldots, \beta_n$ and λ are positive coefficients.

6.3   Deterministic Shifts

A possible way around the curve fitting problem is to use deterministic shifts for the fitting of a single forward curve, usually at $t = 0$. Here, we consider again a Vasicek model of forward rates in which we let $a = 0$ for simplicity. According to (6.2), consider the instantaneous forward rate given by

$$f(t, T) := \varphi(T) + X_t\, \mathrm{e}^{-(T-t)b} - \frac{\sigma^2}{2b^2} \left(1 - \mathrm{e}^{-(T-t)b} \right)^2, \qquad (6.5)$$

where $T \mapsto \varphi(T)$ is given deterministic function, and X_t is the solution of

$$dX_t = -bX_t dt + \sigma dB_t$$

with $X_0 = 0$. Given an initial market term structure $T \mapsto f^M(0, T)$, from Relation (6.5) at $t = 0$ we check that choosing $T \mapsto \varphi(T)$ as

$$\varphi(T) := f^M(0, T) + \frac{\sigma^2}{2b^2} \left(1 - \mathrm{e}^{-bT} \right)^2 \qquad (6.6)$$

allows one to match $T \mapsto f^M(0,T)$, or any (fixed) term structure of our choice at a given (unique) time t, by the equality $f(0,T) = f^M(0,T)$, $T \geqslant 0$.

The interest in this method is to achieve consistency with the absence of arbitrage condition (4.2) and Relations (6.3) and (6.4) above by defining the short rate r_t as

$$r_t = f(t,t) = \varphi(t) + X_t.$$

Indeed, in this case we have

$$P(t,T) = \mathbb{E}^* \left[e^{-\int_t^T r_s ds} \,\middle|\, \mathcal{F}_t \right]$$

$$= \mathbb{E}^* \left[\exp\left(-\int_t^T \varphi(s)ds - \int_t^T X_s ds \right) \,\middle|\, \mathcal{F}_t \right]$$

$$= e^{-\int_t^T \varphi(s)ds} \, \mathbb{E}^* \left[e^{-\int_t^T X_s ds} \,\middle|\, \mathcal{F}_t \right]$$

$$= e^{-\int_t^T \varphi(s)ds} \exp\left(-\int_t^T \left(X_t e^{-(s-t)b} - \frac{\sigma^2}{2b^2}\left(1 - e^{-(s-t)b}\right)^2 \right) ds \right)$$

$$= e^{-\int_t^T f(t,s)ds},$$

where we used (6.2) and the relation

$$\mathbb{E}^* \left[e^{-\int_t^T X_s ds} \,\middle|\, \mathcal{F}_t \right] = \exp\left(-\int_t^T f(t,s)ds \right)$$

$$= \exp\left(-\int_t^T \left(X_t e^{-(s-t)b} - \frac{\sigma^2}{2b^2}\left(1 - e^{-(s-t)b}\right)^2 \right) ds \right)$$

defining the bond price as the exponential of an integral of the instantaneous forward rate in a Vasicek model with $a = 0$, see Proposition 5.2. Note however that this argument is restricted to the fitting of the single (initial) market curve $T \mapsto f^M(0,T)$.

6.4 The Correlation Problem

The correlation problem is another issue of concern when implementing the affine models considered so far. Let us compare three bond price simulations with maturity $T_1 = 10$, $T_2 = 20$, and $T_3 = 30$ based on the same Brownian path, as given in Figure 6.9. Clearly, the bond prices $F(r_t, T_1) = P(t,T_1)$ and $F(r_t, T_2) = P(t,T_2)$ with maturities T_1 and T_2 are linked by the relation

$$P(t,T_2) = P(t,T_1) \exp\left(A(t,T_2) - A(t,T_1) + r_t(C(t,T_2) - C(t,T_1)) \right), \quad (6.7)$$

meaning that bond prices with different maturities could be deduced from each other, which is unrealistic.

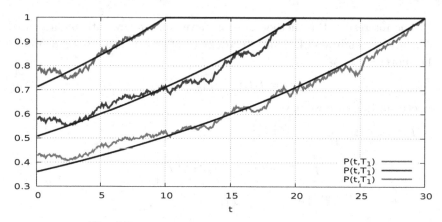

Fig. 6.9: Graphs of $t \mapsto P(t,T_1), P(t,T_2), P(t,T_3)$.

Note that if X and Y are two linear combinations of a same random variable Z:

$$X = a + \alpha Z, \qquad Y = b + \beta Z,$$

then their covariance (see the appendix) equals

$$\begin{aligned}
\operatorname{Cov}(X,Y) &= \operatorname{Cov}(a + \alpha Z, b + \beta Z) \\
&= \operatorname{Cov}(\alpha Z, \beta Z) \\
&= \alpha\beta \operatorname{Cov}(Z,Z) \\
&= \alpha\beta \operatorname{Var}(Z),
\end{aligned}$$

and

$$\operatorname{Var} X = \alpha^2 \operatorname{Var} Z, \qquad \operatorname{Var} Y = \beta^2 \operatorname{Var} Z,$$

hence

$$|\operatorname{Cor}(X,Y)| = \frac{|\operatorname{Cov}(X,Y)|}{\sqrt{\operatorname{Var} X}\sqrt{\operatorname{Var} Y}} = 1,$$

i.e. X and Y are completely correlated or anticorrelated.

In affine short rate models, by (6.7), $\log P(t,T_1)$ and $\log P(t,T_2)$ are linked by the affine relationship

$$\log P(t,T_2) = \log P(t,T_1) + A(t,T_2) - A(t,T_1) + r_t(C(t,T_2) - C(t,T_1))$$

$$= \log P(t, T_1) + A(t, T_2) - A(t, T_1)$$

$$+ (C(t, T_2) - C(t, T_1)) \frac{\log P(t, T_1) - A(t, T_1)}{C(t, T_1)}$$

$$= \left(1 + \frac{C(t, T_2) - C(t, T_1)}{A(t, T_1)} \right) \log P(t, T_1) + A(t, T_2) - A(t, T_1) \frac{C(t, T_2)}{C(t, T_1)}$$

with constant coefficients, which yields the perfect correlation or anticorrelation

$$\mathrm{Cor}(\log P(t, T_1), \log P(t, T_2)) = \pm 1,$$

depending on the sign of the coefficient $1 + (C(t, T_2) - C(t, T_1))/A(t, T_1)$.

6.5 Two-Factor Model

The short rate models considered in the previous chapters are one-factor models in the sense that their stochastic evolution is driven by a single Brownian motion. As such, they pose several restriction to the fitting of forward curves, and as noted above, they induce undesirable correlations in the price of zero-coupon bonds with different maturities. In this section we study a two-factor model which uses two sources of randomness, and allows for a wider choice of parameters for the fitting of forward curves, as a solution to the correlation problem.

This two-factor model is based on two control processes $(X_t)_{t \in \mathbb{R}_+}$, $(Y_t)_{t \in \mathbb{R}_+}$ which are solution of the stochastic differential equations

$$\begin{cases} dX_t = \mu_1(t, X_t)dt + \sigma_1(t, X_t)dB_t^{(1)} \\ \\ dY_t = \mu_2(t, Y_t)dt + \sigma_2(t, Y_t)dB_t^{(2)}, \end{cases} \tag{6.8}$$

where $\left(B_t^{(1)}\right)_{t \in \mathbb{R}_+}$, $\left(B_t^{(2)}\right)_{t \in \mathbb{R}_+}$ are correlated Brownian motion under $\mathbb{P}^*$, with

$$\mathrm{Cov}\left(B_s^{(1)}, B_t^{(2)}\right) = \rho \min(s, t), \qquad s, t \geqslant 0, \tag{6.9}$$

and

$$dB_t^{(1)} \cdot dB_t^{(2)} = \rho dt,$$

for some correlation parameter $\rho \in [-1, 1]$, and to define the associated bond prices as

$$P(t, T_1) = \mathbb{E}^* \left[e^{- \int_t^{T_1} X_s ds} \, \middle| \, \mathcal{F}_t \right]$$

and

$$P(t, T_2) = \mathbb{E}^* \left[e^{- \int_t^{T_2} Y_s ds} \,\middle|\, \mathcal{F}_t \right].$$

This type of approach would nevertheless lead to other difficulties, namely:

- the presence of two candidate processes X_t and Y_t to represent the interest rate,

- the need to introduce a new control process for each additional maturity T_n, $n \geqslant 3$.

In this two-factor model, one chooses to build the short-term interest rate process r_t as

$$r_t = \varphi(t) + X_t + Y_t, \qquad t \geqslant 0,$$

where the function $\varphi(T)$ can be chosen to fit the initial forward curve as in (6.6). Applying previous standard arbitrage arguments as in Chapter 4, we define the price of a bond with maturity T as

$$P(t, T) := \mathbb{E}^* \left[e^{- \int_t^T r_s ds} \,\middle|\, \mathcal{F}_t \right],$$

which however does no longer directly depend on the short rate r_t. Nevertheless, being a made of solutions to the stochastic differential equations (6.8), the couple $(X_t, Y_t)_{t \in \mathbb{R}_+}$ is a two-dimensional Markov process, see Property 4.1, and we can write

$$
\begin{aligned}
P(t, T) : &= \mathbb{E}^* \left[e^{- \int_t^T r_s ds} \,\middle|\, \mathcal{F}_t \right] \\
&= \mathbb{E}^* \left[e^{- \int_t^T r_s ds} \,\middle|\, X_t, \, Y_t \right] \\
&= e^{- \int_t^T \varphi(s) ds} \, \mathbb{E}^* \left[e^{- \int_t^T (X_s + Y_s) ds} \,\middle|\, X_t, \, Y_t \right] \qquad (6.10) \\
&= F(t, X_t, Y_t), \qquad 0 \leqslant t \leqslant T,
\end{aligned}
$$

i.e. $P(t, T)$ can be written as a function

$$P(t, T) = F(t, X_t, Y_t)$$

of t, X_t and Y_t, using the Markov property of the process $(X_t, Y_t)_{t \in \mathbb{R}_+}$.

In the sequel, we will assume that $\left(B_t^{(1)} \right)_{t \in \mathbb{R}_+}$ and $\left(B_t^{(2)} \right)_{t \in \mathbb{R}_+}$ have correlation $\rho \in [-1, 1]$, that is

$$\mathrm{Cov}\left(B_s^{(1)}, B_t^{(2)} \right) = \rho \min(s, t), \qquad s, t \geqslant 0.$$

Note that in terms of stochastic differentials, this implies the product rule

$$dB_t^{(1)} \cdot dB_t^{(2)} = \rho dt. \tag{6.11}$$

In practice, $(B^{(1)})_{t\in\mathbb{R}_+}$ and $(B^{(2)})_{t\in\mathbb{R}_+}$ can be constructed from two independent Brownian motions $(W^{(1)})_{t\in\mathbb{R}_+}$ and $(W^{(2)})_{t\in\mathbb{R}_+}$, by letting

$$\begin{cases} B_t^{(1)} = W_t^{(1)}, \\ \\ B_t^{(2)} = \rho W_t^{(1)} + \sqrt{1-\rho^2}W_t^{(2)}, \qquad t \geqslant 0, \end{cases}$$

and Relations (6.9) and (6.11) are easily satisfied from this construction. In the next proposition we make use of the Itô formula in two variables in order to derive a PDE on $\mathbb{R}^2$ for the bond price $P(t,T) = F(t,X_t,Y_t)$, $t \in [0,T]$.

Proposition 6.1. *In the two-factor model (6.8), the pricing PDE for the bond price $P(t,T) = F(t,X_t,Y_t)$ is given by*

$$-(\varphi(t) + x + y)F(t,x,y) + \mu_1(t,x)\frac{\partial F}{\partial x}(t,x,y) + \mu_2(t,y)\frac{\partial F}{\partial y}(t,x,y)$$

$$+\frac{1}{2}\sigma_1^2(t,x)\frac{\partial^2 F}{\partial x^2}(t,x,y) + \frac{1}{2}\sigma_2^2(t,y)\frac{\partial^2 F}{\partial y^2}(t,x,y)$$

$$+\rho\sigma_1(t,x)\sigma_2(t,y)\frac{\partial^2 F}{\partial x \partial y}(t,x,y) + \frac{\partial F}{\partial t}(t,x,y) = 0, \quad 0 \leqslant t \leqslant T. \tag{6.12}$$

Proof. Applying the Itô formula in two variables (1.12) to

$$t \mapsto F(t,X_t,Y_t) = P(t,T) = \mathbb{E}^*\left[e^{-\int_t^T r_s ds} \,\Big|\, \mathcal{F}_t \right],$$

we have

$$d\left(e^{-\int_0^t r_s ds} P(t,T) \right) = -r_t e^{-\int_0^t r_s ds} P(t,T)dt + e^{-\int_0^t r_s ds}dP(t,T) \tag{6.13}$$

$$= -r_t e^{-\int_0^t r_s ds} P(t,T)dt + e^{-\int_0^t r_s ds}dF(t,X_t,Y_t)$$

$$= -r_t e^{-\int_0^t r_s ds} P(t,T)dt + e^{-\int_0^t r_s ds}\frac{\partial F}{\partial x}(t,X_t,Y_t)dX_t + \frac{\partial F}{\partial t}(t,X_t,Y_t)dt$$

$$+e^{-\int_0^t r_s ds}\frac{\partial F}{\partial y}(t,X_t,Y_t)dY_t + \frac{1}{2}\sigma_1^2(t,X_t)e^{-\int_0^t r_s ds}\frac{\partial^2 F}{\partial x^2}(t,X_t,Y_t)dt$$

$$+\frac{1}{2}\sigma_2^2(t,Y_t)e^{-\int_0^t r_s ds}\frac{\partial^2 F}{\partial y^2}(t,X_t,Y_t)dt$$

$$+\rho e^{-\int_0^t r_s ds}\sigma_1(t,X_t)\sigma_2(t,Y_t)\frac{\partial^2 F}{\partial x \partial y}(t,X_t,Y_t)dt$$

$$= \sigma_1(t,X_t)e^{-\int_0^t r_s ds}\frac{\partial F}{\partial x}(t,X_t,Y_t)dB_t^{(1)}$$

$$+\sigma_2(t,Y_t)\,e^{-\int_0^t r_s ds}\frac{\partial F}{\partial y}(t,X_t,Y_t)dB_t^{(2)}$$

$$e^{-\int_0^t r_s ds}\left(-r_tP(t,T)+\frac{\partial F}{\partial x}(t,X_t,Y_t)\mu_1(t,X_t)+\frac{\partial F}{\partial y}(t,X_t,Y_t)\mu_2(t,Y_t)\right.$$

$$+\frac{1}{2}\sigma_1^2(t,X_t)\frac{\partial^2 F}{\partial x^2}(t,X_t,Y_t)+\frac{1}{2}\sigma_2^2(t,Y_t)\frac{\partial^2 F}{\partial y^2}(t,X_t,Y_t)$$

$$\left.+\rho\sigma_1(t,X_t)\sigma_2(t,Y_t)\frac{\partial^2 F}{\partial x\partial y}(t,X_t,Y_t)+\frac{\partial F}{\partial t}(t,X_t,Y_t)\right)dt,$$

which yields (6.12) since the discounted process

$$t\mapsto e^{-\int_0^t r_s ds}P(t,T)=\mathbb{E}^*\left[e^{-\int_0^T r_s ds}\,\bigg|\,\mathcal{F}_t\right]$$

is an $(\mathcal{F}_t)_{t\in[0,T]}$-martingale under $\mathbb{P}^*$. $\square$

Next, we consider another Vasicek-type example where

$$\begin{cases} dX_t=-aX_tdt+\sigma dB_t^{(1)}, \\[2mm] dY_t=-bY_tdt+\eta dB_t^{(2)}. \end{cases} \tag{6.14}$$

Here, instead of attempting to solve the 2-dimensional PDE (6.12) we choose to compute $P(t,T)=F(t,X_t,Y_t)$ from its expression (6.10) as a conditional expectation. This yields the solution $F(t,x,y)$ of (6.12) as

$$P(t,T)=F(t,X_t,Y_t)=e^{-\int_t^T \varphi(s)ds}F_1(t,X_t)F_2(t,Y_t)e^{U_\rho(t,T)},$$

where $F_1(t,X_t)$ and $F_2(t,Y_t)$ are the bond prices associated to X_t and Y_t in the Vasicek model, and

$$U_\rho(t,T):=$$

$$\rho\frac{\sigma\eta}{ab}\left(T-t+\frac{e^{-(T-t)a}-1}{a}+\frac{e^{-(T-t)b}-1}{b}+\frac{1-e^{-(a+b)(T-t)}}{a+b}\right)$$

is a correlation term which vanishes when $(B_t^{(1)})_{t\in\mathbb{R}_+}$ and $(B_t^{(2)})_{t\in\mathbb{R}_+}$ are independent, *i.e.* when $\rho=0$.

Proposition 6.2. *The bond price $P(t,T)$ in the two-factor model (6.8) is given by*

$$P(t,T)=\exp\left(-\int_t^T\varphi(s)ds-\frac{1}{a}\left(1-e^{-(T-t)a}\right)X_t-\frac{1}{b}\left(1-e^{-(T-t)b}\right)Y_t\right)$$

$$\times\exp\left(\frac{\sigma^2}{2a^2}\int_t^T\left(1-e^{-(T-s)a}\right)^2ds+\frac{\eta^2}{2b^2}\int_t^T\left(1-e^{-(T-s)b}\right)^2ds\right)$$

$$\times\exp\left(\rho\frac{\sigma\eta}{ab}\int_t^T\left(1-e^{-(T-s)a}\right)\left(1-e^{-(T-s)b}\right)ds\right),\qquad 0\leqslant t\leqslant T.$$

Proof. From (3.2) and (6.14), we have

$$\int_0^t X_s ds = \frac{1}{a}\left(\sigma B_t^{(1)} - X_t\right)$$

$$= \frac{\sigma}{a}\left(B_t^{(1)} - \int_0^t e^{-(t-s)a} dB_s^{(1)}\right)$$

$$= \frac{\sigma}{a}\int_0^t \left(1 - e^{-(t-s)a}\right) dB_s^{(1)},$$

hence

$$\int_t^T X_s ds = \int_0^T X_s ds - \int_0^t X_s ds$$

$$= \frac{\sigma}{a}\int_0^T \left(1 - e^{-(T-s)a}\right) dB_s^{(1)} - \frac{\sigma}{a}\int_0^t \left(1 - e^{-(t-s)a}\right) dB_s^{(1)}$$

$$= -\frac{\sigma}{a}\left(\int_0^t \left(e^{-(T-s)a} - e^{-(t-s)a}\right) dB_s^{(1)} - \int_t^T \left(1 - e^{-(T-s)a}\right) dB_s^{(1)}\right)$$

$$= \frac{\sigma}{a}\left(1 - e^{-(T-t)a}\right)\int_0^t e^{-(t-s)a} dB_s^{(1)} + \frac{\sigma}{a}\int_t^T \left(1 - e^{-(T-s)a}\right) dB_s^{(1)}$$

$$= \frac{1}{a}\left(1 - e^{-(T-t)a}\right) X_t + \frac{\sigma}{a}\int_t^T \left(1 - e^{-(T-s)a}\right) dB_s^{(1)},$$

and similarly,

$$\int_t^T Y_s ds = \frac{1}{b}\left(1 - e^{-(T-t)b}\right) Y_t + \frac{\eta}{b}\int_t^T \left(1 - e^{-(T-s)b}\right) dB_s^{(2)}.$$

Hence, conditionally to $\mathcal{F}_t$, the random vector $\left(\int_t^T X_s ds, \int_t^T Y_s ds\right)$ is Gaussian with mean

$$\begin{pmatrix} \mathbb{E}^*\left[\int_t^T X_s ds \,\middle|\, \mathcal{F}_t\right] \\[2mm] \mathbb{E}^*\left[\int_t^T Y_s ds \,\middle|\, \mathcal{F}_t\right] \end{pmatrix} = \begin{pmatrix} \frac{1}{a}\left(1 - e^{-(T-t)a}\right) X_t \\[2mm] \frac{1}{b}\left(1 - e^{-(T-t)b}\right) Y_t \end{pmatrix}$$

and conditional covariance matrix

$$\mathrm{Cov}\left[\int_t^T X_s ds, \int_t^T Y_s ds \,\middle|\, \mathcal{F}_t\right] =$$

$$\begin{pmatrix} \frac{\sigma^2}{a^2}\int_t^T \left(1 - e^{-(T-s)a}\right)^2 ds & \rho\frac{\sigma\eta}{ab}\int_t^T \left(1 - e^{-(T-s)a}\right)\left(1 - e^{-(T-s)b}\right) ds \\[3mm] \rho\frac{\sigma\eta}{ab}\int_t^T \left(1 - e^{-(T-s)a}\right)\left(1 - e^{-(T-s)b}\right) ds & \frac{\eta^2}{b^2}\int_t^T \left(1 - e^{-(T-s)b}\right)^2 ds \end{pmatrix}$$

obtained by the Itô isometry (1.4). The variance of $\int_t^T X_s ds + \int_t^T Y_s ds$ given $\mathcal{F}_t$ equals

$$\left\langle \operatorname{Cov}\left[\int_t^T X_s ds, \int_t^T Y_s ds \,\Big|\, \mathcal{F}_t\right]\begin{bmatrix}1\\1\end{bmatrix}, \begin{bmatrix}1\\1\end{bmatrix}\right\rangle_{\mathbb{R}^2}$$

$$= \operatorname{Var}\left[\int_t^T X_s ds \,\Big|\, \mathcal{F}_t\right] + \operatorname{Var}\left[\int_t^T Y_s ds \,\Big|\, \mathcal{F}_t\right]$$

$$+ 2\operatorname{Cov}\left[\int_t^T X_s ds, \int_t^T Y_s ds \,\Big|\, \mathcal{F}_t\right],$$

and we have

$$P(t,T) = \mathbb{E}^*\left[e^{-\int_t^T r_s ds} \,\Big|\, \mathcal{F}_t\right]$$

$$= e^{-\int_t^T \varphi(s)ds}\, \mathbb{E}^*\left[\exp\left(-\int_t^T X_s ds - \int_t^T Y_s ds\right)\,\Big|\, \mathcal{F}_t\right]$$

$$= \exp\left(-\int_t^T \varphi(s)ds - \mathbb{E}^*\left[\int_t^T X_s ds \,\Big|\, \mathcal{F}_t\right] - \mathbb{E}^*\left[\int_t^T Y_s ds \,\Big|\, \mathcal{F}_t\right]\right)$$

$$\times \exp\left(\frac{1}{2}\left\langle \operatorname{Cov}\left[\int_t^T X_s ds, \int_t^T Y_s ds \,\Big|\, \mathcal{F}_t\right]\begin{bmatrix}1\\1\end{bmatrix}, \begin{bmatrix}1\\1\end{bmatrix}\right\rangle_{\mathbb{R}^2}\right)$$

$$= \exp\left(-\int_t^T \varphi(s)ds - \frac{1}{a}\left(1 - e^{-(T-t)a}\right)X_t - \frac{1}{b}\left(1 - e^{-(T-t)b}\right)Y_t\right)$$

$$\times \exp\left(\frac{\sigma^2}{2a^2}\int_t^T \left(1 - e^{-(T-s)a}\right)^2 ds + \frac{\eta^2}{2b^2}\int_t^T \left(1 - e^{-(T-s)b}\right)^2 ds\right)$$

$$\times \exp\left(\rho\frac{\sigma\eta}{ab}\int_t^T \left(1 - e^{-(T-s)a}\right)\left(1 - e^{-(T-s)b}\right)ds\right),$$

see the appendix for details on the Laplace transform (11.3) of Gaussian random vectors. $\qquad\square$

The above proposition shows in particular that the solution $F(t, x, y)$ to the 2-dimensional PDE (6.12) is given by

$$F(t, x, y) = \exp\left(-\int_t^T \varphi(s)ds - \frac{x}{a}\left(1 - e^{-(T-t)a}\right) - \frac{y}{b}\left(1 - e^{-(T-t)b}\right)\right)$$

$$\times \exp\left(\frac{\sigma^2}{2a^2}\int_t^T \left(1 - e^{-(T-s)a}\right)^2 ds + \frac{\eta^2}{2b^2}\int_t^T \left(1 - e^{-(T-s)b}\right)^2 ds\right)$$

$$\times \exp\left(\rho\frac{\sigma\eta}{ab}\int_t^T \left(1 - e^{-(T-s)a}\right)\left(1 - e^{-(T-s)b}\right)ds\right).$$

The bond price $P(t,T)$ can also be written as

$$P(t,T) = F_1(t,X_t)F_2(t,Y_t)\exp\left(-\int_t^T \varphi(s)ds + U(t,T)\right), \qquad (6.15)$$

where $F_1(t,X_t)$ and $F_2(t,Y_t)$ are the bond prices associated to X_t and Y_t in the Vasicek model:

$$F_1(t,X_t) = \mathbb{E}^*\left[e^{-\int_t^T X_s ds}\,\Big|\,X_t\right]$$

$$= \exp\left(\frac{\sigma^2}{a^2}\left(T - t + \frac{2}{a}e^{-(T-t)a} - \frac{e^{-2(T-t)a}}{2a} - \frac{3}{2a}\right) - \frac{1 - e^{-(T-t)a}}{a}X_t\right),$$

$$F_2(t,Y_t) = \mathbb{E}^*\left[e^{-\int_t^T Y_s ds}\,\Big|\,Y_t\right]$$

$$= \exp\left(\frac{\eta^2}{b^2}\left(T - t + \frac{2}{b}e^{-(T-t)b} - \frac{e^{-2(T-t)b}}{2b} - \frac{3}{2b}\right) - \frac{1 - e^{-(T-t)b}}{b}Y_t\right),$$

and

$$U(t,T) =$$
$$\rho\frac{\sigma\eta}{ab}\left(T - t + \frac{e^{-(T-t)a} - 1}{a} + \frac{e^{-(T-t)b} - 1}{b} - \frac{e^{-(a+b)(T-t)} - 1}{a+b}\right),$$

$0 \leqslant t \leqslant T$, is a correlation term which vanishes when $\left(B_t^{(1)}\right)_{t\in\mathbb{R}_+}$ and $\left(B_t^{(2)}\right)_{t\in\mathbb{R}_+}$ are independent, *i.e.* when $\rho = 0$.

By Proposition 5.1, partial differentiation of $\log P(t,T)$ with respect to T yields the instantaneous forward rate

$$f(t,T) = -\frac{\partial}{\partial T}\log P(t,T)$$

$$= \varphi(T) + f_1(t,T) + f_2(t,T) - \rho\frac{\sigma\eta}{ab}\left(1 - e^{-(T-t)a}\right)\left(1 - e^{-(T-t)b}\right)$$

$$= \varphi(T) + X_t e^{-(T-t)a} - \frac{\sigma^2}{2a^2}\left(1 - e^{-(T-t)a}\right)^2 + Y_t e^{-(T-t)b} \qquad (6.16)$$

$$- \frac{\eta^2}{2b^2}\left(1 - e^{-(T-t)b}\right)^2 - \rho\frac{\sigma\eta}{ab}\left(1 - e^{-(T-t)a}\right)\left(1 - e^{-(T-t)b}\right),$$

where $f_1(t,T)$, $f_2(t,T)$ are the instantaneous forward rates corresponding to X_t and Y_t respectively, *i.e.*

$$f_1(t,T) = X_t e^{-(T-t)a} - \frac{\sigma^2}{2a^2}\left(1 - e^{-(T-t)a}\right)^2$$

and

$$f_2(t, T) = Y_t \, e^{-(T-t)b} - \frac{\eta^2}{2b^2} \left(1 - e^{-(T-t)b}\right)^2.$$

Clearly, the instantaneous forward rate now depends on a larger number of degrees of freedom, in particular one is now allowed to choose independently the coefficients a and b appearing in the exponentials in (6.16).

Next, in Figure 6.10 we present a graph of the evolution of instantaneous forward rate curves in a two-factor model.

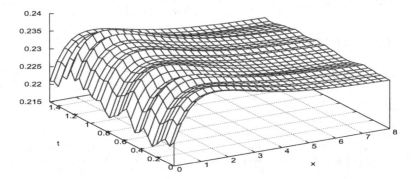

Fig. 6.10: Random evolution of forward rates in a two-factor model.

An example of a single instantaneous forward rate curve obtained in this two-factor model is given in Figure 6.11.

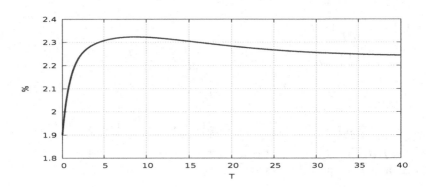

Fig. 6.11: Graph of instantaneous forward curve in a two-factor model.

Exercises

Exercise 6.1. Find the stochastic differential equation satisfied by $P(t, T)$ defined in (6.15).

Exercise 6.2. Consider the two-factor Vasicek model
$$\begin{cases} dX_t = -bX_t dt + \sigma dB_t^{(1)}, \\ dY_t = -bY_t dt + \sigma dB_t^{(2)}, \end{cases}$$
where $\left(B_t^{(1)}\right)_{t\in\mathbb{R}_+}$, $\left(B_t^{(2)}\right)_{t\in\mathbb{R}_+}$ are correlated Brownian motion such that $dB_t^{(1)} \cdot dB_t^{(2)} = \rho dt$, for $\rho \in [-1, 1]$.

(1) Write down the expressions of the short rates X_t and Y_t.

   Hint: Such expressions can be found in Section 3.1.

(2) Compute the variances $\text{Var}[X_t]$, $\text{Var}[Y_t]$, and the covariance $\text{Cov}(X_t, Y_t)$.

   Hint: The expressions of $\text{Var}[X_t]$ and $\text{Var}[Y_t]$ can be found in Section 3.1.

(3) Compute the covariance $\text{Cov}(\log P(t, T_1), \log P(t, T_2))$ for the two-factor bond prices
$$P(t, T_1) = F_1(t, X_t, T_1) F_2(t, Y_t, T_1) e^{\rho U(t, T_1)}$$
   and
$$P(t, T_2) = F_1(t, X_t, T_2) F_2(t, Y_t, T_2) e^{\rho U(t, T_2)},$$
   where
$$\log F_1(t, x, T) = C_1^T + x A_1^T \quad \text{and} \quad \log F_2(t, x) = C_2^T + x A_2^T.$$
   Hint: We have $\text{Cov}(X + Y, Z) = \text{Cov}(X, Z) + \text{Cov}(Y, Z)$ and $\text{Cov}(c, X) = 0$ when c is a constant.

Exercise 6.3. Consider the Hull-White model in which the short-term interest rate process $(r_t)_{t\in\mathbb{R}_+}$ satisfies the equation
$$dr_t = (\theta(t) - ar_t)dt + \sigma dB_t,$$
where $a \in \mathbb{R}$, $\theta(t)$ is a deterministic function of t, the initial condition r_0 is deterministic, and $(B_t)_{t\in\mathbb{R}_+}$ is a standard Brownian motion under $\mathbb{P}$, generating the filtration $(\mathcal{F}_t)_{t\in\mathbb{R}_+}$. Let the bond price $P(t, T)$ be defined under the absence of arbitrage hypothesis as
$$P(t, T) = \mathbb{E}^* \left[e^{-\int_t^T r_s ds} \,\Big|\, \mathcal{F}_t \right], \qquad 0 \leqslant t \leqslant T.$$

Recall that from the Markov property of $(r_t)_{t \in \mathbb{R}_+}$, there exists a function

$$(t, x) \mapsto F(t, x)$$

such that

$$F(t, r_t) = P(t, T), \qquad 0 \leqslant t \leqslant T.$$

(1) Let $(X_t)_{t \in \mathbb{R}_+}$ denote the solution of the stochastic differential equation

$$\begin{cases} dX_t = -aX_t dt + \sigma dB_t, & t \geqslant 0, \\ \\ X_0 = 0. \end{cases}$$

Show that

$$r_t = r_0 e^{-at} + \varphi(t) + X_t, \qquad t \geqslant 0,$$

where

$$\varphi(t) = \int_0^t \theta(u) e^{-(t-u)a} du, \qquad t \geqslant 0.$$

(2) Using Itô's calculus, derive the PDE satisfied by the function $(t, x) \mapsto F(t, x)$.

(3) Recall (cf. Problem 4.8) that $\int_t^T X_s ds$ has a Gaussian distribution given $\mathcal{F}_t$, with

$$\mathbb{E}^* \left[\int_t^T X_s ds \,\Big|\, \mathcal{F}_t \right] = \frac{X_t}{a} \left(1 - e^{-(T-t)a}\right)$$

and

$$\mathrm{Var} \left[\int_t^T X_s ds \,\Big|\, \mathcal{F}_t \right] = \frac{\sigma^2}{a^2} \int_t^T \left(1 - e^{-(T-s)a}\right)^2 ds.$$

Show that the bond price $P(t, T)$ can be written as

$$P(t, T) = e^{A(t,T) + X_t C(t,T)},$$

where $A(t, T)$ and $C(t, T)$ are functions to be determined.

(4) Show that in this model, the instantaneous forward rate

$$f(t, T) = -\frac{\partial}{\partial T} \log P(t, T)$$

is given by

$$f(t, T) = r_0 e^{-aT} + \varphi(T) + X_t e^{-(T-t)a} - \frac{\sigma^2}{2a^2} \left(1 - e^{-(T-t)a}\right)^2, \quad 0 \leqslant t \leqslant T. \tag{6.17}$$

(5) Let the market data of an initial interest rate curve be given by a function

$$T \mapsto f^M(0, T).$$

Show that the function $\varphi(t)$ can be chosen in such a way that the theoretical value $f(0, T)$ matches the market data $f^M(0, T)$, *i.e.*

$$f(0, T) = f^M(0, T), \qquad T \geqslant 0.$$

(6) Show that choosing $\theta(t)$ equal to

$$\theta(t) = a f^M(0, t) + \frac{\partial f^M}{\partial t}(0, t) + \frac{\sigma^2}{2a}\left(1 - e^{-2at}\right), \qquad t \geqslant 0,$$

entails

$$f(0, T) = f^M(0, T), \qquad T \geqslant 0.$$

(7) Show that we have

$$\frac{dP(t, T)}{P(t, T)} = r_t dt + \zeta_t dB_t,$$

and

$$d\left(e^{-\int_0^t r_s ds} P(t, T)\right) = \zeta_t e^{-\int_0^t r_s ds} P(t, T) dB_t, \qquad (6.18)$$

where $(\zeta_t)_{t \in [0, T]}$ is a process to be determined.

Chapter 7

Forward Rate Modeling

In this chapter, we present the Heath, Jarrow and Morton (Heath *et al.* (1992)) and the Brace, Gatarek and Musiela (Brace *et al.* (1997)) frameworks for the modeling of forward rates. In this infinite-dimensional setting, forward rates can be viewed as stochastic processes taking values in a function space. We also cover the HJM absence of arbitrage condition, which is shown to be satisfied in classical time-dependent short-term interest rate models such as the Hull-White model.

7.1 The HJM Model

In previous chapters we started from the modeling of the short rate $(r_t)_{t \in \mathbb{R}_+}$, followed by its consequences on the pricing of bonds $P(t, T)$, and on the expressions of the forward rates $f(t, T, S)$ and $L(t, T, S)$.

In this chapter we choose a different starting point, and consider the problem of directly modeling the instantaneous forward rate $f(t, T)$. The graph given in Figure 6.7 presents a possible random evolution of a forward interest rate curve using the Musiela convention, *i.e.* we will write

$$g(x) = f(t, t + x) = f(t, T), \tag{7.1}$$

under the substitution $x = T - t$, $0 \leqslant x \leqslant T$, and represent a sample of the instantaneous forward curve $x \mapsto f(t, t + x)$ for each $t \in \mathbb{R}_+$.

Definition 7.1. *In the Heath-Jarrow-Morton (HJM) model, the instantaneous forward rate $f(t, T)$ is modeled under $\mathbb{P}^*$ by a Stochastic Differential Equation (SDE) of the form*

$$d_t f(t, T) = \alpha(t, T)dt + \sigma(t, T)dB_t, \qquad 0 \leqslant t \leqslant T, \tag{7.2}$$

where $t \mapsto \alpha(t, T)$ and $t \mapsto \sigma(t, T)$, $0 \leqslant t \leqslant T$, are allowed to be random (adapted) processes.

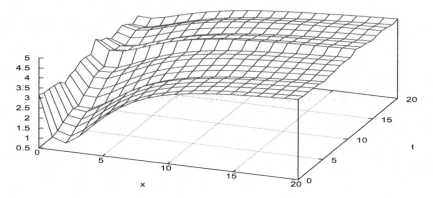

Fig. 7.1: Stochastic process of forward curves.

In the above equation (7.2), the maturity date T is fixed and the differential d_t is with respect to t. Next, we determine the dynamics of the spot forward rate.

Proposition 7.1. *The dynamics of the spot forward rate $f(t, t, T)$ in the HJM model is given by*

$$d_t f(t, t, T) = \frac{f(t, t, T)}{T - t} dt - \frac{r_t}{T - t} dt$$
$$+ \frac{1}{T - t} \int_t^T \alpha(t, s) ds dt + \frac{1}{T - t} \int_t^T \sigma(t, s) ds dB_t, \quad 0 \leqslant t < T.$$

Proof. We have

$$f(t, t, T) = \frac{1}{T - t} \int_t^T f(t, s) ds = \frac{X_t}{T - t}, \qquad 0 \leqslant t < T, \qquad (7.3)$$

under the condition (7.2), where

$$X_t = \int_t^T f(t, s) ds = -\log P(t, T), \qquad 0 \leqslant t \leqslant T$$

and the dynamics of $t \mapsto f(t, s)$ is given by (7.2). We have

$$d_t X_t = -f(t, t) dt + \int_t^T d_t f(t, s) ds$$
$$= -f(t, t) dt + \int_t^T \alpha(t, s) ds dt + \int_t^T \sigma(t, s) ds dB_t$$
$$= -r_t dt + \left(\int_t^T \alpha(t, s) ds \right) dt + \left(\int_t^T \sigma(t, s) ds \right) dB_t,$$

hence

$$d_t X_t \cdot d_t X_t = \left(\int_t^T \sigma(t,s) ds \right)^2 dt,$$

and Itô's formula yields, for h a $\mathcal{C}^2$ function,

$$d_t h(t, X_t) = \frac{\partial h}{\partial t}(t, X_t) dt - r_t \frac{\partial h}{\partial x}(t, X_t) dt + \left(\int_t^T \alpha(t,s) ds \right) \frac{\partial h}{\partial x}(t, X_t) dt$$

$$+ \int_t^T \sigma(t,s) ds \frac{\partial h}{\partial x}(t, X_t) dB_t + \frac{1}{2} \left(\int_t^T \sigma(t,s) ds \right)^2 \frac{\partial^2 h}{\partial x^2}(t, X_t) dt,$$

which gives

$$d_t f(t, t, T) = \frac{X_t dt}{(T-t)^2} + \frac{d_t X_t}{T-t}, \qquad 0 \leqslant t < T.$$

$\square$

7.2 Absence of Arbitrage

An important question is to determine under which conditions the equation (7.2) makes sense in a financial context, and in particular how to "encode" absence of arbitrage in the defining HJM Equation (7.2).

Recall that under absence of arbitrage, the bond price $P(t, T)$ has been constructed (see Chapter 4) as

$$P(t, T) = \mathbb{E}^* \left[e^{-\int_t^T r_s ds} \,\Big|\, \mathcal{F}_t \right] = e^{-\int_t^T f(t,s) ds}, \qquad (7.4)$$

cf. Proposition 5.2. In this framework, we have

$$e^{-\int_0^t r_s ds} P(t, T) = e^{-\int_0^t r_s ds} \mathbb{E}^* \left[e^{-\int_t^T r_s ds} \,\Big|\, \mathcal{F}_t \right]$$

$$= \mathbb{E}^* \left[e^{-\int_0^t r_s ds} e^{-\int_t^T r_s ds} \,\Big|\, \mathcal{F}_t \right]$$

$$= \mathbb{E}^* \left[e^{-\int_0^T r_s ds} \,\Big|\, \mathcal{F}_t \right], \qquad 0 \leqslant t \leqslant T,$$

which is a martingale under $\mathbb{P}^*$, as

$$\mathbb{E}^* \left[\mathbb{E}^* \left[e^{-\int_0^T r_s ds} \,\Big|\, \mathcal{F}_t \right] \,\Big|\, \mathcal{F}_u \right] = \mathbb{E}^* \left[e^{-\int_0^T r_s ds} \,\Big|\, \mathcal{F}_u \right], \qquad 0 \leqslant u \leqslant t,$$

by the tower property (11.7) of conditional expectations in the appendix, see also Proposition 4.1.

Hence, the discounted bond price process given by

$$e^{-\int_0^t r_s ds} P(t,T) = \exp\left(-\int_0^t r_s ds - \int_t^T f(t,s)ds\right)$$

$$= \exp\left(-\int_0^t r_s ds - X_t\right), \qquad 0 \leqslant t \leqslant T, \qquad (7.5)$$

is a martingale under $\mathbb{P}^*$ by Proposition 5.2. In other words, $\mathbb{P}^*$ is a risk-neutral measure, and by the first fundamental theorem of asset pricing, see Harrison and Pliska (1981) and Chapter VII-4a of Shiryaev (1999), we can conclude that the market is without arbitrage opportunities.

Assuming that the short rate $(r_t)_{t \in \mathbb{R}_+}$ has the Markov property, the expression (7.4) can be rewritten as

$$P(t,T) = \mathbb{E}^*\left[e^{-\int_t^T r_s ds} \,\Big|\, r_t\right]$$

$$= F(t, r_t).$$

Using Itô calculus and the martingale property, the above expression lead us in Chapter 4 to the PDE satisfied by $F(t,x)$.

Here, we again apply the same strategy:

(1) Apply Itô's calculus to differentiate (7.5).

(2) Since (7.5) is a martingale under $\mathbb{P}^*$ under absence of arbitrage, we can equate the sum of its dt drift terms to zero.

Proposition 7.2. *(HJM Condition Heath et al. (1992)). Under the condition*

$$\alpha(t,T) = \sigma(t,T) \int_t^T \sigma(t,s)ds, \qquad 0 \leqslant t \leqslant T, \qquad (7.6)$$

which is known as the HJM *absence of arbitrage condition, the discounted bond price process (7.5) is a martingale, and the probability measure $\mathbb{P}^*$ is risk-neutral.*

Proof. Using the process $(X_t)_{t \in [0,T]}$ defined as

$$X_t := \int_t^T f(t,s)ds = -\log P(t,T), \qquad 0 \leqslant t \leqslant T,$$

such that $P(t,T) = e^{-X_t}$, we rewrite the spot forward rate, or yield

$$f(t,t,T) = \frac{1}{T-t} \int_t^T f(t,s)ds,$$

see (5.6) or (7.3), as

$$f(t,t,T) = \frac{1}{T-t} \int_t^T f(t,s)ds = \frac{X_t}{T-t}, \qquad 0 \leqslant t < T,$$

where the dynamics of $t \mapsto f(t,s)$ is given by (7.2). We also use the extended Leibniz integral rule

$$d_t \int_t^T f(t,s)ds = -f(t,t)dt + \int_t^T d_t f(t,s)ds = -r_t dt + \int_t^T d_t f(t,s)ds,$$

which can be checked in the particular case where $f(t,s) = g(t)h(s)$ is a smooth function that satisfies the separation of variables property, as

$$d_t\left(\int_t^T g(t)h(s)ds\right) = d_t\left(g(t)\int_t^T h(s)ds\right)$$

$$= \int_t^T h(s)dsdg(t) + g(t)d_t \int_t^T h(s)ds$$

$$= g'(t)\left(\int_t^T h(s)ds\right)dt - g(t)h(t)dt.$$

We have

$$d_t X_t = d_t \int_t^T f(t,s)ds$$

$$= -f(t,t)dt + \int_t^T d_t f(t,s)ds$$

$$= -f(t,t)dt + \int_t^T \alpha(t,s)dsdt + \int_t^T \sigma(t,s)dsdB_t$$

$$= -r_t dt + \left(\int_t^T \alpha(t,s)ds\right)dt + \left(\int_t^T \sigma(t,s)ds\right)dB_t,$$

hence

$$d_t X_t \cdot d_t X_t = \left(\int_t^T \sigma(t,s)ds\right)^2 dt.$$

Therefore, by Itô's calculus we find

$$d_t P(t,T) = d_t e^{-X_t}$$

$$= -e^{-X_t}d_t X_t + \frac{1}{2}e^{-X_t}(d_t X_t)^2$$

$$= -e^{-X_t} d_t X_t + \frac{1}{2} e^{-X_t} \left(\int_t^T \sigma(t,s)ds \right)^2 dt$$

$$= -e^{-X_t} \left(-r_t dt + \int_t^T \alpha(t,s)dsdt + \int_t^T \sigma(t,s)dsdB_t \right)$$

$$+ \frac{1}{2} e^{-X_t} \left(\int_t^T \sigma(t,s)ds \right)^2 dt,$$

and the discounted bond price process satisfies

$$d_t \left(e^{-\int_0^t r_s ds} P(t,T) \right)$$

$$= -r_t \exp \left(-\int_0^t r_s ds - X_t \right) dt + e^{-\int_0^t r_s ds} d_t P(t,T)$$

$$= -r_t \exp \left(-\int_0^t r_s ds - X_t \right) dt - \exp \left(-\int_0^t r_s ds - X_t \right) d_t X_t$$

$$+ \frac{1}{2} \exp \left(-\int_0^t r_s ds - X_t \right) \left(\int_t^T \sigma(t,s)ds \right)^2 dt$$

$$= -r_t \exp \left(-\int_0^t r_s ds - X_t \right) dt$$

$$- \exp \left(-\int_0^t r_s ds - X_t \right) \left(-r_t dt + \int_t^T \alpha(t,s)dsdt + \int_t^T \sigma(t,s)dsdB_t \right)$$

$$+ \frac{1}{2} \exp \left(-\int_0^t r_s ds - X_t \right) \left(\int_t^T \sigma(t,s)ds \right)^2 dt$$

$$= -\exp \left(-\int_0^t r_s ds - X_t \right) \int_t^T \sigma(t,s)dsdB_t$$

$$- \exp \left(-\int_0^t r_s ds - X_t \right) \left(\int_t^T \alpha(t,s)ds - \frac{1}{2} \left(\int_t^T \sigma(t,s)ds \right)^2 \right) dt.$$

Thus, the discounted bond price process

$$t \mapsto e^{-\int_0^t r_s ds} P(t,T)$$

will be a martingale provided that

$$\int_t^T \alpha(t,s)ds - \frac{1}{2} \left(\int_t^T \sigma(t,s)ds \right)^2 = 0, \qquad 0 \leqslant t \leqslant T. \qquad (7.7)$$

Differentiating the above relation with respect to T, we get

$$\alpha(t,T) = \sigma(t,T) \int_t^T \sigma(t,s)ds,$$

which is in fact equivalent to (7.7).   □

As a consequence of Relation (7.6), the stochastic differential equation defining the instantaneous forward rate $f(t, T)$ rewrites as

$$d_t f(t, T) = \sigma(t, T) \left(\int_t^T \sigma(t, s) ds \right) dt + \sigma(t, T) dB_t,$$

and in integral form this gives

$$f(t, T) = f(0, T) + \int_0^t \alpha(s, T) ds + \int_0^t \sigma(s, T) dB_s \qquad (7.8)$$

$$= f(0, T) + \int_0^t \sigma(s, T) \int_s^T \sigma(s, u) du ds + \int_0^t \sigma(s, T) dB_s,$$

$0 \leqslant t \leqslant T$.

7.3 HJM-Vasicek Forward Rates

In the Vasicek model, the short rate process $(r_t)_{t \in \mathbb{R}_+}$ is solution of

$$dr_t = (a - br_t) dt + \sigma dB_t, \qquad (7.9)$$

see (3.1)-(3.2), as illustrated in Figure 4.4b, with

$$r_t = e^{-bt} r_0 + \frac{a}{b} (1 - e^{-bt}) + \sigma \int_0^t e^{-(t-s)b} dB_s$$

$$= f(0, t) + \frac{\sigma^2}{2b^2} (1 - e^{-bt})^2 + \sigma \int_0^t e^{-(t-s)b} dB_s,$$

where

$$f(0, t) = e^{-bt} r_0 + \frac{a}{b} (1 - e^{-bt}) - \frac{\sigma^2}{2b^2} (1 - e^{-bt})^2, \qquad t \geqslant 0,$$

is deterministic. By (6.2), the instantaneous forward rate process is given by

$$f(t, T) = r_t e^{-(T-t)b} + \frac{a}{b} (1 - e^{-(T-t)b}) - \frac{\sigma^2}{2b^2} (1 - e^{-(T-t)b})^2$$

$$= r_t e^{-(T-t)b} - aC(T - t) - \frac{\sigma^2}{2} C^2(T - t), \qquad (7.10)$$

where $C(x) = -(1 - e^{-bx})/b$, $x > 0$, or, in the Musiela notation, taking $x := T - t$,

$$f(t, T) = f(t, t + x)$$

$$= r_t e^{-bx} + \frac{a}{b} (1 - e^{-bx}) - \frac{\sigma^2}{2b^2} (1 - e^{-bx})^2$$

$$= \frac{a}{b} - \frac{\sigma^2}{2b^2} + \left(r_t - \frac{a}{b} + \frac{\sigma^2}{b^2} \right) e^{-bx} - \frac{\sigma^2}{2b^2} e^{-2bx}, \quad x \geqslant 0.$$

Let us determine the dynamics of the instantaneous forward rate process $(f(t,T))_{t \in [0,T]}$ in the Vasicek model. By (7.10), we have

$$
\begin{aligned}
d_t f(t,T) \\
&= e^{-(T-t)b} dr_t + b e^{-(T-t)b} r_t dt + aC'(T-t)dt + \sigma^2 C(T-t)C'(T-t)dt \\
&= (a - br_t) e^{-(T-t)b} dt + \sigma e^{-(T-t)b} dB_t + b e^{-(T-t)b} r_t dt \\
&\quad + aC'(T-t)dt + \sigma^2 C(T-t)C'(T-t)dt \\
&= -\sigma^2 C(T-t) e^{-(T-t)b} dt + \sigma e^{-(T-t)b} dB_t \\
&= \frac{\sigma^2}{b} e^{-(T-t)b} \left(1 - e^{-(T-t)b} \right) dt + \sigma e^{-(T-t)b} dB_t \\
&= \sigma^2 e^{-(T-t)b} \left(\int_t^T e^{(t-s)b} ds \right) dt + \sigma e^{-(T-t)b} dB_t.
\end{aligned}
$$

Hence, $d_t f(t,T)$ can be written as

$$d_t f(t,T) = \alpha(t,T)dt + \sigma(t,T)dB_t,$$

with

$$\alpha(t,T) = \sigma^2 e^{-(T-t)b} \int_t^T e^{(t-s)b} ds = \frac{\sigma^2}{b} e^{-(T-t)b} \left(1 - e^{-(T-t)b} \right),$$

and $\sigma(t,T) = \sigma e^{-(T-t)b}$, $t \in [0,T]$. Thus, the HJM coefficients in the Vasicek model are deterministic functions of time t and maturity T, with the relation

$$\alpha(t,T) = \sigma^2 e^{-(T-t)b} \int_t^T e^{-(s-t)b} ds$$

$$= \sigma(t,T) \int_t^T \sigma(t,s)ds, \quad 0 \leqslant t \leqslant T, \tag{7.11}$$

which recovers the HJM condition (7.6) between $\alpha(t,T)$ and $\sigma(t,T)$, as the Vasicek model is consistent with the absence of arbitrage hypothesis. Note also that the coefficient a present in (7.9) is not present in (7.11).

Random simulations of the Vasicek instantaneous forward rates are presented in Figures 7.2, 7.3 and 7.4.

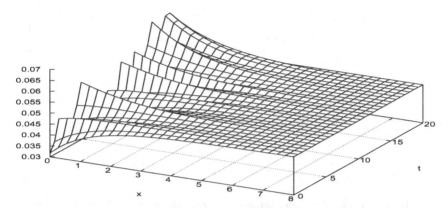

Fig. 7.2: Forward instantaneous curves $(t, x) \mapsto f(t, t + x)$ in the Vasicek model.

Regarding the spot forward rates, we find the dynamics

$$d_t f(t, t, T) = \frac{f(t, t, T)}{T - t} dt - \frac{r_t}{T - t} dt$$

$$+ \frac{\sigma^2}{T - t} \left(\int_t^T e^{-(s-t)b} \int_t^s e^{-(t-u)b} du\, ds \right) dt + \frac{\sigma}{T - t} \left(\int_t^T e^{-(s-t)b} ds \right) dB_t.$$

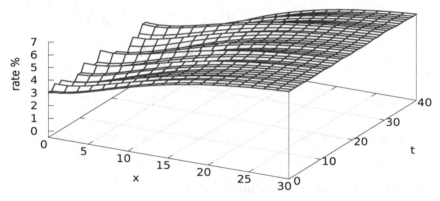

Fig. 7.3: Forward instantaneous curves $(t, x) \mapsto f(t, t + x)$ in the Vasicek model.

We check that at fixed time t, all Vasicek instantaneous forward curves converge to the "long rate"

$$\lim_{x \to \infty} f(t, t + x) = \lim_{T \to \infty} f(t, T) = \frac{a}{b} - \frac{\sigma^2}{2b^2}$$

as x goes to infinity.

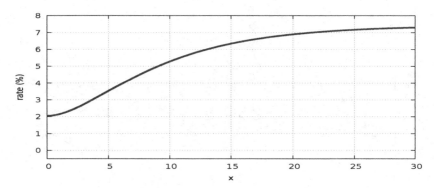

Fig. 7.4: Forward instantaneous curve $x \mapsto f(0, x)$ in the Vasicek model.

Note that in Figure 7.3, the first "slice" for $x = 0$ of the surface is actually the short rate Vasicek process $r_t = f(t, t) = f(t, t+0)$, which is represented in Figure 7.5 using another discretization.

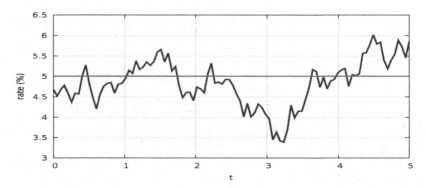

Fig. 7.5: Short-term interest rate curve $t \mapsto r_t$ in the Vasicek model.

7.4   Markov Property of Short Rates

As noted above, the Markov property of the short rate is of capital importance when deriving the pricing PDE for $P(t, T) = F(t, r_t)$. Thus, a natural question is:

- When does the short-term interest rate process have the Markov property in the HJM model?

Recall that from Relations (7.6) and (7.8), in the HJM model the short rate process is given by

$$r_t = f(t,t) = f(0,t) + \int_0^t \sigma(s,t) \int_s^t \sigma(s,u) du ds + \int_0^t \sigma(s,t) dB_s.$$

In general, a process of the form

$$t \mapsto Z_t := \int_0^t \sigma(s,t) dB_s, \qquad t \geqslant 0, \tag{7.12}$$

where $s \mapsto \sigma(s,t)$ is $(\mathcal{F}_s)_{s \in \mathbb{R}_+}$-adapted, may *not* be Markovian due to the dependence of $\sigma(s,t)$ in the time variable t. In fact, we have

$$
\begin{aligned}
\mathbb{E}[Z_t \mid \mathcal{F}_u] &= \mathbb{E}\left[\int_0^t \sigma(s,t) dB_s \,\middle|\, \mathcal{F}_u\right] \\
&= \mathbb{E}\left[\int_0^u \sigma(s,t) dB_s + \int_u^t \sigma(s,t) dB_s \,\middle|\, \mathcal{F}_u\right] \\
&= \mathbb{E}\left[\int_0^u \sigma(s,t) dB_s \,\middle|\, \mathcal{F}_u\right] + \mathbb{E}\left[\int_u^t \sigma(s,t) dB_s \,\middle|\, \mathcal{F}_u\right] \\
&= \mathbb{E}\left[\int_0^u \sigma(s,t) dB_s \,\middle|\, \mathcal{F}_u\right] \\
&= \int_0^u \sigma(s,t) dB_s, \tag{7.13}
\end{aligned}
$$

where we applied Relation (1.7) of Chapter 1.

According to the Markov property the above quantity should depend only on u, t, and

$$Z_u = \int_0^u \sigma(s,u) dB_s,$$

and there is a priori no reason for this property to hold here, as follows by inspection of (7.13).

Nevertheless, the Markov property of the process $(Z_t)_{t \in \mathbb{R}_+}$ defined as in (7.12) does hold for some particular choices of $\sigma(t,T)$. For example, in case

$$\sigma(s,t) = e^{-(t-s)b}, \qquad 0 \leqslant s \leqslant t,$$

we have

$$\mathbb{E}[Z_t \mid \mathcal{F}_u] = \int_0^u \sigma(s,t) dB_s$$

$$= \int_0^u e^{-(t-s)b} dB_s$$

$$= e^{-(t-u)b} \int_0^u e^{-(u-s)b} dB_s$$

$$= e^{-(t-u)b} Z_u, \qquad 0 \leqslant u \leqslant t,$$

as known from the explicit solution to the Vasicek model, cf. (3.2). More generally, the Markov property does hold for a stochastic integral process of the form

$$t \mapsto \int_0^t \sigma(s,t) dB_s,$$

under the product condition

$$\sigma(s,t) = \xi(s)\psi(t), \qquad 0 \leqslant s \leqslant t, \tag{7.14}$$

where $\xi(s)$ and $\psi(t)$ are two deterministic functions. Indeed, we have

$$\mathbb{E}[Z_t \mid \mathcal{F}_u] = \int_0^u \sigma(s,t) dB_s$$

$$= \psi(t) \int_0^u \xi(s) dB_s$$

$$= \frac{\psi(t)}{\psi(u)} \int_0^u \psi(u)\xi(s) dB_s$$

$$= \frac{\psi(t)}{\psi(u)} Z_u, \qquad 0 \leqslant u \leqslant t.$$

Recall that in the Vasicek model we have $\sigma(s,t) = \sigma e^{-(t-s)b}$, $0 \leqslant s \leqslant t$, hence condition (7.14) holds and the short rate is indeed a Markov process.

Hull-White model

Our goal is now to derive a stochastic differential equation satisfied by the short rate process $(r_t)_{t \in \mathbb{R}_+}$ in the Markovian HJM model under the product assumption (7.14) on the volatility coefficient $\sigma(s,t)$. In this way, we will recover the time-dependent Hull-White short rate model described in Section 3.3.

By (7.2) and the HJM Condition (7.6), or directly by (7.8), using (7.14) we have

$$r_t = f(t,t)$$

$$= f(0,t) + \int_0^t \sigma(s,t) \int_s^t \sigma(s,u) du ds + \int_0^t \sigma(s,t) dB_s$$

$$= f(0,t) + \int_0^t \xi(s)\psi(t) \int_s^t \xi(s)\psi(u) du ds + \psi(t) \int_0^t \xi(s) dB_s,$$

hence

$$r_t = U(t) + \psi(t) \int_0^t \xi(s) dB_s, \tag{7.15}$$

where

$$U(t) = f(0,t) + \psi(t) \int_0^t \xi^2(s) \int_s^t \psi(u) du ds.$$

Using the relation

$$\int_0^t \xi(s) dB_s = \frac{r_t - U(t)}{\psi(t)}, \qquad t \geqslant 0,$$

that follows from (7.15), we have

$$dr_t = U'(t)dt + \psi'(t)\left(\int_0^t \xi(s) dB_s \right)dt + \psi(t)\xi(t)dB_t$$

$$= U'(t)dt + (r_t - U(t))\frac{\psi'(t)}{\psi(t)}dt + \psi(t)\xi(t)dB_t,$$

which indeed shows that the short rate process $(r_t)_{t \in \mathbb{R}_+}$ has the Markov property as the solution of a stochastic differential equation, cf. Property 4.1.

The above equation belongs to the class of short rate models of the form Hull and White (1990)

$$dr_t = (a(t) - b(t)r_t)dt + \sigma(t)dB_t,$$

cf. Section 3.3, which can be interpreted as time-dependent Vasicek models with explicit solution

$$r_t = r_s e^{-\int_s^t b(\tau)d\tau} + \int_s^t e^{-\int_u^t b(\tau)d\tau} a(u)du + \int_s^t \sigma(u) e^{-\int_u^t b(\tau)d\tau} dB_u,$$

$0 \leqslant s \leqslant t$.

7.5   The BGM Model

The BGM model has been introduced in Brace *et al.* (1997) for the pricing of interest rate derivatives such as interest rate caps and swaptions on the LIBOR market.

The models considered in the previous sections suffer from the following drawbacks:

- explicitly computable models such as the Vasicek model do not satisfy the positivity of rates property.

- models with positive rates (*e.g.* the CIR model) do not lead to explicit analytical formulas.

- the lack of explicit analytical formulas makes it necessary to use the Monte Carlo method for pricing, which makes model calibration difficult in practice.

- fitting the forward interest rate curves in these models is problematic.

Thus there is a strong interest in models that:

- yield positive interest rates, and

- permit to derive explicit formulas for the computation of prices.

These two goals can be achieved by the BGM model. In the sequel we assume that the bond price $P(t, T_i)$ satisfies
$$\frac{dP(t, T_i)}{P(t, T_i)} = r_t dt + \zeta_i(t) dB_t, \qquad i = 1, 2, \ldots, n,$$
where $(B_t)_{t \in \mathbb{R}_+}$ is a standard Brownian motion under $\mathbb{P}$, and we consider a family $\widehat{\mathbb{P}}_1, \ldots, \widehat{\mathbb{P}}_n$ of probability measures such that the process
$$B_t^{(i)} := B_t - \int_0^t \zeta_i(s) ds, \qquad 0 \leqslant t \leqslant T_i,$$
is a standard valued Brownian motion for $i = 1, 2, \ldots, n$, see Definition 8.1 and Chapter 8 for details.

Definition 7.2. *In the BGM model, we assume that the LIBOR rate $L(t, T_i, T_{i+1})$ is a geometric Brownian motion under $\widehat{\mathbb{P}}_{i+1}$, i.e.*
$$\frac{dL(t, T_i, T_{i+1})}{L(t, T_i, T_{i+1})} = \gamma_i(t) dB_t^{(i+1)}, \tag{7.16}$$

$0 \leqslant t \leqslant T_i$, $i = 1, 2, \ldots, n - 1$, *for some adapted process* $(\gamma_i(t))_{t \in [0, T_i]}$, $i = 1, 2, \ldots, n - 1$.

Equation (7.16) can be solved as

$$L(u, T_i, T_{i+1}) = L(t, T_i, T_{i+1}) \exp\left(\int_t^u \gamma_i(s) dB_s^{(i+1)} - \frac{1}{2} \int_t^u |\gamma_i(s)|^2 ds \right),$$

i.e. for $u = T_i$,

$$L(T_i, T_i, T_{i+1}) = L(t, T_i, T_{i+1}) \exp\left(\int_t^{T_i} \gamma_i(s) dB_s^{(i+1)} - \frac{1}{2} \int_t^{T_i} |\gamma_i(s)|^2 ds \right).$$

Since $L(t, T_i, T_{i+1})$ is a geometric Brownian motion under $\widehat{\mathbb{P}}_{i+1}$, $i = 0, 1, \ldots, n-1$, standard interest rate caplets can be priced at time $t \in [0, T_i]$ from the Black-Scholes formula of Section 2.3, see Section 9.1 below.

In the next proposition we determine the dynamics of $L(t, T_i, T_{i+1})$ under $\widehat{\mathbb{P}}_k$, $1 \leqslant i < k \leqslant n$. Again, we let

$$\delta_k = T_{k+1} - T_k, \qquad k = 1, 2, \ldots, n - 1.$$

Proposition 7.3. *For* $1 \leqslant i < k \leqslant n$ *we have*

$$\frac{dL(t, T_i, T_{i+1})}{L(t, T_i, T_{i+1})} = -\gamma_i(t) \sum_{j=i+1}^{k-1} \gamma_j(t) \frac{\delta_j L(t, T_j, T_{j+1})}{1 + \delta_j L(t, T_j, T_{j+1})} dt + \gamma_i(t) dB_t^{(k)},$$

$0 \leqslant t \leqslant T_i$, *where* $(\gamma_i(t))_{t \in [0, T_i]}$ *is an adapted process,* $\left(B_t^{(k)} \right)_{t \in \mathbb{R}_+}$ *is a standard Brownian motion under* $\widehat{\mathbb{P}}_k$, *and* $L(t, T_i, T_{i+1})$, $0 \leqslant t \leqslant T_i$, *is a martingale under* $\widehat{\mathbb{P}}_{i+1}$, $i = 1, 2, \ldots, n - 1$.

Proof. By (7.16), we have

$$d\left(\frac{P(t, T_i)}{P(t, T_{i+1})} \right) = d(1 + \delta_i L(t, T_i, T_{i+1}))$$

$$= \delta_i \gamma_i(t) L(t, T_i, T_{i+1}) dB_t^{(i+1)}$$

$$= \gamma_i(t) \frac{P(t, T_i)}{P(t, T_{i+1})} \frac{\delta_i L(t, T_i, T_{i+1})}{1 + \delta_i L(t, T_i, T_{i+1})} dB_t^{(i+1)}. \qquad (7.17)$$

On the other hand, using the dynamics

$$\frac{dP(t, T_i)}{P(t, T_i)} = r_t dt + \zeta_i(t) dB_t, \qquad i = 1, 2, \ldots, n,$$

and Itô's calculus, we have

$$d\left(\frac{P(t, T_i)}{P(t, T_{i+1})} \right) = \frac{P(t, T_i)}{P(t, T_{i+1})} (\zeta_i(t) - \zeta_{i+1}(t))(dB_t - \zeta_{i+1}(t) dt)$$

$$= \frac{P(t,T_i)}{P(t,T_{i+1})}(\zeta_i(t) - \zeta_{i+1}(t))dB_t^{(i+1)}. \tag{7.18}$$

By identification of (7.17) with (7.18), we get

$$\zeta_{i+1}(t) - \zeta_i(t) = -\frac{\delta_i L(t,T_i,T_{i+1})}{1 + \delta_i L(t,T_i,T_{i+1})}\gamma_i(t), \tag{7.19}$$

$0 \leqslant t \leqslant T_i$, $i = 1, 2, \ldots, n - 1$, hence

$$\zeta_k(t) = \zeta_i(t) - \sum_{j=i}^{k-1} \frac{\delta_j L(t,T_j,T_{j+1})}{1 + \delta_j L(t,T_j,T_{j+1})}\gamma_j(t), \tag{7.20}$$

$0 \leqslant t \leqslant T_i$, $1 \leqslant i < k \leqslant n$. Since $dB_t^{(i+1)} = dB_t - \zeta_{i+1}(t)dt$ and

$$dB_t^{(k)} = dB_t^{(i+1)} - (\zeta_k(t) - \zeta_{i+1}(t))dt, \quad 1 \leqslant i < k \leqslant n,$$

for $k = i + 1, \ldots, n$, we have

$$\frac{dL(t,T_i,T_{i+1})}{L(t,T_i,T_{i+1})} = \gamma_i(t)dB_t^{(i+1)}$$

$$= \gamma_i(t)dB_t^{(k)} - \gamma_i(t)\big(dB_t^{(k)} - dB_t^{(i+1)}\big)$$

$$= \gamma_i(t)dB_t^{(k)} + \gamma_i(t)(\zeta_k(t) - \zeta_{i+1}(t))dt$$

$$= \gamma_i(t)dB_t^{(k)} - \gamma_i(t)\sum_{j=i+1}^{k-1}(\zeta_j(t) - \zeta_{j+1}(t))dt$$

$$= -\sum_{j=i+1}^{k-1}\frac{\delta_j L(t,T_j,T_{j+1})}{1 + \delta_j L(t,T_j,T_{j+1})}\gamma_i(t)\gamma_j(t)dt + \gamma_i(t)dB_t^{(k)},$$

$0 \leqslant t \leqslant T_i$, $k = i + 1, \ldots, n$. $\qquad\square$

Similarly, for $1 \leqslant k \leqslant i < n$, we have:

$$\frac{dL(t,T_i,T_{i+1})}{L(t,T_i,T_{i+1})} = \gamma_i(t)dB_t^{(k)} + \gamma_i(t)\sum_{j=k}^{i}(\zeta_j(t) - \zeta_{j+1}(t))dt$$

$$= \gamma_i(t)dB_t^{(k)} + \gamma_i(t)\sum_{j=k}^{i}\frac{\delta_j L(t,T_j,T_{j+1})}{1 + \delta_j L(t,T_j,T_{j+1})}\gamma_j(t)dt,$$

and

$$\frac{dL(t,T_i,T_{i+1})}{L(t,T_i,T_{i+1})} = \gamma_i(t)dB_t^{(i+1)}$$

$$= \gamma_i(t)dB_t^{(i)} + \frac{\delta_i L(t,T_i,T_{i+1})}{1 + \delta_i L(t,T_i,T_{i+1})}|\gamma_i(t)|^2 dt.$$

In Table 7.1, we summarize some stochastic models used for interest rates.

	Model
Short rate r_t	Mean reverting SDEs
Instantaneous forward rate $f(t, s)$	HJM model
Forward rate $f(t, T, S)$	BGM model

Table 7.1: Stochastic interest rate models.

Exercises

Exercise 7.1. (Exercise 5.1 continued).

(1) Derive the stochastic equation satisfied by the instantaneous forward rate $f(t, T)$ in the Ho-Lee model (3.8).
(2) Check that the HJM absence of arbitrage Condition (7.6) is satisfied in this equation.

Exercise 7.2. (Exercise 5.2 continued).

(1) Derive the stochastic equation satisfied by the instantaneous forward rate $f(t, T)$ in the Vasicek model (4.25).
(2) Check that the HJM absence of arbitrage condition is satisfied in the equation of Question (1).

Exercise 7.3. (Exercise 5.4 continued). Check that the HJM absence of arbitrage condition is satisfied.

Exercise 7.4. (Exercise 6.3 continued).

(1) Compute $d_t f(t, T)$ from (6.17), and derive the stochastic equation satisfied by the instantaneous forward rate $f(t, T)$.
(2) Check that the HJM absence of arbitrage condition is satisfied in this equation.

Exercise 7.5. Given $(B_t)_{t \in \mathbb{R}_+}$ a standard Brownian motion, consider a HJM model given by

$$d_t f(t, T) = \frac{\sigma^2}{2} T(T^2 - t^2) dt + \sigma T dB_t. \tag{7.21}$$

(1) Show that the HJM condition (7.6) is satisfied by (7.21).
(2) Compute $f(t, T)$ by solving (7.21).

   Hint: We have $f(t, T) = f(0, T) + \int_0^t d_s f(s, T) = \cdots$
(3) Compute the short rate $r_t = f(t, t)$ from the result of Question (2).
(4) Show that the short rate process $(r_t)_{t \in \mathbb{R}_+}$ satisfies a stochastic differential equation of the form

$$dr_t = \eta(t)dt + (r_t - f(0, t))\psi(t)dt + \xi(t)dB_t,$$

where $\eta(t)$, $\psi(t)$, $\xi(t)$ are deterministic functions to be determined.

Exercise 7.6. Stochastic string model (Santa-Clara and Sornette (2001)). Consider an instantaneous forward rate $f(t, x)$ solution of

$$d_t f(t, x) = \alpha x^2 dt + \sigma d_t B(t, x), \qquad (7.22)$$

with a flat initial curve $f(0, x) = r$, where x represents the time to maturity, and $(B(t, x))_{(t,x) \in \mathbb{R}_+^2}$ is a standard *Brownian sheet* with covariance

$$\mathbb{E}[B(s, x)B(t, y)] = (\min(s, t))(\min(x, y)), \qquad s, t, x, y \in \mathbb{R}_+, \qquad (7.23)$$

and initial conditions $B(t, 0) = B(0, x) = 0$ for all $t, x \in \mathbb{R}_+$.

(1) Solve the equation (7.22) for $f(t, x)$.
(2) Compute the short-term interest rate $r_t = f(t, 0)$.
(3) Compute the value at time $t \in [0, T]$ of the bond price

$$P(t, T) = \exp\left(-\int_0^{T-t} f(t, x)dx \right)$$

with maturity T.
(4) Compute the variance $\mathbb{E}\left[\left(\int_0^{T-t} B(t, x)dx \right)^2 \right]$ of the centered Gaussian random variable $\int_0^{T-t} B(t, x)dx$.
(5) Compute the expected value $\mathbb{E}^*[P(t, T)]$.
(6) Find the value of α such that the discounted bond price

$$e^{-rt} P(t, T) = \exp\left(-rT - \frac{\alpha}{3}t(T - t)^3 - \sigma \int_0^{T-t} B(t, x)dx \right),$$

$0 \leqslant t \leqslant T$, satisfies $\mathbb{E}^*[P(t, T)] = e^{-(T-t)r}$.

Chapter 8

Forward Measures and Derivative Pricing

In this chapter we introduce the notion of forward rate and forward swap measures for the pricing of interest rate derivatives. The dynamics of interest rate and asset prices processes under forward measures are determined using the Girsanov theorem. Applications are given to the pricing of bond options, with explicit calculations in the Vasicek model.

8.1 Forward Rate Measures

We consider a sequence of maturity dates arranged according to a discrete *tenor structure*

$$\{0 = T_0 < T_1 < T_2 < \cdots < T_n\}.$$

Maturity	2D	1W	1M	2M	3M	1Y	2Y	3Y	4Y	5Y	6Y	7Y
Rate (%)	2.55	2.53	2.56	2.52	2.48	2.34	2.49	2.79	3.07	3.31	3.52	3.71
Maturity	8Y	9Y	10Y	11Y	12Y	13Y	14Y	15Y	20Y	25Y	30Y	
Rate (%)	3.88	4.02	4.14	4.23	4.33	4.40	4.47	4.54	4.74	4.83	4.86	

Table 8.1: Forward rates arranged according to a tenor structure.

A sample of data used to generate the forward interest rate curve of Table 8.2 is given in Table 8.1, which contains the values of $(T_1, T_2, \ldots, T_{23})$ and of $\{f(t, t + T_i, t + T_i + \delta)\}_{i=1,2,\ldots,23}$, with $t = 07/05/2003$ and $\delta = $ six months.

TimeSerieNb	505
AsOfDate	7-mai-03
2D	2,55
1W	2,53
1M	2,56
2M	2,52
3M	2,48
1Y	2,34
2Y	2,49
3Y	2,79
4Y	3,07
5Y	3,31
6Y	3,52
7Y	3,71
8Y	3,88
9Y	4,02
10Y	4,14
11Y	4,23
12Y	4,33
13Y	4,4
14Y	4,47
15Y	4,54
20Y	4,74
25Y	4,83
30Y	4,86

Table 8.2: Forward rates according to a tenor structure.

Given a money market account earning interest at the instantaneous short rate r_t, the price at time t of a contingent claim with payoff F at maturity time T_i is determined by the conditional expectation

$$\mathbb{E}^* \left[e^{- \int_t^{T_i} r_s ds} F \,\middle|\, \mathcal{F}_t \right]$$

under a risk neutral probability measure $\mathbb{P}^*$, see (2.17). When the interest rate process $(r(t))_{t \in \mathbb{R}_+}$ is a deterministic function of time, this expression becomes

$$e^{- \int_t^{T_i} r(s) ds} \, \mathbb{E}^*[F \mid \mathcal{F}_t],$$

and when $(r(t))_{t \in \mathbb{R}_+}$ equals a deterministic constant r, we find the well-known expression

$$e^{-(T_i - t)r} \, \mathbb{E}^*[F \mid \mathcal{F}_t], \qquad 0 \leqslant t \leqslant T.$$

In most interest rate models, the short-term interest rate $(r_t)_{t \in \mathbb{R}_+}$ is a random process and the above manipulation will not be allowed, meaning that we will have to evaluate expressions of the form

$$\mathbb{E}^* \left[e^{- \int_t^{T_i} r_s ds} F \,\middle|\, \mathcal{F}_t \right] \tag{8.1}$$

where $(r_t)_{t \in \mathbb{R}_+}$ will be a random process, adding another level of complexity in comparison with the standard Black-Scholes framework of Chapter 2.

In the case of a constant payoff $F = \$1$, such expressions have been evaluated as solutions of PDEs for the computation of bond prices in Chapter 4.

When the payoff F is random, *e.g.* of the form $F = h(f(T, T, S))$ for an option on the spot forward rate $f(T, T, S)$, the computation of (8.1) would require the knowledge of the joint distribution of $\int_t^T r_s ds$ and $f(T, T, S)$, which can lead to complex computations.

The forward measure defined in Definition 8.1 will be our main tool for the evaluations of expressions of the form (8.1).

Definition 8.1. *The T_i-forward probability measure $\widehat{\mathbb{P}}_i$ is defined as*

$$\frac{d\widehat{\mathbb{P}}_i}{d\mathbb{P}^*} = \frac{1}{P(0, T_i)} e^{-\int_0^{T_i} r_s ds}, \qquad i = 1, 2, \ldots, n. \tag{8.2}$$

In the sequel, the expectation under $\widehat{\mathbb{P}}_i$ will be denoted by $\widehat{\mathbb{E}}_i$, $i = 1, 2, \ldots, n$. Relation (8.2) can be equivalently stated as

$$\begin{aligned}
\widehat{\mathbb{E}}_i[F] &= \int_\Omega F(\omega) d\widehat{\mathbb{P}}_i(\omega) \\
&= \frac{1}{P(0, T_i)} \int_\Omega e^{-\int_0^{T_i} r_s ds} F(\omega) d\mathbb{P}^*(\omega) \\
&= \frac{1}{P(0, T_i)} \mathbb{E}^* \left[e^{-\int_0^{T_i} r_s ds} F \right],
\end{aligned}$$

for all integrable random variables F. Note that for $i = 1, 2, \ldots, n$, we also have

$$\begin{aligned}
\mathbb{E}^* \left[\frac{d\widehat{\mathbb{P}}_i}{d\mathbb{P}^*} \;\middle|\; \mathcal{F}_t \right] &= \frac{1}{P(0, T_i)} \mathbb{E}^* \left[e^{-\int_0^{T_i} r_s ds} \;\middle|\; \mathcal{F}_t \right] \\
&= \frac{e^{-\int_0^t r_s ds}}{P(0, T_i)} \mathbb{E}^* \left[e^{-\int_t^{T_i} r_s ds} \;\middle|\; \mathcal{F}_t \right] \\
&= \frac{P(t, T_i)}{P(0, T_i)} e^{-\int_0^t r_s ds}, \qquad 0 \leqslant t \leqslant T_i.
\end{aligned}$$

Proposition 8.1 will allow us to price contingent claims using the forward measure $\widehat{\mathbb{P}}_i$ with maturity T_i, $i = 1, 2, \ldots, n$.

Proposition 8.1. *For all sufficiently integrable random variables F, we have the pricing relation*

$$\mathbb{E}^* \left[F e^{-\int_t^{T_i} r_s ds} \;\middle|\; \mathcal{F}_t \right] = P(t, T_i) \widehat{\mathbb{E}}_i[F \mid \mathcal{F}_t], \qquad 0 \leqslant t \leqslant T_i, \tag{8.3}$$

$i = 1, 2, \ldots, n.$

Proof. For all bounded and $\mathcal{F}_t$-measurable random variables G, we have

$$\mathbb{E}^*\left[GF\,\mathrm{e}^{-\int_t^{T_i} r_s ds}\right] = P(0,T_i)\widehat{\mathbb{E}}_i\left[G\,\mathrm{e}^{\int_0^t r_s ds}F\right]$$

$$= P(0,T_i)\widehat{\mathbb{E}}_i\left[G\,\mathrm{e}^{\int_0^t r_s ds}\widehat{\mathbb{E}}_i[F\mid\mathcal{F}_t]\right]$$

$$= \mathbb{E}^*\left[G\,\mathrm{e}^{-\int_t^{T_i} r_s ds}\widehat{\mathbb{E}}_i[F\mid\mathcal{F}_t]\right]$$

$$= \mathbb{E}^*[GP(t,T_i)\widehat{\mathbb{E}}_i[F\mid\mathcal{F}_t]],$$

hence

$$\mathbb{E}^*\left[F\,\mathrm{e}^{-\int_t^{T_i} r_s ds}\,\Big|\,\mathcal{F}_t\right] = P(t,T_i)\widehat{\mathbb{E}}_i[F\mid\mathcal{F}_t],\qquad 0\leqslant t\leqslant T_i,$$

for all integrable random variables F, which shows (8.3). In other words, we have

$$\mathbb{E}^*\left[GF\,\mathrm{e}^{-\int_t^{T_i} r_s ds}\right] = P(0,T_i)\widehat{\mathbb{E}}_i\left[G\,\mathrm{e}^{\int_0^t r_s ds}F\right]$$

$$= P(0,T_i)\widehat{\mathbb{E}}_i\left[G\,\mathrm{e}^{\int_0^t r_s ds}\widehat{\mathbb{E}}_i[F\mid\mathcal{F}_t]\right]$$

$$= P(0,T_i)\,\mathbb{E}^*\left[\frac{d\widehat{\mathbb{P}}_i}{d\mathbb{P}^*}G\,\mathrm{e}^{\int_0^t r_s ds}\widehat{\mathbb{E}}_i[F\mid\mathcal{F}_t]\right]$$

$$= P(0,T_i)\,\mathbb{E}^*\left[\frac{1}{P(0,T_i)}\mathrm{e}^{-\int_0^{T_i} r_s ds}G\,\mathrm{e}^{\int_0^t r_s ds}\widehat{\mathbb{E}}_i[F\mid\mathcal{F}_t]\right]$$

$$= \widehat{\mathbb{E}}_i\left[G\,\mathrm{e}^{-\int_t^{T_i} r_s ds}\widehat{\mathbb{E}}_i[F\mid\mathcal{F}_t]\right]$$

$$= \widehat{\mathbb{E}}_i[GP(t,T_i)\widehat{\mathbb{E}}_i[F\mid\mathcal{F}_t]].$$

On the last line we used the characterization of conditional expectation $X = \mathbb{E}^*[F\mid\mathcal{F}_t] \iff \mathbb{E}[GX] = \mathbb{E}[GF]$ for all G bounded and $\mathcal{F}_t$-measurable, see Relation (11.4) in the appendix, the bond pricing relation

$$P(t,T_i) = \mathbb{E}^*\left[\mathrm{e}^{-\int_t^{T_i} r_s ds}\,\Big|\,\mathcal{F}_t\right],$$

and the fact that G and $\widehat{\mathbb{E}}_i[F\mid\mathcal{F}_t]$ are $\mathcal{F}_t$-measurable random variables.□

As a consequence of Proposition 8.1, the computation of the claim price

$$\mathbb{E}^*\left[F\,\mathrm{e}^{-\int_t^{T_i} r_s ds}\,\Big|\,\mathcal{F}_t\right]$$

can be replaced by that of $P(t,T_i)\widehat{\mathbb{E}}_i[F\mid\mathcal{F}_t]$ under the *forward* measure $\widehat{\mathbb{P}}_i$.

As a consequence of Proposition 8.1, the next Proposition 8.2 tells us how the Radon-Nikodym density $d\widehat{\mathbb{P}}_i/d\mathbb{P}^*$ behaves under conditioning with respect to $\mathcal{F}_t$. Recall that by definition,

$$\Lambda_t := \frac{d\widehat{\mathbb{P}}_{i|\mathcal{F}_t}}{d\mathbb{P}^*_{|\mathcal{F}_t}}, \qquad 0 \leqslant t \leqslant T_i,$$

is the only random variable satisfying

$$\widehat{\mathbb{E}}_i[F \mid \mathcal{F}_t] = \mathbb{E}^*[F\Lambda_t \mid \mathcal{F}_t], \qquad 0 \leqslant t \leqslant T_i,$$

i.e.

$$\int_\Omega F(\omega) d\widehat{\mathbb{P}}_{i|\mathcal{F}_t}(\omega) = \int_\Omega F(\omega)\Lambda_t(\omega) d\mathbb{P}^*_{|\mathcal{F}_t}(\omega), \qquad 0 \leqslant t \leqslant T_i,$$

for all bounded random variables F.

Proposition 8.2. *We have*

$$\frac{d\widehat{\mathbb{P}}_{i|\mathcal{F}_t}}{d\mathbb{P}^*_{|\mathcal{F}_t}} = \frac{e^{-\int_t^{T_i} r_s ds}}{P(t,T_i)}, \qquad 0 \leqslant t \leqslant T_i. \tag{8.4}$$

Proof. Rewrite (8.3) as

$$\widehat{\mathbb{E}}_i[F \mid \mathcal{F}_t] = \mathbb{E}^* \left[F \frac{e^{-\int_t^{T_i} r_s ds}}{P(t,T_i)} \,\middle|\, \mathcal{F}_t \right], \qquad 0 \leqslant t \leqslant T_i,$$

for all F bounded and measurable, which implies (8.4). □

Note that $\dfrac{d\widehat{\mathbb{P}}_{i|\mathcal{F}_t}}{d\mathbb{P}^*_{|\mathcal{F}_t}}$ is *not* equal to $\mathbb{E}^* \left[\dfrac{d\widehat{\mathbb{P}}_i}{d\mathbb{P}^*} \,\middle|\, \mathcal{F}_t \right]$, in fact we have

$$\mathbb{E}^* \left[\frac{d\widehat{\mathbb{P}}_i}{d\mathbb{P}^*} \,\middle|\, \mathcal{F}_t \right] = \frac{1}{P(0,T_i)} \mathbb{E}^* \left[e^{-\int_0^{T_i} r_s ds} \,\middle|\, \mathcal{F}_t \right] \tag{8.5}$$

$$= \frac{P(t,T_i)}{P(0,T_i)} e^{-\int_0^t r_s ds}, \qquad 0 \leqslant t \leqslant T_i,$$

since the discounted bond price process (8.6) is an $(\mathcal{F}_t)_{t\in[0,T_i]}$-martingale under $\mathbb{P}^* = \mathbb{P}$. In addition, we have the following result, which states a martingale property for the deflated processes $P(t,T_j)/P(t,T_i)$.

Proposition 8.3. *For all $T_i, T_j \geqslant 0$, the process*

$$t \mapsto \frac{P(t,T_j)}{P(t,T_i)}, \qquad 0 \leqslant t \leqslant \min(T_i, T_i),$$

is an $(\mathcal{F}_t)_{t\in[0,\min(T_i,T_j)]}$-martingale under $\widehat{\mathbb{P}}_i$, provided that it is integrable.

Proof. For all bounded and $\mathcal{F}_s$-measurable random variables F, from Relation (8.5) we have

$$
\begin{aligned}
\widehat{\mathbb{E}}_i \left[F \frac{P(t, T_j)}{P(t, T_i)} \right] &= \mathbb{E}^* \left[F \frac{e^{-\int_0^{T_i} r_u du}}{P(0, T_i)} \frac{P(t, T_j)}{P(t, T_i)} \right] \\
&= \frac{1}{P(0, T_i)} \mathbb{E}^* \left[F e^{-\int_0^t r_u du} P(t, T_j) \right] \\
&= \frac{1}{P(0, T_i)} \mathbb{E}^* \left[F e^{-\int_0^s r_u du} P(s, T_j) \right] \\
&= \mathbb{E}^* \left[F \frac{e^{-\int_0^{T_i} r_u du}}{P(0, T_i)} \frac{P(s, T_j)}{P(s, T_i)} \right] \\
&= \widehat{\mathbb{E}}_i \left[F \frac{P(s, T_j)}{P(s, T_i)} \right],
\end{aligned}
$$

where we used the characterization of conditional expectation $X = \mathbb{E}^*[F \mid \mathcal{F}_t] \iff \mathbb{E}[GX] = \mathbb{E}[GF]$ for all G bounded and $\mathcal{F}_t$-measurable, see Relation (11.4) in the appendix. Hence, we have

$$
\widehat{\mathbb{E}}_i \left[\frac{P(t, T_j)}{P(t, T_i)} \,\Big|\, \mathcal{F}_s \right] = \frac{P(s, T_j)}{P(s, T_i)}, \qquad 0 \leqslant t \leqslant \min(T_i, T_j). \qquad \square
$$

Note that Proposition 8.3 can also be recovered from Proposition 8.2, as follows:

$$
\begin{aligned}
\widehat{\mathbb{E}}_i \left[\frac{P(t, T_j)}{P(t, T_i)} \,\Big|\, \mathcal{F}_s \right] &= \mathbb{E}^* \left[\frac{P(t, T_j)}{P(t, T_i)} \frac{d\widehat{\mathbb{P}}_{i|\mathcal{F}_s}}{d\mathbb{P}^*_{|\mathcal{F}_s}} \,\Big|\, \mathcal{F}_s \right] \\
&= \mathbb{E}^* \left[\frac{P(t, T_j)}{P(t, T_i)} \frac{e^{-\int_s^{T_i} r_u du}}{P(s, T_i)} \,\Big|\, \mathcal{F}_s \right] \\
&= \frac{1}{P(s, T_i)} \mathbb{E}^* \left[\frac{P(t, T_j)}{P(t, T_i)} e^{-\int_s^{T_i} r_u du} \,\Big|\, \mathcal{F}_s \right] \\
&= \frac{1}{P(s, T_i)} \mathbb{E}^* \left[\mathbb{E}^* \left[\frac{P(t, T_j)}{P(t, T_i)} e^{-\int_s^{T_i} r_u du} \,\Big|\, \mathcal{F}_t \right] \,\Big|\, \mathcal{F}_s \right] \\
&= \frac{1}{P(s, T_i)} \mathbb{E}^* \left[e^{-\int_s^t r_u du} \frac{P(t, T_j)}{P(t, T_i)} \mathbb{E}^* \left[e^{-\int_t^{T_i} r_u du} \,\Big|\, \mathcal{F}_t \right] \,\Big|\, \mathcal{F}_s \right] \\
&= \frac{1}{P(s, T_i)} \mathbb{E}^* \left[e^{-\int_s^t r_u du} P(t, T_j) \,\Big|\, \mathcal{F}_s \right] \\
&= \frac{e^{\int_0^s r_u du}}{P(s, T_i)} \mathbb{E}^* \left[e^{-\int_0^t r_u du} P(t, T_j) \,\Big|\, \mathcal{F}_s \right] \\
&= \frac{P(s, T_j)}{P(s, T_i)},
\end{aligned}
$$

$0 \leqslant s \leqslant t \leqslant T_i \leqslant T_j$, since the discounted bond price process (8.6) is a martingale under $\mathbb{P}$ by Proposition 4.1.

We work under the assumption of absence of arbitrage under $\mathbb{P}^*$, which states that the discounted bond price process

$$t \mapsto e^{-\int_0^t r_s ds} P(t, T_i), \quad 0 \leqslant t \leqslant T_i, \quad i = 1, 2, \ldots, n, \qquad (8.6)$$

is an $(\mathcal{F}_t)_{t \in [0, T_i]}$-martingale under $\mathbb{P}^* = \mathbb{P}$ by Proposition 4.1. The next result is related to Proposition 8.3.

Proposition 8.4. *For all $1 \leqslant i, j \leqslant n$ we have*

$$\widehat{\mathbb{E}}_i \left[\frac{d\widehat{\mathbb{P}}_j}{d\widehat{\mathbb{P}}_i} \,\bigg|\, \mathcal{F}_t \right] = \frac{P(0, T_i)}{P(0, T_j)} \frac{P(t, T_j)}{P(t, T_i)}, \qquad 0 \leqslant t \leqslant \min(T_i, T_j), \qquad (8.7)$$

and in particular the process

$$t \mapsto \frac{P(t, T_j)}{P(t, T_i)}, \qquad 0 \leqslant t \leqslant \min(T_i, T_j),$$

is an $(\mathcal{F}_t)_{t \in [0, \min(T_i, T_j)]}$-martingale under $\widehat{\mathbb{P}}_i$, $1 \leqslant i, j \leqslant n$.

Proof. For all bounded and $\mathcal{F}_t$-measurable random variables F we have[1]

$$\begin{aligned}
\widehat{\mathbb{E}}_i \left[F \frac{d\widehat{\mathbb{P}}_j}{d\widehat{\mathbb{P}}_i} \right] &= \mathbb{E}^* \left[F \frac{d\widehat{\mathbb{P}}_j}{d\mathbb{P}^*} \right] \\
&= \frac{1}{P(0, T_j)} \mathbb{E}^* \left[F e^{-\int_0^{T_j} r_\tau d\tau} \right] \\
&= \frac{1}{P(0, T_j)} \mathbb{E}^* \left[F e^{-\int_0^t r_\tau d\tau} P(t, T_j) \right] \\
&= \frac{1}{P(0, T_j)} \mathbb{E}^* \left[F e^{-\int_0^{T_i} r_\tau d\tau} \frac{P(t, T_j)}{P(t, T_i)} \right] \\
&= \widehat{\mathbb{E}}_i \left[F \frac{P(0, T_i)}{P(0, T_j)} \frac{P(t, T_j)}{P(t, T_i)} \right],
\end{aligned}$$

which shows (8.7). $\qquad \square$

---

[1] We repeatedly use the characterization of conditional expectation $X = \mathbb{E}^*[F \mid \mathcal{F}_t] \Leftrightarrow \mathbb{E}[GX] = \mathbb{E}[GF]$ for all bounded and $\mathcal{F}_t$-measurable random variables G, see Relation (11.4) in the appendix.

8.2   Dynamics under the Forward Measure

In order to apply Proposition 8.1 and to compute the price

$$\mathbb{E}^* \left[e^{-\int_t^{T_i} r_s ds} C \,\middle|\, \mathcal{F}_t \right] = P(t, T_i) \widehat{\mathbb{E}}_i[C \mid \mathcal{F}_t],$$

of a random claim payoff C from the evaluation of

$$P(t, T) \widehat{\mathbb{E}}_i[C \mid \mathcal{F}_t],$$

we need to determine the dynamics of the underlying processes $(r_t)_{t \in \mathbb{R}_+}$, $(f(t, T_i, T_j))_{t \in [0, T_i]}$, and $(P(t, T_i))_{t \in [0, T_i]}$ via their stochastic differential equations written under the forward measure $\widehat{\mathbb{P}}_i$. Recall that by Proposition 8.3, the deflated process

$$t \mapsto \frac{P(t, T_j)}{P(t, T_i)}, \qquad 0 \leqslant t \leqslant \min(T_i, T_j),$$

is an $(\mathcal{F}_t)_{t \in [0, \min(T_i, T_j)]}$-martingale under $\widehat{\mathbb{P}}_i$ for all $i, j = 1, 2, \ldots, n$.

Forward Brownian motions

In order to determine the dynamics of the underlying processes under $\widehat{\mathbb{P}}_i$, we will use the following Proposition 8.5 which is obtained from the Girsanov Theorem 2.1.

Proposition 8.5. *For all* $i = 1, 2, \ldots, n$, *the process*

$$\widehat{B}_t^{(i)} := B_t - \int_0^t \zeta_i(s) ds, \qquad 0 \leqslant t \leqslant T_i, \tag{8.8}$$

is a standard Brownian motion under the forward measure $\widehat{\mathbb{P}}_i$, $i = 1, 2, \ldots, n$.

Proof. Letting

$$\Phi_i(t) := \mathbb{E}^* \left[\frac{d\widehat{\mathbb{P}}_i}{d\mathbb{P}^*} \,\middle|\, \mathcal{F}_t \right]$$

$$= \frac{1}{P(0, T_i)} \mathbb{E}^* \left[e^{-\int_0^{T_i} r_s ds} \,\middle|\, \mathcal{F}_t \right]$$

$$= \frac{P(t, T_i)}{P(0, T_i)} e^{-\int_0^t r_s ds}, \qquad 0 \leqslant t \leqslant T_i,$$

Equation (8.14) rewrites as

$$d\Phi_i(t) = \zeta_i(t) \Phi_i(t) dB_t,$$

which is solved as

$$\Phi_i(t) = \exp\left(\int_0^t \zeta_i(s)dB_s - \frac{1}{2}\int_0^t |\zeta_i(s)|^2 ds\right), \qquad 0 \leqslant t \leqslant T_i,$$

hence

$$\mathbb{E}^*\left[\frac{d\widehat{\mathbb{P}}_i}{d\mathbb{P}^*}\,\bigg|\,\mathcal{F}_{T_i}\right] = \Phi_i(T_i) = \exp\left(\int_0^{T_i} \zeta_i(s)dB_s - \frac{1}{2}\int_0^{T_i} |\zeta_i(s)|^2 ds\right),$$

and we conclude by the Girsanov Theorem 2.1, which shows that

$$\widehat{B}_t^{(i)} := B_t - \int_0^t \zeta_i(s)ds,$$

is a standard Brownian motion under the forward measure $\widehat{\mathbb{P}}_i$, $i = 1, 2, \ldots, n$. $\qquad\square$

We also have the relations

$$d\widehat{B}_t^{(i)} = dB_t - \zeta_i(t)dt, \qquad i = 1, 2, \ldots, n, \tag{8.9}$$

and

$$d\widehat{B}_t^{(j)} = dB_t - \zeta_j(t)dt = d\widehat{B}_t^{(i)} + (\zeta_i(t) - \zeta_j(t))dt, \quad i, j = 1, 2, \ldots, n,$$

which shows that $(\widehat{B}_t^{(j)})_{t\in\mathbb{R}_+}$ has drift $(\zeta_i(t) - \zeta_j(t))_{t\in\mathbb{R}_+}$ under $\widehat{\mathbb{P}}_i$.

Short rate dynamics under the forward measure

In case the short rate process $(r_t)_{t\in\mathbb{R}_+}$ is given as the (Markovian) solution to the stochastic differential equation

$$dr_t = \mu(t, r_t)dt + \sigma(t, r_t)dB_t,$$

by (8.9) its dynamics will be given under $\widehat{\mathbb{P}}_i$ by

$$dr_t = \mu(t, r_t)dt + \sigma(t, r_t)\big(\zeta_i(t)dt + d\widehat{B}_t^{(i)}\big)$$
$$= \mu(t, r_t)dt + \sigma(t, r_t)\zeta_i(t)dt + \sigma(t, r_t)d\widehat{B}_t^{(i)}. \tag{8.10}$$

In the case of the Vasicek model, by (4.22) we have

$$dr_t = (a - br_t)dt + \sigma dB_t,$$

and

$$\zeta_i(t) = -\frac{\sigma}{b}\big(1 - e^{-(T_i-t)b}\big), \qquad 0 \leqslant t \leqslant T_i,$$

hence from (8.10) we have

$$d\widehat{B}_t^{(i)} = dB_t - \zeta_i(t)dt = dB_t + \frac{\sigma}{b}\big(1 - e^{-(T_i-t)b}\big)dt, \tag{8.11}$$

and

$$dr_t = (a - br_t)dt - \frac{\sigma^2}{b}\big(1 - e^{-(T_i-t)b}\big)dt + \sigma d\widehat{B}_t^{(i)}. \tag{8.12}$$

Bond price dynamics under the forward measure

In the sequel, we assume that the dynamics of the bond price $P(t, T_i)$ is given by

$$\frac{dP(t, T_i)}{P(t, T_i)} = r_t dt + \zeta_i(t) dB_t, \qquad i = 1, 2, \ldots, n, \qquad (8.13)$$

where $(B_t)_{t \in \mathbb{R}_+}$ is a standard Brownian motion under $\mathbb{P}^*$ and $(r_t)_{t \in \mathbb{R}_+}$ and $(\zeta_i(t))_{t \in \mathbb{R}_+}$ are $(\mathcal{F}_t)_{t \in \mathbb{R}_+}$-adapted processes with respect to the filtration $(\mathcal{F}_t)_{t \in \mathbb{R}_+}$ generated by $(B_t)_{t \in \mathbb{R}_+}$. Equation (8.13) can be solved as

$$P(t, T_i) = P(0, T_i) \exp\left(\int_0^t r_s ds + \int_0^t \zeta_i(s) dB_s - \frac{1}{2} \int_0^t |\zeta_i(s)|^2 ds \right),$$

$0 \leqslant t \leqslant T_i$, $i = 1, 2, \ldots, n$. An application of Itô's calculus to (8.13) shows that

$$d\left(e^{-\int_0^t r_s ds} P(t, T_i) \right) = \zeta_i(t) \left(e^{-\int_0^t r_s ds} P(t, T_i) \right) dB_t, \qquad (8.14)$$

which is consistent with the fact that

$$t \mapsto e^{-\int_0^t r_s ds} P(t, T)$$

is a martingale under $\mathbb{P}^*$ by Proposition 4.1.

As a consequence of Proposition 8.5 and (8.13), the dynamics of $(P(t, T_j))_{t \in [0, T_i]}$ under $\widehat{\mathbb{P}}_i$ is given by

$$\frac{dP(t, T_j)}{P(t, T_j)} = r_t dt + \zeta_i(t) \zeta_j(t) dt + \zeta_j(t) d\widehat{B}_t^{(i)}, \quad i, j = 1, 2, \ldots, n, \quad (8.15)$$

where $\left(\widehat{B}_t^{(i)} \right)_{t \in \mathbb{R}_+}$ is a standard Brownian motion under $\widehat{\mathbb{P}}_i$, and we also have

$$d\left(e^{-\int_0^t r_s ds} P(t, T_j) \right) = |\zeta_t|^2 e^{-\int_0^t r_s ds} P(t, T_j) dt + \zeta_t e^{-\int_0^t r_s ds} P(t, T_j) d\widehat{B}_t.$$

Next, we note that

$$P(t, T_j)$$

$$= e^{\int_0^t r_s ds} P(0, T_j) \exp\left(\int_0^t \zeta_j(s) dB_s - \frac{1}{2} \int_0^t |\zeta_j(s)|^2 ds \right) \qquad \text{[under } \mathbb{P}^*\text{]}$$

$$= e^{\int_0^t r_s ds} P(0, T_j) \exp\left(\int_0^t \zeta_j(s) d\widehat{B}_s^{(j)} + \frac{1}{2} \int_0^t |\zeta_j(s)|^2 ds \right)$$

$$= e^{\int_0^t r_s ds} P(0, T_j)$$

$$\times \exp\left(\int_0^t \zeta_j(s) d\widehat{B}_s^{(i)} + \int_0^t \zeta_j(s)\zeta_i(s) ds - \frac{1}{2} \int_0^t |\zeta_j(s)|^2 ds \right)$$

$$= e^{\int_0^t r_s ds} P(0, T_j)$$

$$\times \exp\left(\int_0^t \zeta_j(s) d\widehat{B}_s^{(i)} - \frac{1}{2} \int_0^t |\zeta_j(s) - \zeta_i(s)|^2 ds + \frac{1}{2} \int_0^t |\zeta_i(s)|^2 ds \right),$$

$t \in [0, T_j]$, $i, j = 1, 2, \ldots, n$. Consequently, the forward price $P(t, T_j)/P(t, T_i)$ can be written as

$$\frac{P(t, T_j)}{P(t, T_i)}$$

$$= \frac{P(0, T_j)}{P(0, T_i)} \exp\left(\int_0^t (\zeta_j(s) - \zeta_i(s)) d\widehat{B}_s^{(j)} + \frac{1}{2} \int_0^t |\zeta_j(s) - \zeta_i(s)|^2 ds \right)$$

$$[\text{under } \widehat{\mathbb{P}}_j]$$

$$= \frac{P(0, T_j)}{P(0, T_i)} \exp\left(\int_0^t (\zeta_j(s) - \zeta_i(s)) d\widehat{B}_s^{(i)} - \frac{1}{2} \int_0^t |\zeta_i(s) - \zeta_j(s)|^2 ds \right),$$

$$[\text{under } \widehat{\mathbb{P}}_i]$$

$$(8.16)$$

$0 \leqslant t \leqslant \min(T_i, T_j)$, $i, j = 1, 2, \ldots, n$.

By Itô's calculus and (8.13), for any $1 \leqslant i < j \leqslant n$ we can also obtain the relation

$$d\left(\frac{P(t, T_j)}{P(t, T_i)} \right) = \frac{P(t, T_j)}{P(t, T_i)} (\zeta_j(t) - \zeta_i(t))(dB_t - \zeta_i(t) dt)$$

$$= \frac{P(t, T_j)}{P(t, T_i)} (\zeta_j(t) - \zeta_i(t)) d\widehat{B}_t^{(i)},$$

where $\left(\widehat{B}_t^{(i)} \right)_{t \in \mathbb{R}_+}$ is a standard Brownian motion under $\widehat{\mathbb{P}}_i$, see Exercise 8.6-(1) for details, which recovers the martingale property of $P(t, T_j)/P(t, T_i)$ stated in Proposition 8.3 and in the second part of Proposition 8.4.

Next, assume as above that the short rate process $(r_t)_{t \in \mathbb{R}_+}$ is Markovian, and solution of

$$dr_t = \mu(t, r_t) dt + \sigma(t, r_t) dB_t,$$

with dynamics under $\widehat{\mathbb{P}}_i$ given by

$$dr_t = \mu(t, r_t) dt + \sigma(t, r_t)\zeta_i(t) dt + \sigma(t, r_t) d\widehat{B}_t^{(i)}.$$

Recall that in this Markovian setting, the bond price $P(t, T_i)$ is expressed as

$$P(t, T_i) = \mathbb{E}^* \left[e^{-\int_t^{T_i} r_s ds} \,\middle|\, \mathcal{F}_t \right]$$

$$= \mathbb{E}^* \left[e^{-\int_t^{T_i} r_s ds} \,\middle|\, r_t \right]$$

$$= F_i(t, r_t), \qquad 0 \leqslant t \leqslant T_i,$$

i.e. it becomes a function $F_i(t, r_t)$ of t and r_t. Itô's formula then shows that

$$d\left(e^{-\int_0^t r_s ds} P(t, T_i) \right) = e^{-\int_0^t r_s ds} \sigma(t, r_t) \frac{\partial F_i}{\partial x}(t, r_t) dB_t,$$

since, by Corollary II-1 of Protter (2004), the sum of all terms in dt vanish in the above expression because

$$t \mapsto e^{-\int_0^t r_s ds} P(t, T_i) = e^{-\int_0^t r_s ds} F_i(t, r_t),$$

is a martingale under $\mathbb{P}^*$ by Proposition 4.1. Hence, the dynamics of $(P(t, T_i))_{t \in [0, T_i]}$ is given by

$$\frac{dP(t, T_i)}{P(t, T_i)} = r_t dt + \frac{\sigma(t, r_t)}{P(t, T_i)} \frac{\partial F_i}{\partial x}(t, r_t) dB_t$$

$$= r_t dt + \frac{\sigma(t, r_t)}{F_i(t, r_t)} \frac{\partial F_i}{\partial x}(t, r_t) dB_t$$

$$= r_t dt + \sigma(t, r_t) \frac{\partial}{\partial x} \log F_i(t, r_t) dB_t,$$

and the process $(\zeta_i(t))_{t \in \mathbb{R}_+}$ in (8.13) is given by

$$\zeta_i(t) = \sigma(t, r_t) \frac{\partial}{\partial x} \log F_i(t, r_t), \qquad 0 \leqslant t \leqslant T_i.$$

As an example, in the Vasicek model, where $\sigma(t, x)$ is constant equal to σ, the price $P(t, T_i)$ has the form

$$P(t, T_i) = F_i(t, r_t) = e^{C(T_i - t) r_t + A(T_i - t)},$$

where

$$C(T_i - t) = -\frac{1}{b}(1 - e^{-(T_i - t)b}),$$

hence

$$\log F_i(t, r_t) = C(T_i - t) r_t + A(T_i - t),$$

and

$$\zeta_i(t) = \sigma C(T_i - t) = -\frac{\sigma}{b}(1 - e^{-(T_i - t)b}), \qquad 0 \leqslant t \leqslant T_i,$$

which recovers (4.22). From (8.11), we obtain the dynamics

$$\frac{dP(t, T_i)}{P(t, T_i)} = r_t dt + \frac{\sigma^2}{b^2}(1 - e^{-(T_i - t)b})^2 dt - \frac{\sigma}{b}(1 - e^{-(T_i - t)b}) d\widehat{B}_t^{(i)}.$$

of the Vasicek bond price $(P(t, T_i))_{t \in [0, T_i]}$ under the forward measure $\widehat{\mathbb{P}}_i$.

8.3 Bond Options

Restatement of objectives

Before proceeding further we would like to recall our general objectives, which are:

(1) to find a stochastic model for the underlying (forward) interest rate processes.

(2) to derive option pricing formulas as functions of the model parameters.

(3) to calibrate the parameters of the model by matching computed prices to market data.

(4) to compute "new" prices using the calibrated pricing formulas.

Caps and floors are standard examples of options on interest rates. An interest rate cap can protect a borrower against interest rates going above a certain level κ. As an example, a cap on the underlying short rate at time T yields a payoff equal to

$$r_T - \min(\kappa, r_T) = (r_T - \kappa)^+$$

expressed in interest rate (base) points. However, this type of cap makes little sense in practice, since

a) the rate r_T is not a tradable asset,

b) r_T is an instantaneous rate which makes sense only over an infinitesimally short period of time $[T, T + dt]$.

With reference to point (a) above, the bond price $P(T, S)$ is a tradable asset, and in a Markovian setting it can be written as a function

$$P(T, S) = F(T, r_T)$$

of r_T, therefore writing a put option with payoff

$$(K - P(T, S))^+ = (K - F(T, r_T))^+ \tag{8.17}$$

certainly makes sense.

Referring to point (b) above, interest rate option contracts are built on forward rates rather than on the short rate process $(r_t)_{t \in \mathbb{R}_+}$. Since

$f(T, T, S)$ is part of the data known at time T (*i.e.*, it is $\mathcal{F}_T$-measurable), it makes sense to consider an option on the *forward spot rate*

$$f(T, T, S) = \frac{1}{S - T} \int_T^S f(T, t)dt,$$

which appears as being random at time $t \in [0, T]$ (precisely, it is $\mathcal{F}_T$-measurable but not $\mathcal{F}_t$-measurable). For example, an interest rate floor contract on $f(T, T, S)$ generates a payoff of the form

$$\text{Max}\,(\kappa, f(T, T, S)) - f(T, T, S) = (\kappa - f(T, T, S))^+.$$

Under a different choice of payoff function on the same asset, the payoff of the contract can take the form

$$\left(\kappa - e^{-(S-T)f(T,T,S)}\right)^+ = \left(\kappa - e^{-\int_T^S f(T,s)ds}\right)^+$$
$$= (\kappa - P(T, S))^+,$$

which coincides with the payoff (8.17) of a put option on the bond price $P(T, S) = e^{-(S-T)f(T,T,S)}$.

Using the framework of Section 8.2, we are now able to compute the price at time t of a contingent claim with random payoff F and maturity time T_i from the relation

$$\mathbb{E}^*\left[e^{-\int_t^{T_i} r_s ds} F \,\Big|\, \mathcal{F}_t\right] = P(t, T_i)\widehat{\mathbb{E}}_i[F \mid \mathcal{F}_t]$$

and the knowledge of the dynamics of the process $(r_t)_{t\in\mathbb{R}_+}$ under the probability measure $\widehat{\mathbb{P}}_i$. Next, we compute the price

$$\mathbb{E}^*\left[e^{-\int_t^{T_i} r_s ds}(P(T_i, T_j) - K)^+ \,\Big|\, \mathcal{F}_t\right] = P(t, T_i)\widehat{\mathbb{E}}_i[(P(T_i, T_j) - K)^+ \mid \mathcal{F}_t]$$

of a bond call option on $P(T_i, T_j)$ with payoff $F = (P(T_i, T_j) - K)^+$, when the dynamics of $P(t, T_i)$ takes the form

$$\frac{dP(t, T_i)}{P(t, T_i)} = r_t dt + \zeta_i(t)dB_t, \qquad 0 \leqslant t \leqslant T_i,$$

see Exercises 8.6 and 8.11 for the pricing of bond put options.

Proposition 8.6. *Let $0 \leqslant T_i \leqslant T_j$ and assume that the dynamics of the bond prices $P(t, T_i)$, $P(t, T_j)$ under $\mathbb{P}^*$ are given by*

$$\frac{dP(t, T_i)}{P(t, T_i)} = r_t dt + \zeta_i(t)dB_t, \qquad \frac{dP(t, T_j)}{P(t, T_j)} = r_t dt + \zeta_j(t)dB_t, \qquad (8.18)$$

where $(\zeta_i(t))_{t\in\mathbb{R}_+}$ and $(\zeta_j(t))_{t\in\mathbb{R}_+}$ are deterministic volatility functions. Then the price of a bond call option on $P(T_i, T_j)$ with payoff

$$C := (P(T_i, T_j) - K)^+$$

can be written as

$$
\mathbb{E}^*\left[e^{-\int_t^{T_i} r_s ds}(P(T_i, T_j) - K)^+ \,\Big|\, \mathcal{F}_t\right] \tag{8.19}
$$
$$
= P(t, T_j)\Phi\left(\frac{v(t, T_i)}{2} + \frac{1}{v(t, T_i)}\log\frac{P(t, T_j)}{KP(t, T_i)}\right)
$$
$$
- KP(t, T_i)\Phi\left(-\frac{v(t, T_i)}{2} + \frac{1}{v(t, T_i)}\log\frac{P(t, T_j)}{KP(t, T_i)}\right), \quad 0 \leqslant t < T_i,
$$

where $v^2(t, T_i) := \displaystyle\int_t^{T_i} |\zeta_j(s) - \zeta_i(s)|^2 ds$ *and*

$$
\Phi(x) := \frac{1}{\sqrt{2\pi}}\int_{-\infty}^x e^{-y^2/2}dy, \qquad x \in \mathbb{R},
$$

is the standard normal cumulative distribution function.

Proof. First, we note that by Proposition 8.1, the price of the bond call option on $P(T_i, T_j)$ with payoff $F = (P(T_i, T_j) - K)^+$ can be written using the forward measure $\widehat{\mathbb{P}}_i$ as

$$
\mathbb{E}^*\left[e^{-\int_t^{T_i} r_s ds}(P(T_i, T_j) - K)^+ \,\Big|\, \mathcal{F}_t\right] = P(t, T_i)\widehat{\mathbb{E}}_i\left[(P(T_i, T_j) - K)^+ \,\big|\, \mathcal{F}_t\right]. \tag{8.20}
$$

Next, by (8.16) we can write $P(T_i, T_j)$ as the geometric Brownian motion

$$
P(T_i, T_j) =
$$
$$
\frac{P(t, T_j)}{P(t, T_i)}\exp\left(\int_t^{T_i}(\zeta_j(s) - \zeta_i(s))d\widehat{B}_s^{(i)} - \frac{1}{2}\int_t^{T_i}|\zeta_j(s) - \zeta_i(s)|^2 ds\right),
$$

under the forward measure $\widehat{\mathbb{P}}_i$, and rewrite (8.20) as

$$
\mathbb{E}^*\left[e^{-\int_t^{T_i} r_s ds}(P(T_i, T_j) - K)^+ \,\Big|\, \mathcal{F}_t\right]
$$
$$
= P(t, T_i)
$$
$$
\times \widehat{\mathbb{E}}_i\left[\left(\frac{P(t, T_j)}{P(t, T_i)}e^{\int_t^{T_i}(\zeta_j(s) - \zeta_i(s))d\widehat{B}_s^{(i)} - \int_t^{T_i}|\zeta_j(s) - \zeta_i(s)|^2 ds/2} - K\right)^+ \,\Big|\, \mathcal{F}_t\right]
$$
$$
= \widehat{\mathbb{E}}_i\left[\left(P(t, T_j)e^{\int_t^{T_i}(\zeta_j(s) - \zeta_i(s))d\widehat{B}_s^{(i)} - \int_t^{T_i}|\zeta_j(s) - \zeta_i(s)|^2 ds/2} - KP(t, T_i)\right)^+ \,\Big|\, \mathcal{F}_t\right].
$$

Since $(\zeta_i(s))_{s\in[0,T_i]}$ and $(\zeta_j(s))_{s\in[0,T_j]}$ in (8.13) are deterministic volatility functions, $P(T_i, T_j)$ is a lognormal random variable given $\mathcal{F}_t$ under $\widehat{\mathbb{P}}_i$ and we can use Lemma 2.3 to price the bond call option by the Black-Scholes formula

$$
\mathrm{Bl}\big(P(t, T_j), KP(0, T_i), v(t, T_i)/\sqrt{T_i - t}, 0, T_i - t\big)
$$

with underlying asset $P(t, T_j)$, strike price $KP(t, T_i)$, volatility parameter $v(t, T_i)/\sqrt{T_i - t}$, time to maturity $T_i - t$ and zero interest rate, which yields (8.19).

In other words, we have

$$
\mathbb{E}^* \left[e^{-\int_t^T r_s ds} (P(T_i, T_j) - K)^+ \,\Big|\, \mathcal{F}_t \right]
$$

$$
= P(t, T_i) \widehat{\mathbb{E}}_i \left[\left(\frac{P(t, T_j)}{P(t, T_i)} e^{X - \int_t^T |\zeta_j(s) - \zeta_i(s)|^2 ds/2} - K \right)^+ \,\Big|\, \mathcal{F}_t \right]
$$

$$
= P(t, T_i) \widehat{\mathbb{E}}_i \left[\left(e^{X + m(t, T_i)} - K \right)^+ \,\Big|\, \mathcal{F}_t \right],
$$

where

$$
m(t, T_i) := -\frac{1}{2} v^2(t, T_i) + \log \frac{P(t, T_j)}{P(t, T_i)},
$$

and X is a centered Gaussian random variable with variance

$$
v^2(t, T_i) := \int_t^{T_i} |\zeta_j(s) - \zeta_i(s)|^2 ds,
$$

given $\mathcal{F}_t$. Recall that by Lemma 2.3, when X is a centered Gaussian random variable with variance $v^2 > 0$, the expectation of $\left(e^{m+X} - K \right)^+$ is given, as in the standard Black-Scholes formula, by

$$
\mathbb{E}^* \left[\left(e^{m+X} - K \right)^+ \right] = e^{m + v^2/2} \Phi \left(v + \frac{m - \log K}{v} \right) - K \Phi \left(\frac{m - \log K}{v} \right),
$$

where

$$
\Phi(z) = \int_{-\infty}^z e^{-y^2/2} \frac{dy}{\sqrt{2\pi}}, \qquad z \in \mathbb{R},
$$

denotes the standard normal cumulative distribution function and for simplicity of notation we dropped the indices t, T_i in $m(t, T_i)$ and $v^2(t, T_i)$.

$\square$

Note that (8.19) also yields the decomposition of the corresponding self-financing hedging portfolio in the assets $P(t, T_i)$ and $P(t, T_j)$ for the claim with payoff $(P(T_i, T_j) - K)^+$, see *e.g.* Corollary 10.1 in Privault (2014).

8.4   Vasicek Bond Option Pricing

In this section we show how the bond option pricing formula (8.19) can be recovered in the Vasicek model by an independent computation based on the dynamics of $(r_t)_{t \in \mathbb{R}_+}$ and $(P(t,T_i))_{t \in [0,T_i]}$ under $\widehat{\mathbb{P}}_i$, see also Jamshidian (1989). In the Vasicek model, we have

$$dr_t = (a - br_t)dt + \sigma dB_t,$$

and

$$\zeta_t = \sigma C(T_i - t) = -\frac{\sigma}{b}\left(1 - e^{-(T_i - t)b}\right),$$

hence

$$v^2(t,T_i) = \frac{\sigma^2}{b^2} \int_t^{T_i} \left(e^{-(T_i - s)b} - e^{-(T_j - s)b}\right)^2 ds, \qquad 0 \leqslant t \leqslant T_i.$$

Under the forward measure $\widehat{\mathbb{P}}_i$, we have

$$dr_t = (a - br_t)dt - \frac{\sigma^2}{b}\left(1 - e^{-(T_i - t)b}\right)dt + \sigma d\widehat{B}_t^{(i)} \qquad (8.21)$$

and

$$\frac{dP(t,T_i)}{P(t,T_i)} = r_t dt + \frac{\sigma^2}{b^2}\left(1 - e^{-(T_i - t)b}\right)^2 dt - \frac{\sigma}{b}\left(1 - e^{-(T_i - t)b}\right)d\widehat{B}_t^{(i)}.$$

Equation (8.21) can be solved as

$$r_t = r_s e^{-(t-s)b} + \int_s^t e^{-(t-u)b}\left(a + \sigma^2 C(T_i - u)\right)du + \sigma \int_s^t e^{-(t-u)b}d\widehat{B}_u^{(i)}$$

$$= \widehat{\mathbb{E}}_i[r_t \mid \mathcal{F}_s] + \sigma \int_s^t e^{-(t-u)b}d\widehat{B}_u^{(i)},$$

hence the conditional mean and variance of r_t under $\widehat{\mathbb{P}}_i$ are given by

$$\widehat{\mathbb{E}}_i[r_t \mid \mathcal{F}_s] = r_s e^{-(t-s)b} + \int_s^t e^{-(t-u)b}\left(a + \sigma^2 C(T_i - u)\right)du,$$

and

$$\mathrm{Var}_i[r_t \mid \mathcal{F}_s] = \widehat{\mathbb{E}}_i\left[\left(r_t - \widehat{\mathbb{E}}_i[r_t \mid \mathcal{F}_s]\right)^2 \mid \mathcal{F}_s\right]$$

$$= \frac{\sigma^2}{2b}\left(1 - e^{-2(t-s)b}\right).$$

Therefore, the price the bond call option on $P(T_i, T_j)$ is given by

$$\mathbb{E}^*\left[e^{-\int_t^{T_i} r_s ds}(P(T_i, T_j) - K)^+ \mid \mathcal{F}_t\right] = P(t, T_i)\widehat{\mathbb{E}}_i\left[(P(T_i, T_j) - K)^+ \mid \mathcal{F}_t\right]$$

$$= P(t, T_i)\widehat{\mathbb{E}}_i\left[(F(T_i, r_{T_i}) - K)^+ \mid \mathcal{F}_t\right]$$

$$= P(t, T_i)\widehat{\mathbb{E}}_i\left[\left(e^{A(T_j - T_i) + r_{T_i} C(T_j - T_i)} - K\right)^+ \mid \mathcal{F}_t\right]$$

$$= P(t, T_i)\widehat{\mathbb{E}}_i\left[\left(e^{m(t, T_i) + X} - K\right)^+ \mid r_t\right],$$

where X is a centered Gaussian random variable with variance

$$
\begin{aligned}
v^2(t, T_i) &:= \operatorname{Var}_i[C(T_j - T_i)r_{T_i} \mid \mathcal{F}_t] \\
&= C^2(T_j - T_i)\operatorname{Var}_i[r_{T_i} \mid r_t] \\
&= \frac{\sigma^2}{2b}C^2(T_j - T_i)\left(1 - e^{-2(T_i - t)b}\right)
\end{aligned}
\tag{8.22}
$$

given $\mathcal{F}_t$, and

$$
\begin{aligned}
\widetilde{m}(t, T_i) &:= A(T_j - T_i) + C(T_j - T_i)\,\mathbb{E}^*[r_{T_i} \mid \mathcal{F}_t] \\
&= A(T_j - T_i) \\
&\quad + C(T_j - T_i)\left(r_t\, e^{-(T_i - t)b} + \int_t^{T_i} e^{-(T_i - u)b}(a + \sigma C(T_i - u))du\right),
\end{aligned}
$$

where

$$
\begin{aligned}
A(T_j - T_i) &= \frac{4ab - 3\sigma^2}{4b^3} + \frac{\sigma^2 - 2ab}{2b^2}(T_j - T_i) \\
&\quad + \frac{\sigma^2 - ab}{b^3}\, e^{-(T_j - T_i)b} - \frac{\sigma^2}{4b^3}\, e^{-2(T_j - T_i)b}.
\end{aligned}
$$

Since from Proposition 8.3,

$$
t \mapsto \frac{P(t, T_j)}{P(t, T_i)}, \qquad 0 \leqslant t \leqslant T_i < T_j,
$$

is a martingale under $\widehat{\mathbb{P}}_i$, we have the relation

$$
\begin{aligned}
\frac{P(t, T_j)}{P(t, T_i)} &= \widehat{\mathbb{E}}_i[P(T_i, T_j) \mid \mathcal{F}_t] \\
&= \widehat{\mathbb{E}}_i\left[e^{A(T_j - T_i) + r_{T_i} C(T_j - T_i)} \mid \mathcal{F}_t\right] \\
&= \exp\!\left(A(T_j - T_i) + C(T_j - T_i)\,\mathbb{E}^*[r_{T_i} \mid \mathcal{F}_t] + \frac{1}{2}C^2(T_j - T_i)\operatorname{Var}_i[r_{T_i} \mid \mathcal{F}_t]\right) \\
&= e^{\widetilde{m}(t, T_i) + v^2(t, T_i)/2}.
\end{aligned}
\tag{8.23}
$$

The above non-trivial Relation (8.23) can actually be checked by hand calculations, as follows:

$$
\begin{aligned}
-\frac{1}{2}v^2(t, T_i) + \log\frac{P(t, T_j)}{P(t, T_i)} &= -\frac{1}{2}v^2(t, T_i) + \log P(t, T_j) - \log P(t, T_i) \\
&= -\frac{1}{2}v^2(t, T_i) + A(T_j - t) + r_t C(T_j - t) - (A(T_i - t) + r_t C(T_i - t)) \\
&= -\frac{1}{2}v^2(t, T_i) + A(T_j - t) - A(T_i - t) + r_t(C(T_j - t) - C(T_i - t)) \\
&= -\frac{\sigma^2}{4b}C^2(T_j - T_i)\left(1 - e^{-2(T_i - t)b}\right)
\end{aligned}
$$

$$+ A(T_j - t) - A(T_i - t) + r_t C(T_j - T_i) e^{-(T_i - t)b}$$
$$= A(T_j - T_i)$$
$$+ C(T_j - T_i) \left(r_t e^{-(T_i - t)b} + \int_t^{T_i} e^{-(T_i - u)b} (a + \sigma^2 C(T_i - u)) du \right)$$
$$= \widetilde{m}(t, T_i).$$

Using (8.22) and (8.23) we finally recover (8.19), *i.e.*

$$\mathbb{E}^* \left[e^{-\int_t^{T_i} r_s ds} (P(T_i, T_j) - K)^+ \,\Big|\, \mathcal{F}_t \right] = P(t, T_i) \widehat{\mathbb{E}}_i \left[(e^{\widetilde{m}(t, T_i) + X} - K)^+ \,\Big|\, r_t \right]$$

$$= P(t, T_i) e^{\widetilde{m}(t, T_i) + v^2(t, T_i)/2} \Phi \left(v(t, T_i) + \frac{\widetilde{m}(t, T_i) - \log K}{v(t, T_i)} \right)$$

$$- K P(t, T_i) \Phi \left(\frac{\widetilde{m}(t, T_i) - \log K}{v(t, T_i)} \right)$$

$$= P(t, T_j) \Phi \left(v(t, T_i) + \frac{\widetilde{m}(t, T_i) - \log K}{v(t, T_i)} \right)$$

$$- K P(t, T_i) \Phi \left(\frac{\widetilde{m}(t, T_i) - \log K}{v(t, T_i)} \right)$$

$$= P(t, T_j) \Phi \left(\frac{v(t, T_i)}{2} + \frac{1}{v(t, T_i)} \log \frac{P(t, T_j)}{K P(t, T_i)} \right)$$

$$- K P(t, T_i) \Phi \left(-\frac{v(t, T_i)}{2} + \frac{1}{v(t, T_i)} \log \frac{P(t, T_j)}{K P(t, T_i)} \right).$$

Note that bond option prices in the Vasicek model could also be computed from the joint distribution of $(r_{T_i}, \int_t^{T_i} r_s ds)$, which is jointly Gaussian in the Vasicek case, or from the dynamics (8.15)-(8.12) of $(P(t, T_i))_{t \in [0, T_i]}$ and $(r_t)_{t \in [0, T_i]}$ under $\widehat{\mathbb{P}}_i$, see *e.g.* Kim (2002) for the CIR and other short rate models with correlated Brownian motions.

8.5 Forward Swap Measures

In this section we present the construction of forward swap measures, for use in Chapter 9 on the pricing of swaptions. This construction is based on the *annuity numeraire*

$$P(t, T_i, T_j) = \sum_{k=i}^{j-1} (T_{k+1} - T_k) P(t, T_{k+1}), \quad 0 \leqslant t \leqslant T_{i+1}, \qquad (8.24)$$

$1 \leqslant i < j \leqslant n$, which can be viewed as the value at time t of a sequence of \$1 payments received at the times $T_{i+1}, \ldots, T_j$, weighted by the time interval lengths $\delta_k = T_{k+1} - T_k$, $k = i, \ldots, j - 1$. When $j = i + 1$, the *annuity numeraire* (8.24) satisfies

$$P(t, T_i, T_{i+1}) = (T_{i+1} - T_i)P(t, T_{i+1}), \quad 0 \leqslant t \leqslant T_{i+1},$$

$1 \leqslant i \leqslant n - 1$. The annuity numeraire can be also used to price a *bond ladder*. The discounted annuity numeraire satisfies the following martingale property, which can be proved by linearity and the fact that $t \mapsto e^{-\int_0^t r_s ds} P(t, T_k)$ is a martingale for all $k = 1, 2, \ldots, n$ by Proposition 4.1.

Proposition 8.7. *The discounted annuity numeraire*

$$e^{-\int_0^t r_s ds} P(t, T_i, T_j) = e^{-\int_0^t r_s ds} \sum_{k=i}^{j-1} (T_{k+1} - T_k) P(t, T_{k+1}), \quad 0 \leqslant t \leqslant T_{i+1},$$

is a martingale under $\mathbb{P}^*$, $1 \leqslant i < j \leqslant n$.

Proof. This result can be recovered by linearity and the fact that $t \mapsto e^{-\int_0^t r_s ds} P(t, T_k)$ is a martingale for all $k = i, \ldots, j$. Alternatively, by standard arguments, given that $\delta_k = T_{k+1} - T_k$, $k = i, \ldots, j - 1$, we obtain

$$\mathbb{E}^* \left[e^{-\int_0^T r_s ds} P(T, T_i, T_j) \,\middle|\, \mathcal{F}_t \right] = \sum_{k=i}^{j-1} \delta_k \, \mathbb{E}^* \left[e^{-\int_0^T r_s ds} P(T, T_{k+1}) \,\middle|\, \mathcal{F}_t \right]$$

$$= \sum_{k=i}^{j-1} \delta_k \, \mathbb{E}^* \left[e^{-\int_0^T r_s ds} \, \mathbb{E}^* \left[e^{-\int_T^{T_{k+1}} r_s ds} \,\middle|\, \mathcal{F}_T \right] \,\middle|\, \mathcal{F}_t \right]$$

$$= \sum_{k=i}^{j-1} \delta_k \, \mathbb{E}^* \left[\mathbb{E}^* \left[e^{-\int_0^T r_s ds} e^{-\int_T^{T_{k+1}} r_s ds} \,\middle|\, \mathcal{F}_T \right] \,\middle|\, \mathcal{F}_t \right]$$

$$= \sum_{k=i}^{j-1} \delta_k \, \mathbb{E}^* \left[\mathbb{E}^* \left[e^{-\int_0^{T_{k+1}} r_s ds} \,\middle|\, \mathcal{F}_T \right] \,\middle|\, \mathcal{F}_t \right]$$

$$= \sum_{k=i}^{j-1} \delta_k \, \mathbb{E}^* \left[e^{-\int_0^{T_{k+1}} r_s ds} \,\middle|\, \mathcal{F}_t \right]$$

$$= \sum_{k=i}^{j-1} \delta_k e^{-\int_0^t r_s ds} P(t, T_{k+1})$$

$$= e^{-\int_0^t r_s ds} P(t, T_i, T_j),$$

for $0 \leqslant t \leqslant T_{i+1}$. $\qquad\qquad\square$

Note that by (8.18) we have

$$d\left(e^{-\int_0^t r_s ds} P(t, T_i, T_j)\right) = \sum_{k=i}^{j-1} d\left(e^{-\int_0^t r_s ds} P(t, T_{k+1})\right)$$

$$= e^{-\int_0^t r_s ds} \sum_{k=i}^{j-1} \zeta_{k+1}(t) P(t, T_{k+1}) dB_t$$

$$= e^{-\int_0^t r_s ds} P(t, T_i, T_j) \sum_{k=i}^{j-1} v_{k+1}^{i,j}(t) \zeta_{k+1}(t) dB_t,$$

where

$$v_k^{i,j}(t) := \frac{P(t, T_k)}{P(t, T_i, T_j)}, \qquad 0 \leqslant t \leqslant T_{i+1}, \quad 1 \leqslant i < k \leqslant j \leqslant n,$$

which also recovers the result of Proposition 8.7. Next, we introduce the forward measures to be used for the pricing of swaptions in Chapter 9.

Definition 8.2. *The forward swap measure* $\widehat{\mathbb{P}}_{i,j}$ *is defined by*

$$\frac{d\widehat{\mathbb{P}}_{i,j}}{d\mathbb{P}^*} = e^{-\int_0^{T_i} r_s ds} \frac{P(T_i, T_i, T_j)}{P(0, T_i, T_j)}, \qquad 1 \leqslant i < j \leqslant n. \tag{8.25}$$

By Remark 8.7, we have

$$\mathbb{E}^* \left[\frac{d\widehat{\mathbb{P}}_{i,j}}{d\mathbb{P}^*} \,\middle|\, \mathcal{F}_t \right] = \frac{1}{P(0, T_i, T_j)} \mathbb{E}^* \left[e^{-\int_0^{T_i} r_s ds} P(T_i, T_i, T_j) \,\middle|\, \mathcal{F}_t \right]$$

$$= \frac{1}{P(0, T_i, T_j)} \mathbb{E}^* \left[e^{-\int_0^{T_i} r_s ds} \sum_{k=i}^{j-1} (T_{k+1} - T_k) P(T_i, T_{k+1}) \,\middle|\, \mathcal{F}_t \right]$$

$$= \frac{1}{P(0, T_i, T_j)} \sum_{k=i}^{j-1} (T_{k+1} - T_k) \mathbb{E}^* \left[e^{-\int_0^{T_i} r_s ds} P(T_i, T_{k+1}) \,\middle|\, \mathcal{F}_t \right]$$

$$= \frac{1}{P(0, T_i, T_j)} e^{-\int_0^t r_s ds} \sum_{k=i}^{j-1} (T_{k+1} - T_k) P(t, T_{k+1})$$

$$= \frac{P(t, T_i, T_j)}{P(0, T_i, T_j)} e^{-\int_0^t r_s ds}, \qquad 0 \leqslant t \leqslant T_{i+1},$$

and

$$\frac{d\widehat{\mathbb{P}}_{i,j|\mathcal{F}_t}}{d\mathbb{P}^*_{|\mathcal{F}_t}} = e^{-\int_t^{T_i} r_s ds} \frac{P(T_i, T_i, T_j)}{P(t, T_i, T_j)}, \qquad 0 \leqslant t \leqslant T_{i+1}, \tag{8.26}$$

see *e.g.* Relation (10.3) in Privault (2014).

Proposition 8.8. *We have*

$$\frac{d\widehat{\mathbb{P}}_{i,j|\mathcal{F}_t}}{d\mathbb{P}^*_{|\mathcal{F}_t}} = e^{-\int_t^{T_i} r_s ds} \frac{P(T_i, T_i, T_j)}{P(t, T_i, T_j)}, \qquad 0 \leqslant t \leqslant T_{i+1}. \tag{8.27}$$

Proof. It suffices to show that

$$\widehat{\mathbb{E}}_{i,j}[F \mid \mathcal{F}_t] = \mathbb{E}^* \left[F e^{-\int_t^T r_s ds} \frac{P(T_i, T_i, T_j)}{P(t, T_i, T_j)} \,\middle|\, \mathcal{F}_t \right] \tag{8.28}$$

for all integrable random variables F. Now, for any random variable G bounded and $\mathcal{F}_t$-measurable, we have

$$\mathbb{E}^* \left[GFP(T_i, T_i, T_j) e^{-\int_t^T r_s ds} \right] = P(0, T_i, T_j) \widehat{\mathbb{E}}_{i,j} \left[GF e^{\int_0^t r_s ds} \right]$$

$$= P(0, T_i, T_j) \widehat{\mathbb{E}}_{i,j} \left[G e^{\int_0^t r_s ds} \widehat{\mathbb{E}}_{i,j}[F \mid \mathcal{F}_t] \right]$$

$$= \mathbb{E}^* \left[GP(T_i, T_i, T_j) e^{-\int_0^{T_i} r_s ds} e^{\int_0^t r_s ds} \widehat{\mathbb{E}}_{i,j}[F \mid \mathcal{F}_t] \right]$$

$$= \mathbb{E}^* \left[GP(T_i, T_i, T_j) e^{-\int_t^{T_i} r_s ds} \widehat{\mathbb{E}}_{i,j}[F \mid \mathcal{F}_t] \right]$$

$$= \mathbb{E}^* \left[G \mathbb{E}^* \left[P(T_i, T_i, T_j) e^{-\int_t^{T_i} r_s ds} \,\middle|\, \mathcal{F}_t \right] \widehat{\mathbb{E}}_{i,j}[F \mid \mathcal{F}_t] \right]$$

$$= \mathbb{E}^* \left[GP(t, T_i, T_j) \widehat{\mathbb{E}}_{i,j}[F \mid \mathcal{F}_t] \right],$$

where we used Proposition 8.7, which proves (8.28) by the characterization (11.4) of conditional expectations. □

As a consequence of Propositions 8.2 and 8.8, we also have

$$\frac{d\widehat{\mathbb{P}}_{i,j|\mathcal{F}_t}}{d\widehat{\mathbb{P}}_{k|\mathcal{F}_t}} = P(t, T_k) e^{-\int_{T_k}^{T_i} r_s ds} \frac{P(T_i, T_i, T_j)}{P(t, T_i, T_j)}, \qquad 0 \leqslant t \leqslant \min(T_{i+1}, T_k),$$

hence at $t = 0$ we find

$$\frac{d\widehat{\mathbb{P}}_{i,j}}{d\widehat{\mathbb{P}}_k} = \frac{P(0, T_k)}{P(0, T_i, T_j)} e^{-\int_{T_k}^{T_i} r_s ds} P(T_i, T_i, T_j), \tag{8.29}$$

$1 \leqslant i < j \leqslant n$. Using the forward swap measure $\widehat{\mathbb{P}}_{i,j}$, we obtain the following pricing formula for a given integrable claim with payoff of the form $P(T_i, T_i, T_j)F$:

$$\mathbb{E}^* \left[e^{-\int_t^{T_i} r_s ds} P(T_i, T_i, T_j) F \,\middle|\, \mathcal{F}_t \right] = P(t, T_i, T_j) \mathbb{E}^* \left[F \frac{d\widehat{\mathbb{P}}_{i,j|\mathcal{F}_t}}{d\mathbb{P}^*_{|\mathcal{F}_t}} \,\middle|\, \mathcal{F}_t \right]$$

$$= P(t, T_i, T_j) \widehat{\mathbb{E}}_{i,j}[F \mid \mathcal{F}_t], \tag{8.30}$$

after applying (8.25) and (8.27) on the last line, see also Proposition 10.1 in Privault (2014). The following result extends Propositions 8.3 and 8.4 to the setting of forward swap measures.

Proposition 8.9. *For all* $1 \leqslant i < j \leqslant n$ *and* $1 \leqslant k \leqslant n$ *we have*

$$\widehat{\mathbb{E}}_{i,j}\left[\frac{d\widehat{\mathbb{P}}_k}{d\widehat{\mathbb{P}}_{i,j}}\;\middle|\;\mathcal{F}_t\right] = \frac{P(0,T_i,T_j)}{P(0,T_k)}\frac{P(t,T_k)}{P(t,T_i,T_j)}, \quad 0 \leqslant t \leqslant \min(T_{i+1},T_k),$$

$$(8.31)$$

and

$$\widehat{\mathbb{E}}_k\left[\frac{d\widehat{\mathbb{P}}_{i,j}}{d\widehat{\mathbb{P}}_k}\;\middle|\;\mathcal{F}_t\right] = \frac{P(0,T_k)}{P(0,T_i,T_j)}\frac{P(t,T_i,T_j)}{P(t,T_k)}, \quad 0 \leqslant t \leqslant \min(T_{i+1},T_k).$$

$$(8.32)$$

In particular, the process

$$t \mapsto v_k^{i,j}(t) := \frac{P(t,T_k)}{P(t,T_i,T_j)}$$

is an $(\mathcal{F}_t)_{t\in[0,\min(T_k,T_{i+1})]}$-*martingale under* $\widehat{\mathbb{P}}_{i,j}$, *and*

$$t \mapsto v_{i,j}^k(t) := \frac{P(t,T_i,T_j)}{P(t,T_k)}$$

is an $(\mathcal{F}_t)_{t\in[0,\min(T_k,T_{i+1})]}$-*martingale under* $\widehat{\mathbb{P}}_k$.

Proof. For all bounded and $\mathcal{F}_t$-measurable random variables F, we have

$$\widehat{\mathbb{E}}_{i,j}\left[F\frac{d\widehat{\mathbb{P}}_k}{d\widehat{\mathbb{P}}_{i,j}}\right] = \mathbb{E}^*\left[F\frac{d\widehat{\mathbb{P}}_k}{d\mathbb{P}^*}\right]$$

$$= \frac{1}{P(0,T_k)}\,\mathbb{E}^*\left[F\,e^{-\int_0^{T_k} r_u du}\right]$$

$$= \frac{1}{P(0,T_k)}\,\mathbb{E}^*\left[F\,e^{-\int_0^{t} r_u du}P(t,T_k)\right]$$

$$= \frac{1}{P(0,T_k)}\,\mathbb{E}^*\left[F\,e^{-\int_0^{T_i} r_u du}P(T_i,T_i,T_j)\frac{P(t,T_k)}{P(t,T_i,T_j)}\right]$$

$$= \frac{P(0,T_i,T_j)}{P(0,T_k)}\widehat{\mathbb{E}}_{i,j}\left[F\frac{P(t,T_k)}{P(t,T_i,T_j)}\right],$$

which shows (8.31). Similarly, we have

$$\widehat{\mathbb{E}}_k\left[F\frac{d\widehat{\mathbb{P}}_{i,j}}{d\widehat{\mathbb{P}}_k}\right] = \mathbb{E}^*\left[F\frac{d\widehat{\mathbb{P}}_{i,j}}{d\mathbb{P}^*}\right]$$

$$= \frac{1}{P(0,T_i,T_j)}\,\mathbb{E}^*\left[F\,e^{-\int_0^{T_i} r_u du}P(T_i,T_i,T_j)\right]$$

$$= \frac{1}{P(0,T_i,T_j)}\,\mathbb{E}^*\left[F\,e^{-\int_0^{t} r_u du}P(t,T_i,T_j)\right]$$

$$= \frac{1}{P(0,T_i,T_j)} \mathbb{E}^* \left[F e^{-\int_0^{T_k} r_u du} \frac{P(T_k,T_k)}{P(t,T_k)} P(t,T_i,T_j) \right]$$

$$= \frac{P(0,T_k)}{P(0,T_i,T_j)} \widehat{\mathbb{E}}_k \left[F \frac{P(t,T_i,T_j)}{P(t,T_k)} \right],$$

which shows (8.32). The fact that $t \mapsto v_k^{i,j}(t)$ and $t \mapsto v_{i,j}^k(t)$ are martingales respectively under $\widehat{\mathbb{P}}_{i,j}$ and $\widehat{\mathbb{P}}_k$ then follows from (8.31)-(8.32) and the remark after Proposition 11.2 in the appendix. $\qquad \square$

It follows from Proposition 8.9 that the LIBOR swap rate $S(t,T_i,T_j)$ defined in Proposition 5.4 and Corollary 5.1 is a martingale under $\widehat{\mathbb{P}}_{i,j}$, using *e.g.* the tower property (11.7) in the appendix.

Proposition 8.10. *The LIBOR swap rate process*

$$S(t,T_i,T_j) = \frac{P(t,T_i) - P(t,T_j)}{P(t,T_i,T_j)} = v_i^{i,j}(t) - v_j^{i,j}(t), \quad 0 \leqslant t \leqslant T_i,$$

is a martingale under the forward swap measure $\widehat{\mathbb{P}}_{i,j}$.

Proof. This follows from the fact that the deflated process

$$t \mapsto v_k^{i,j}(t) := \frac{P(t,T_k)}{P(t,T_i,T_j)}, \quad 1 \leqslant i < j \leqslant n, \quad 1 \leqslant k \leqslant n,$$

is an $(\mathcal{F}_t)_{t \in [0,\min(T_k,T_{i+1})]}$-martingale under $\widehat{\mathbb{P}}_{i,j}$ by Proposition 8.9. $\qquad \square$

More precisely, in the next Proposition 8.11 we construct a standard Brownian motion $\left(\widehat{B}_t^{i,j} \right)_{t \in \mathbb{R}_+}$ under $\widehat{\mathbb{P}}_{i,j}$, which is driving the stochastic evolution of $S(t,T_i,T_j)$ in Proposition 8.12 below. Recall that we let $\delta_k = T_{k+1} - T_k$, $k = 1, 2, \ldots, n-1$.

Proposition 8.11. *For all* $1 \leqslant i < j \leqslant n$, *the process*

$$\widehat{B}_t^{i,j} := B_t - \sum_{l=i}^{j-1} \delta_l \int_0^t v_{l+1}^{i,j}(s) \zeta_{l+1}(s) ds, \quad 0 \leqslant t \leqslant T_{i+1}, \qquad (8.33)$$

is a standard Brownian motion under $\widehat{\mathbb{P}}_{i,j}$.

Proof. By Itô's calculus we have, for any $1 \leqslant i < j \leqslant n$ and $1 \leqslant k \leqslant n$,

$$dv_k^{i,j}(t) = d \left(\frac{P(t,T_k)}{P(t,T_i,T_j)} \right)$$

$$= \frac{dP(t,T_k)}{P(t,T_i,T_j)} - \frac{P(t,T_k)}{P(t,T_i,T_j)^2} dP(t,T_i,T_j)$$

$$+ \frac{P(t,T_k)}{P(t,T_i,T_j)^3} dP(t,T_i,T_j) \cdot dP(t,T_i,T_j)$$

$$- \frac{1}{P(t,T_i,T_j)^2} dP(t,T_k) \cdot dP(t,T_i,T_j)$$

$$= \frac{dP(t,T_k)}{P(t,T_i,T_j)} - \frac{P(t,T_k)}{P(t,T_i,T_j)^2} \sum_{l=i}^{j-1} \delta_l dP(t,T_{l+1})$$

$$+ \frac{P(t,T_k)}{P(t,T_i,T_j)^3} dP(t,T_i,T_j) \cdot dP(t,T_i,T_j)$$

$$- \frac{1}{P(t,T_i,T_j)^2} dP(t,T_k) \cdot dP(t,T_i,T_j)$$

$$= \frac{P(t,T_k)}{P(t,T_i,T_j)} (r_t dt + \zeta_k(t) dB_t)$$

$$- \frac{P(t,T_k)}{P(t,T_i,T_j)^2} \sum_{l=i}^{j-1} \delta_l P(t,T_{l+1})(r_t dt + \zeta_{l+1}(t) dB_t)$$

$$+ \frac{P(t,T_k)}{P(t,T_i,T_j)^3} \sum_{l,l'=i}^{j-1} \delta_l \delta_{l'} P(t,T_{l+1}) P(t,T_{l'+1}) \zeta_{l+1}(t) \zeta_{l'+1}(t) dt$$

$$- \zeta_k(t) \frac{P(t,T_k)}{P(t,T_i,T_j)^2} \sum_{l=i}^{j-1} \delta_l \zeta_{l+1}(t) P(t,T_{l+1}) dt$$

$$= v_k^{i,j}(t) \left(\zeta_k(t) dB_t - \sum_{l=i}^{j-1} \delta_l v_{l+1}^{i,j}(t) \zeta_{l+1}(t) dB_t \right.$$

$$\left. + \sum_{l,l'=i}^{j-1} \delta_l \delta_{l'} v_{l+1}^{i,j}(t) v_{l'+1}^{i,j}(t) \zeta_{l+1}(t) \zeta_{l'+1}(t) dt - \zeta_k(t) \sum_{l=i}^{j-1} \delta_l \zeta_{l+1}(t) v_{l+1}^{i,j}(t) dt \right)$$

$$= v_k^{i,j}(t) \left(\sum_{l=i}^{j-1} \delta_l v_{l+1}^{i,j}(t)(\zeta_k(t) - \zeta_{l+1}(t)) dB_t \right.$$

$$\left. + \sum_{l,l'=i}^{j-1} \delta_l \delta_{l'} v_{l+1}^{i,j}(t) v_{l'+1}^{i,j}(t)(\zeta_{l+1}(t) - \zeta_k(t)) \zeta_{l'+1}(t) dt \right)$$

$$= v_k^{i,j}(t) \sum_{l=i}^{j-1} \delta_l v_{l+1}^{i,j}(t)(\zeta_k(t) - \zeta_{l+1}(t)) \left(dB_t - \sum_{l'=i}^{j-1} \delta_{l'} v_{l'+1}^{i,j}(t) \zeta_{l'+1}(t) dt \right)$$

$$= v_k^{i,j}(t) \left(\sum_{l=i}^{j-1} \delta_l v_{l+1}^{i,j}(t)(\zeta_k(t) - \zeta_{l+1}(t)) \right) d\widehat{B}_t^{i,j}.$$

Since the process $\left(v_k^{i,j}(t)\right)_{t\in[0,\min(T_k,T_{i+1})]}$ is a continuous martingale by Proposition 8.9, the same holds for the process $(\widehat{B}_t^{i,j})_{t\in[0,T_{i+1}]}$ defined in (8.33), which becomes a standard Brownian motion under $\widehat{\mathbb{P}}_{i,j}$ by the Lévy characterization of Brownian motion, see Section 1.4.   □

When $j = i + 1$, Relation (8.33) reads

$$\widehat{B}^{i,i+1}(t) = B_t - \int_0^t \zeta_{i+1}(s)ds, \qquad 0 \leqslant t \leqslant T_{i+1},$$

since $v_{i+1}^{i,i+1}(t) = 1/\delta_i$, hence from Relation (8.8), we have

$$\left(\widehat{B}_t^{i,i+1}\right)_{t\in[0,T_i]} = \left(\widehat{B}_t^{i+1}\right)_{t\in[0,T_i]}. \tag{8.34}$$

The equality is valid up to time T_i as by Proposition 8.9 we have

$$\widehat{\mathbb{E}}_{i,i+1}\left[\frac{d\widehat{\mathbb{P}}_{i,i+1}}{d\widehat{\mathbb{P}}_{i+1}}\,\bigg|\,\mathcal{F}_t\right] = 1, \qquad 0 \leqslant t \leqslant T_{i+1}, \tag{8.35}$$

although by (8.29) we have

$$\frac{d\widehat{\mathbb{P}}_{i,i+1}}{d\widehat{\mathbb{P}}_{i+1}} = P(T_i, T_{i+1})e^{-\int_{T_i}^{T_{i+1}} r_s ds},$$

hence $\widehat{\mathbb{P}}_{i,i+1} \neq \widehat{\mathbb{P}}_{i+1}$.

We can now compute the dynamics of the swap rate $S(t, T_i, T_j)$ under $\widehat{\mathbb{P}}_{i,j}$ using the Brownian process $\left(\widehat{B}_t^{i,j}\right)_{t\in\mathbb{R}_+}$, as on page 17 of Schoenmakers (2005).

Proposition 8.12. *The dynamics of the swap rate $S(t, T_i, T_j)$ process under $\widehat{\mathbb{P}}_{i,j}$ is given by*

$$dS(t, T_i, T_j) = \sigma_{i,j}(t)S(t, T_i, T_j)d\widehat{B}_t^{i,j}, \qquad 0 \leqslant t \leqslant T_i,$$

where the swap rate volatility is given by

$$\sigma_{i,j}(t) = \frac{P(t, T_j)}{P(t, T_i) - P(t, T_j)}(\zeta_i(t) - \zeta_j(t)) + \sum_{l=i}^{j-1} \delta_l v_{l+1}^{i,j}(t)(\zeta_i(t) - \zeta_{l+1}(t)),$$

$1 \leqslant i < j \leqslant n$.

Proof. From the proof of Proposition 8.11 we have

$$dv_k^{i,j}(t) = v_k^{i,j}(t)\left(\sum_{l=i}^{j-1} \delta_l v_{l+1}^{i,j}(t)(\zeta_k(t) - \zeta_{l+1}(t))\right)d\widehat{B}_t^{i,j},$$

hence

$$dS(t, T_i, T_j) = d\left(\frac{P(t, T_i) - P(t, T_j)}{P(t, T_i, T_j)}\right)$$

$$= dv_i^{i,j}(t) - dv_j^{i,j}(t)$$

$$= \left(\sum_{l=i}^{j-1} \delta_l v_{l+1}^{i,j}(t)(v_i^{i,j}(t)(\zeta_i(t) - \zeta_{l+1}(t)) - v_j^{i,j}(t)(\zeta_j(t) - \zeta_{l+1}(t)))\right) d\widehat{B}_t^{i,j}$$

$$= \left(\sum_{l=i}^{j-1} \zeta_{l+1}(t)\delta_l v_{l+1}^{i,j}(t)(v_j^{i,j}(t) - v_i^{i,j}(t))\right) d\widehat{B}_t^{i,j}$$

$$+ \left(v_i^{i,j}(t)\zeta_i(t) - v_j^{i,j}(t)\zeta_j(t)\right) d\widehat{B}_t^{i,j}$$

$$= \left(\sum_{l=i}^{j-1} (\zeta_{l+1}(t) - \zeta_i(t))\delta_l v_{l+1}^{i,j}(t)(v_j^{i,j}(t) - v_i^{i,j}(t))\right) d\widehat{B}_t^{i,j}$$

$$+ v_j^{i,j}(t)\left(\zeta_i(t) - \zeta_j(t)\right) d\widehat{B}_t^{i,j}$$

$$= S(t, T_i, T_j)\left(\frac{(\zeta_i(t) - \zeta_j(t))P(t, T_j)}{P(t, T_i) - P(t, T_j)} + \sum_{l=i}^{j-1} \delta_l v_{l+1}^{i,j}(t)(\zeta_i(t) - \zeta_{l+1}(t))\right) d\widehat{B}_t^{i,j}$$

$$= S(t, T_i, T_j)\sigma_{i,j}(t)d\widehat{B}_t^{i,j}.$$

$\square$

As a consequence of Corollary 1.3 and Proposition 8.12, we recover the fact that the swap rate $S(t, T_i, T_j)$ is a martingale under the forward swap measure $\widehat{\mathbb{P}}_{i,j}$.

Exercises

Exercise 8.1. We consider a bond with maturity T, priced

$$P(t, T) = \mathbb{E}^*\left[e^{-\int_t^T r_s ds} \,\middle|\, \mathcal{F}_t\right]$$

at time $t \in [0, T]$.

(1) Apply the change of numeraire formula (8.3) to compute the derivative $\dfrac{\partial P}{\partial T}(t, T)$.

(2) Using Relation (5.4), find an expression of the instantaneous forward rate $f(t, T)$ using the short rate r_T and the forward expectation $\widehat{\mathbb{E}}$.

Exercise 8.2. Bond call options (1). (Exercise 7.1 continued).

(1) Derive a stochastic differential equation satisfied by $t \mapsto P(t, T)$.
(2) Derive a stochastic differential equation satisfied by

$$t \mapsto e^{-\int_0^t r_s ds} P(t, T), \qquad 0 \leqslant t \leqslant T.$$

(3) Express the conditional expectation

$$\mathbb{E}^* \left[\frac{d\widehat{\mathbb{P}}_i}{d\mathbb{P}^*} \,\middle|\, \mathcal{F}_t \right]$$

in terms of $P(t, T)$, $P(0, T)$ and $e^{-\int_0^t r_s ds}$, $0 \leqslant t \leqslant T$.
(4) Find a stochastic differential equation satisfied by

$$t \mapsto \mathbb{E}^* \left[\frac{d\widehat{\mathbb{P}}_i}{d\mathbb{P}^*} \,\middle|\, \mathcal{F}_t \right].$$

(5) Compute the Radon-Nikodym density $d\widehat{\mathbb{P}}_i/d\mathbb{P}^*$ of the forward measure with respect to $\mathbb{P}^*$ by solving the stochastic differential equation of question (4).
(6) Using the Girsanov Theorem 2.1, compute the dynamics of $(r_t)_{t \in \mathbb{R}_+}$ under the forward measure.
(7) Compute the price

$$\mathbb{E}^* \left[e^{-\int_t^T r_s ds} (P(T, S) - K)^+ \,\middle|\, \mathcal{F}_t \right] = P(t, T) \widehat{\mathbb{E}}_i \left[(P(T, S) - K)^+ \,\middle|\, \mathcal{F}_t \right]$$

of a bond call option at time $t \in [0, T]$.

Exercise 8.3. Bond call options (2). (Exercise 7.2 continued).

(1) Compute the Radon-Nikodym density

$$\frac{d\widehat{\mathbb{P}}_i}{d\mathbb{P}^*} = \frac{1}{P(0, T)} e^{-\int_0^T r_t dt}$$

of the forward measure $\widehat{\mathbb{P}}_i$ with respect to $\mathbb{P}^*$.
(2) Using the Girsanov Theorem 2.1, compute the dynamics of $(r_t)_{t \in [0, T_i]}$ under the forward measure.
(3) Assuming for simplicity that $b = 0$, compute the price

$$\mathbb{E}^* \left[e^{-\int_0^T r_s ds} (P(T, S) - K)^+ \right] = P(0, T) \widehat{\mathbb{E}}_i \left[(P(T, S) - K)^+ \right]$$

of a bond call option at time $t = 0$.

Exercise 8.4. Stochastic string model (Exercise 7.6 continued). Compute the bond call option price $\mathbb{E}^*\left[e^{-\int_0^T r_s ds}(P(T,S) - K)^+\right]$ by the Black-Scholes formula, knowing that for any centered Gaussian random variable $X \simeq \mathcal{N}(0, v^2)$ with variance $v^2 > 0$, we have

$$\mathbb{E}^*\left[\left(x\,e^{m+X} - K\right)^+\right]$$
$$= x\,e^{m+v^2/2}\Phi\left(v + \frac{m + \log(x/K)}{v}\right) - K\Phi\left(\frac{m + \log(x/K)}{v}\right).$$

Exercise 8.5. (Exercise 5.5 continued).

(1) Compute the conditional Radon-Nikodym density

$$\mathbb{E}^*\left[\frac{d\widehat{\mathbb{P}}_T}{d\mathbb{P}^*}\,\middle|\,\mathcal{F}_t\right] = \frac{P(t,T)}{P(0,T)}\,e^{-\int_0^t r_s^T ds}$$

of the forward measure $\widehat{\mathbb{P}}_T$ with respect to $\mathbb{P}^*$.
(2) Show that the process
$$\widehat{B}_t := B_t - \sigma t, \qquad 0 \leqslant t \leqslant T,$$
is a standard Brownian motion under $\widehat{\mathbb{P}}_T$.
(3) Compute the dynamics of $(X_t^S)_{t\in[0,T]}$ and $(P(t,S))_{t\in[0,T]}$ under $\widehat{\mathbb{P}}_T$.
Hint: Show that

$$-\mu(S - T) + \sigma(S - T)\int_0^t \frac{dB_s}{S - s} = \frac{S - T}{S - t}\log P(t,S).$$

(4) Compute the bond call option price

$$\mathbb{E}^*\left[e^{-\int_t^T r_s^T ds}(P(T,S) - K)^+\,\middle|\,\mathcal{F}_t\right] = P(t,T)\widehat{\mathbb{E}}\left[(P(T,S) - K)^+\,\middle|\,\mathcal{F}_t\right],$$
$$0 \leqslant t \leqslant T < S.$$

Hint: Given X a Gaussian random variable with mean m and variance $v^2 > 0$ given $\mathcal{F}_t$, we have:

$$\mathbb{E}^*\left[\left(e^X - K\right)^+\,\middle|\,\mathcal{F}_t\right] = e^{m+v^2/2}\Phi\left(\frac{v}{2} + \frac{1}{v}\left(m + \frac{v^2}{2} - \log K\right)\right)$$
$$-K\Phi\left(-\frac{v}{2} + \frac{1}{v}\left(m + \frac{v^2}{2} - \log K\right)\right).$$

Exercise 8.6. Bond put options (1). Consider two bonds with maturities T_1 and T_2, $T_1 < T_2$, which follow the stochastic differential equations
$$dP(t,T_1) = r_t P(t,T_1)dt + \zeta_1(t)P(t,T_1)dB_t$$
and
$$dP(t,T_2) = r_t P(t,T_2)dt + \zeta_2(t)P(t,T_2)dB_t.$$

(1) Using Itô calculus, show that the forward process $P(t, T_2)/P(t, T_1)$ is a driftless geometric Brownian motion driven by $d\widehat{B}_t := dB_t - \zeta_1(t)dt$ under the T_1-forward measure $\widehat{\mathbb{P}}$.

(2) Compute the price $\mathbb{E}^* \left[e^{-\int_t^{T_1} r_s ds} (K - P(T_1, T_2))^+ \,\Big|\, \mathcal{F}_t \right]$ of a bond put option at time $t \in [0, T_1]$ using change of numeraire and the Black-Scholes formula.

Hint: Given X a Gaussian random variable with mean m and variance $v^2 > 0$ given $\mathcal{F}_t$, we have:

$$\mathbb{E}^* \left[(K - e^X)^+ \mid \mathcal{F}_t \right] = K\Phi \left(\frac{v}{2} - \frac{1}{v} \left(m + \frac{v^2}{2} - \log K \right) \right) \qquad (8.36)$$

$$- e^{m + v^2/2} \Phi \left(-\frac{v}{2} - \frac{1}{v} \left(m + \frac{v^2}{2} - \log K \right) \right).$$

Exercise 8.7. Consider two zero-coupon bond prices of the form $P(t, T) = F(t, r_t)$ and $P(t, S) = G(t, r_t)$, where $(r_t)_{t \in \mathbb{R}_+}$ is a short-term interest rate process. Compute the dynamics of $(P(t, S))_{t \in [0, T]}$ under the forward measure $\widehat{\mathbb{P}}$, using a standard Brownian motion $\left(\widehat{B}_t \right)_{t \in [0, T]}$ under $\widehat{\mathbb{P}}$.

Exercise 8.8. Forward contracts. Using a change of numeraire argument based on Proposition 8.1, compute the price at time $t \in [0, T]$ of a forward (or future) contract with payoff $P(T, S) - K$ in a bond market with short-term interest rate $(r_t)_{t \in \mathbb{R}_+}$. How would you hedge this forward contract?

Exercise 8.9. Convertible bonds. Consider an underlying stock price process $(S_t)_{t \in \mathbb{R}_+}$ given by

$$dS_t = rS_t dt + \sigma S_t dB_t^{(1)},$$

and a short-term interest rate process $(r_t)_{t \in \mathbb{R}_+}$ given by

$$dr_t = \gamma(t, r_t)dt + \eta(t, r_t)dB_t^{(2)},$$

where $\left(B_t^{(1)} \right)_{t \in \mathbb{R}_+}$ and $\left(B_t^{(2)} \right)_{t \in \mathbb{R}_+}$ are two correlated Brownian motions under the risk-neutral measure $\mathbb{P}^*$, with $dB_t^{(1)} \cdot dB_t^{(2)} = \rho dt$. A convertible bond is made of a corporate bond priced $P(t, T)$ at time $t \in [0, T]$, that can be exchanged into a quantity $\alpha > 0$ of the underlying company's stock S_τ at a future time τ, whichever has a higher value, where α is a *conversion rate*.

(1) Find the payoff of the convertible bond at time τ.
(2) Rewrite the convertible bond payoff at time τ as the linear combination of $P(\tau, T)$ and a call option payoff on S_τ, whose strike price is to be determined.
(3) Write down the convertible bond price at time $t \in [0, \tau]$ as a function $C(t, S_t, r_t)$ of the underlying asset price and interest rate, using a discounted conditional expectation, and show that the discounted corporate bond price

$$\mathrm{e}^{-\int_0^t r_s ds} C(t, S_t, r_t), \qquad 0 \leqslant t \leqslant \tau,$$

is a martingale.
(4) Write down $d\big(\mathrm{e}^{-\int_0^t r_s ds} C(t, S_t, r_t)\big)$ using the Itô formula, and derive the pricing PDE satisfied by the function $C(t, x, y)$ together with its terminal condition.
(5) Taking the bond price $P(t, T)$ as a numeraire, price the convertible bond as a European option with strike price $K = 1$ on an underlying asset priced $Z_t := S_t/P(t, T)$, $t \in [0, \tau]$ under the forward measure $\widehat{\mathbb{P}}$ with maturity T.
(6) Assuming the bond price dynamics

$$dP(t, T) = r_t P(t, T) dt + \sigma_B(t) P(t, T) dB_t,$$

determine the dynamics of the process $(Z_t)_{t \in \mathbb{R}_+}$ under the forward measure $\widehat{\mathbb{P}}$.
(7) Assuming that $(Z_t)_{t \in \mathbb{R}_+}$ can be modeled as a geometric Brownian motion, price the convertible bond using the Black-Scholes formula.

Exercise 8.10. Bond option hedging (1). Let $(r_t)_{t \in \mathbb{R}_+}$ denote a short-term interest rate process. Consider two bonds with maturities T and S, with prices $P(t, T)$ and $P(t, S)$ given by

$$\frac{dP(t, T)}{P(t, T)} = r_t dt + \zeta^T(t) dB_t, \tag{8.37}$$

and

$$\frac{dP(t, S)}{P(t, S)} = r_t dt + \zeta^S(t) dB_t,$$

where $\big(\zeta^T(s)\big)_{s \in [0, T]}$ and $\big(\zeta^S(s)\big)_{s \in [0, S]}$ are deterministic functions.

(1) Solve the stochastic differential equation (8.37).

(2) Derive the stochastic differential equation satisfied by the discounted bond price process

$$t \mapsto e^{-\int_0^t r_s ds} P(t,T), \qquad 0 \leqslant t \leqslant T,$$

and show that it is a martingale under $\mathbb{P}^*$.

(3) Show that

$$\mathbb{E}^* \left[e^{-\int_0^T r_s ds} \, \middle| \, \mathcal{F}_t \right] = e^{-\int_0^t r_s ds} P(t,T), \qquad 0 \leqslant t \leqslant T.$$

(4) Show, using Itô's formula, that

$$d\left(\frac{P(t,S)}{P(t,T)} \right) = \frac{P(t,S)}{P(t,T)} \left(\zeta^S(t) - \zeta^T(t) \right) d\widehat{B}_t,$$

where $\left(\widehat{B}_t \right)_{t \in \mathbb{R}_+}$ the standard Brownian motion under $\widehat{\mathbb{P}}$ defined as

$$\widehat{B}_t := B_t - \int_0^t \zeta^T(s) ds, \qquad 0 \leqslant t \leqslant T.$$

(5) Compute $P(t,S)/P(t,T)$, $0 \leqslant t \leqslant T$, show that it is a martingale under the forward measure $\widehat{\mathbb{P}}$ with maturity T, and that

$$P(T,S)$$
$$= \frac{P(t,S)}{P(t,T)} \exp\left(\int_t^T \left(\zeta^S(s) - \zeta^T(s) \right) d\widehat{B}_s - \frac{1}{2} \int_t^T \left| \zeta^S(s) - \zeta^T(s) \right|^2 ds \right).$$

Let $\widehat{\mathbb{P}}$ denote the forward measure associated to the bond price $P(t,T)$, $0 \leqslant t \leqslant T$, and defined as

$$\mathbb{E}^* \left[\frac{d\widehat{\mathbb{P}}}{d\mathbb{P}^*} \, \middle| \, \mathcal{F}_t \right] = \frac{P(t,T)}{P(0,T)} e^{-\int_0^t r_s ds}, \qquad 0 \leqslant t \leqslant T.$$

(6) Given that $\left(\zeta^T(t) \right)_{t \in [0,T]}$ and $\left(\zeta^S(t) \right)_{t \in [0,S]}$ are deterministic functions, show that for all $S, T > 0$ the price at time t

$$\mathbb{E}^* \left[e^{-\int_t^T r_s ds} (P(T,S) - K)^+ \, \middle| \, \mathcal{F}_t \right] = P(t,T) \widehat{\mathbb{E}} \left[(P(T,S) - K)^+ \, \middle| \, \mathcal{F}_t \right]$$

of a bond call option on $P(T,S)$ with strike price K and payoff $(P(T,S) - K)^+$ is equal to

$$\mathbb{E}^* \left[e^{-\int_t^T r_s ds} (P(T,S) - K)^+ \, \middle| \, \mathcal{F}_t \right]$$
$$= P(t,S) \Phi \left(\frac{v(t,T)}{2} + \frac{1}{v(t,T)} \log \frac{P(t,S)}{KP(t,T)} \right)$$
$$- KP(t,T) \Phi \left(-\frac{v(t,T)}{2} + \frac{1}{v(t,T)} \log \frac{P(t,S)}{KP(t,T)} \right),$$

where

$$v^2(t, T) := \int_t^T |\zeta^S(s) - \zeta^T(s)|^2 ds, \qquad 0 \leqslant t \leqslant T.$$

Recall that if X is a Gaussian random variable with mean $m(t, T)$ and variance $v^2(t, T)$ given $\mathcal{F}_t$, we have

$$\mathbb{E}\left[(e^X - K)^+ \mid \mathcal{F}_t\right]$$

$$= e^{m(t,T)+v^2(t,T)/2} \Phi\left(\frac{v(t, T)}{2} + \frac{1}{v(t, T)}\left(m(t, T) + \frac{v^2(t, T)}{2} - \log K\right)\right)$$

$$-K\Phi\left(-\frac{v(t, T)}{2} + \frac{1}{v(t, T)}\left(m(t, T) + \frac{v^2(t, T)}{2} - \log K\right)\right)$$

where $\Phi(x)$, $x \in \mathbb{R}$, denotes the standard normal cumulative distribution function.

(7) Compute the self-financing hedging strategy that hedges this bond call option using a portfolio based on the assets $P(t, T)$ and $P(t, S)$.

Exercise 8.11. Bond put options (2). (Exercise 6.3 continued).

(1) Let the forward measure $\widehat{\mathbb{P}}$ be defined via its conditional Radon-Nikodym density

$$\mathbb{E}\left[\frac{d\widehat{\mathbb{P}}}{d\mathbb{P}^*} \,\middle|\, \mathcal{F}_t\right] = \frac{P(t, T)}{P(0, T)} e^{-\int_0^t r_s ds}$$

with respect to $\mathbb{P}^*$, $0 \leqslant t \leqslant T$. Compute $d\widehat{\mathbb{P}}/d\mathbb{P}^*$ by solving Equation (6.18).

(2) Using the Girsanov Theorem 2.1, compute the dynamics of $(r_t)_{t \in [0,T]}$ under the forward measure $\widehat{\mathbb{P}}$.

(3) Using Itô's calculus, show that

$$t \mapsto \frac{P(t, S)}{P(t, T)}, \qquad 0 \leqslant t \leqslant T \leqslant S,$$

is a martingale under $\widehat{\mathbb{P}}$.

(4) Show that

$$\widehat{\mathbb{E}}[P(T, S) \mid \mathcal{F}_t] = \frac{P(t, S)}{P(t, T)}, \qquad 0 \leqslant t \leqslant T \leqslant S,$$

and from this identity, deduce the value of

$$A(T, S) + C(T, S)\,\mathbb{E}[X_T \mid \mathcal{F}_t] + \frac{1}{2}|C(T, S)|^2 \operatorname{Var}[X_T \mid \mathcal{F}_t]$$

in terms of $P(t, S)/P(t, T)$, $t \in [0, T]$.

(5) Compute the price

$$\mathbb{E}\left[e^{-\int_t^T r_s ds}(K - P(T,S))^+ \,\middle|\, \mathcal{F}_t\right] = P(t,T)\widehat{\mathbb{E}}\left[(K - P(T,S))^+ \,\middle|\, \mathcal{F}_t\right]$$

at time t of the bond *put* option with payoff $(K - P(T,S))^+$.

Recall that $x - K = (x - K)^+ - (K - x)^+$, and that if X is a Gaussian random variable with mean m_t and variance $v_t^2 > 0$ given $\mathcal{F}_t$, we have

$$\mathbb{E}\left[\left(e^X - K\right)^+ \,\middle|\, \mathcal{F}_t\right] = \Phi\left(\frac{v_t}{2} + \frac{1}{v_t}(m_t + v_t^2/2 - \log K)\right)$$

$$-K\Phi\left(-\frac{v_t}{2} + \frac{1}{v_t}(m_t + v_t^2/2 - \log K)\right),$$

where

$$\Phi(x) := \int_{-\infty}^x e^{-y^2/2}\frac{dy}{\sqrt{2\pi}}, \qquad x \in \mathbb{R},$$

denotes the standard normal cumulative distribution function.

Exercise 8.12. Consider an asset priced S_t at time t, with

$$dS_t = rS_t dt + \sigma^S S_t dB_t^S,$$

and an exchange rate $(R_t)_{t \in \mathbb{R}_+}$ given by

$$dR_t = (r - r^R)R_t dt + \sigma^R R_t dB_t^R,$$

where $(B_t^S)_{t \in \mathbb{R}_+}$ and $(B_t^R)_{t \in \mathbb{R}_+}$ are standard Brownian motions under $\mathbb{P}^*$, with the correlation

$$dB_t^S \cdot dB_t^R = \rho dt,$$

where $\rho \in [-1, 1]$.

(1) Show that $(B_t^R)_{t \in \mathbb{R}_+}$ can be written as

$$B_t^R = \rho B_t^S + \sqrt{1 - \rho^2}B_t, \qquad t \geqslant 0,$$

where $(B_t)_{t \in \mathbb{R}_+}$ is a standard Brownian motion under $\mathbb{P}^*$, independent of $(B_t^S)_{t \in \mathbb{R}_+}$.

(2) Letting $X_t := S_t/R_t$, show that dX_t can be written as

$$dX_t = \left(r - r^R + \left(\sigma^R\right)^2 - \rho\sigma^R\sigma^S\right)X_t dt + \widehat{\sigma}X_t dB_t^X,$$

where $(B_t^X)_{t \in \mathbb{R}_+}$ is a standard Brownian motion under $\mathbb{P}^*$, and $\widehat{\sigma}$ is to be computed.

(3) Let
$$a = r^R + \rho \sigma^R \sigma^S - (\sigma^R)^2$$
and $Y_t := e^{at} S_t / R_t$, $t \in \mathbb{R}_+$. Show that dY_t can be written as
$$dY_t = rY_t dt + \widehat{\sigma} Y_t dB_t^X.$$

(4) Compute the price
$$e^{-(T-t)r} \mathbb{E}^* \left[\left(\frac{S_T}{R_T} - K \right)^+ \Big| \mathcal{F}_t \right]$$
of a quanto option at time $t \in [0, T]$.

Problem 8.13. Bond option hedging (2). Consider a portfolio $(\xi_t^T, \xi_t^S)_{t \in [0,T]}$ made of two bonds with maturities $0 < T < S$, and value
$$V_t = \xi_t^T P(t, T) + \xi_t^S P(t, S), \qquad 0 \leqslant t \leqslant T,$$
at time t. We assume that the portfolio is self-financing, *i.e.*
$$dV_t = \xi_t^T dP(t, T) + \xi_t^S dP(t, S), \qquad 0 \leqslant t \leqslant T, \tag{8.38}$$
and that it *hedges* the claim the bond call option payoff $(P(T, S) - K)^+$, so that we have
$$V_t = \mathbb{E}^* \left[e^{-\int_t^T r_s ds} (P(T, S) - K)^+ \Big| \mathcal{F}_t \right]$$
$$= P(t, T) \widehat{\mathbb{E}} [(P(T, S) - K)^+ | \mathcal{F}_t], \qquad 0 \leqslant t \leqslant T,$$
under the forward measure $\widehat{\mathbb{P}}$ with maturity T.

(1) Show that under the self-financing condition (8.38), we have
$$\mathbb{E}^* \left[e^{-\int_t^T r_s ds} (P(T, S) - K)^+ \Big| \mathcal{F}_t \right]$$
$$= P(0, T) \widehat{\mathbb{E}} [(P(T, S) - K)^+] + \int_0^t \xi_s^T dP(s, T) + \int_0^t \xi_s^S dP(s, S).$$

(2) Show that the discounted portfolio value process $\widetilde{V}_t = e^{-\int_0^t r_s ds} V_t$ satisfies
$$d\widetilde{V}_t = \xi_t^T d\widetilde{P}(t, T) + \xi_t^S d\widetilde{P}(t, S),$$
where
$$\widetilde{P}(t, T) := e^{-\int_0^t r_s ds} P(t, T), \quad t \in [0, T],$$
and
$$\widetilde{P}(t, S) := e^{-\int_0^t r_s ds} P(t, S), \quad t \in [0, S],$$
denote the discounted bond prices.

(3) Assuming that $(\zeta^T(t))_{t\in[0,T]}$ and $(\zeta^S(t))_{t\in[0,S]}$ are deterministic functions, show that the price of a bond call option with strike price K can be written as

$$\mathbb{E}\left[e^{-\int_t^T r_s ds}(P(T,S)-K)^+ \,\Big|\, \mathcal{F}_t\right] = P(t,T)\widehat{\mathbb{E}}[(P(T,S)-K)^+ \,|\, \mathcal{F}_t]$$
$$= P(t,T)C(X_t, v(t,T)),$$

where X_t is the forward price $X_t := P(t,S)/P(t,T)$,

$$v^2(t,T) := \int_t^T \left|\zeta^S(s) - \zeta^T(s)\right|^2 ds, \qquad 0 \leqslant t \leqslant T,$$

and $C(x,\sigma)$ is a function to be determined. Recall that if X is a Gaussian random variable with mean m_t and variance $v_t^2 > 0$ given $\mathcal{F}_t$, we have

$$\mathbb{E}^*\left[(e^X - K)^+ \,|\, \mathcal{F}_t\right] = e^{m_t + v_t^2/2}\Phi\left(\frac{v_t}{2} + \frac{1}{v_t}\left(m_t + \frac{v_t^2}{2} - \log K\right)\right)$$
$$-K\Phi\left(-\frac{v_t}{2} + \frac{1}{v_t}\left(m_t + \frac{v_t^2}{2} - \log K\right)\right)$$

where $\Phi(x)$, $x \in \mathbb{R}$, denotes the standard normal cumulative distribution function, cf. Lemma 2.3.

(4) Letting $X_t := P(t,S)/P(t,T)$, $t \in [0,T]$, show that

$$\widehat{\mathbb{E}}[(P(T,S)-K)^+ \,|\, \mathcal{F}_t] = \widehat{\mathbb{E}}[(P(T,S)-K)^+] + \int_0^t \frac{\partial C}{\partial x}(X_u, v(u,T))dX_u,$$

$0 \leqslant t \leqslant T$.

Hint: Use the martingale property and the Itô formula.

(5) Under the self-financing condition (8.38), show that the forward portfolio value process $\widehat{V}_t := V_t/P(t,T)$, $t \in [0,T]$, satisfies

$$d\widehat{V}_t = \frac{\partial C}{\partial x}(X_t, v(t,T))dX_t \tag{8.39}$$
$$= (\zeta^S(t) - \zeta^T(t))\frac{P(t,S)}{P(t,T)}\frac{\partial C}{\partial x}(X_t, v(t,T))d\widehat{B}_t,$$

where $(\widehat{B}_t)_{t\in[0,T]}$ is the standard Brownian motion under $\widehat{\mathbb{P}}$ given by

$$\widehat{B}_t := B_t - \int_0^t \zeta^T(s)ds, \qquad 0 \leqslant t \leqslant T.$$

(6) Show that the portfolio value process $(V_t)_{t\in[0,T]}$ satisfies

$$dV_t = (\zeta^S(t) - \zeta^T(t))P(t,S)\frac{\partial C}{\partial x}(X_t, v(t,T))dB_t + \widehat{V}_t dP(t,T), \quad 0 \leqslant t \leqslant T.$$

(7) Show that the discounted portfolio value process $\widetilde{V}_t = e^{-\int_0^t r_s ds} V_t$ satisfies

$$d\widetilde{V}_t = \left(\zeta^S(t) - \zeta^T(t)\right) \widetilde{P}(t, S) \frac{\partial C}{\partial x}(X_t, v(t, T)) dB_t + \widehat{V}_t d\widetilde{P}(t, T).$$

(8) Show that the portfolio value process $(V_t)_{t \in [0,T]}$ satisfies

$$dV_t = \left(\widehat{V}_t - \frac{P(t, S)}{P(t, T)} \frac{\partial C}{\partial x}(X_t, v(t, T))\right) dP(t, T) + \frac{\partial C}{\partial x}(X_t, v(t, T)) dP(t, S).$$

(9) Compute the hedging portfolio strategy $(\xi_t^T, \xi_t^S)_{t \in [0,T]}$ of the bond call option on $P(T, S)$.

(10) Show that

$$\frac{\partial C}{\partial x}(x, v) = \Phi\left(\frac{\log(x/K) + \tau v^2/2}{\sqrt{\tau} v}\right),$$

and compute the hedging portfolio strategy $(\xi_t^T, \xi_t^S)_{t \in [0,T]}$ accordingly.

Chapter 9

Pricing of Caps and Swaptions

In this chapter we consider the pricing of interest rate derivatives such as interest rate caplets, caps and swaptions, using change of numeraire and forward rate and swap measures. This approach covers the Black pricing formulas for caps and swaptions, and the pricing of swaptions in the BGM model. We also give an outlook on the calibration of the BGM model from the data of interest rate cap and swaption prices.

9.1 Black Caplet Pricing

We consider two maturity dates $T_i < T_j$ and let $\widehat{\mathbb{E}}_j$ denote the expectation under the forward measure $\widehat{\mathbb{P}}_j$ with maturity T_j and Radon-Nikodym density

$$\frac{\mathrm{d}\widehat{\mathbb{P}}_j}{\mathrm{d}\mathbb{P}^*} = \frac{1}{P(0, T_j)} \mathrm{e}^{-\int_0^{T_j} r_s \, ds}, \tag{9.1}$$

with

$$\frac{\mathrm{d}\widehat{\mathbb{P}}_{T_j \mid \mathcal{F}_t}}{\mathrm{d}\mathbb{P}^*_{\mid \mathcal{F}_t}} = \frac{\mathrm{e}^{-\int_t^{T_j} r_s \, ds}}{P(t, T_j)}, \quad \text{and} \quad \mathbb{E}^* \left[\frac{\mathrm{d}\widehat{\mathbb{P}}_j}{\mathrm{d}\mathbb{P}^*} \, \middle| \, \mathcal{F}_t \right] = \frac{P(t, T_j)}{P(0, T_j)} \mathrm{e}^{-\int_0^t r_s \, ds},$$

$0 \leqslant t \leqslant T_j$.

Proposition 9.1. *The LIBOR rate process*

$$L(t, T_i, T_j) := \frac{1}{T_j - T_i} \left(\frac{P(t, T_i)}{P(t, T_j)} - 1 \right), \qquad 0 \leqslant t \leqslant T_i < T_j,$$

is a martingale under the forward measure $\widehat{\mathbb{P}}_j$ defined in (9.1).

Proof. The LIBOR rate $L(t, T_i, T_j)$ is a deflated process according to the forward numeraire process $(P(t, T_j))_{t \in [0, T_j]}$. Therefore, by Proposition 8.3 it is a martingale under $\widehat{\mathbb{P}}_j$. $\qquad \square$

Interest rate caplets are option contracts that offer protection against the fluctuations of a variable (or floating) rate with respect to a fixed rate κ. The payoff of an interest rate caplet on the yield (or spot forward rate) $L(T_i, T_i, T_j)$ with strike level κ can be written as

$$(L(T_i, T_i, T_j) - \kappa)^+.$$

The interest rate caplet on $L(T_i, T_i, T_j)$ can be priced at time $t \in [0, T_i]$ as

$$\mathbb{E}^* \left[e^{-\int_t^{T_j} r_s ds} (L(T_i, T_i, T_j) - \kappa)^+ \,\Big|\, \mathcal{F}_t \right]$$

$$= \mathbb{E}^* \left[e^{-\int_t^{T_j} r_s ds} \left(\frac{1}{T_j - T_i} \left(\frac{P(t, T_i)}{P(t, T_j)} - 1 \right) - \kappa \right)^+ \,\Big|\, \mathcal{F}_t \right],$$

where the discount factor is counted from the settlement date T_j. More generally, instead of interest rate caplets one can consider caps that are relative to a given tenor structure $\{0 = T_0 < T_1 < T_2 < \cdots < T_n\}$, with discounted payoff

$$\sum_{k=i}^{j-1} (T_{k+1} - T_k) e^{-\int_t^{T_{k+1}} r_s ds} (L(T_k, T_k, T_{k+1}) - \kappa)^+.$$

Using the forward measure $\widehat{\mathbb{P}}_j$, the interest rate caplet on $L(T_i, T_i, T_j)$ can be priced at time $t \in [0, T_i]$

$$\mathbb{E}^* \left[e^{-\int_t^{T_j} r_s ds} (L(T_i, T_i, T_j) - \kappa)^+ \,\Big|\, \mathcal{F}_t \right] \tag{9.2}$$

$$= P(t, T_j) \widehat{\mathbb{E}}_j \left[(L(T_i, T_i, T_j) - \kappa)^+ \,\Big|\, \mathcal{F}_t \right].$$

The next pricing formula (9.4) is known as the *Black caplet formula*. It allows us to price and hedge an interest rate caplet using a portfolio based on the bonds $P(t, T_i)$ and $P(t, T_j)$ when $L(t, T_i, T_j)$ is modeled in the BGM model.

Proposition 9.2. *Assume that $L(t, T_i, T_j)$ is modeled in the BGM model as*

$$\frac{dL(t, T_i, T_j)}{L(t, T_i, T_j)} = \gamma_i(t) d\widehat{B}_t^j, \qquad 0 \leqslant t \leqslant T_i, \tag{9.3}$$

$i = 1, 2, \ldots, n - 1$, *where $\gamma_i(t)$ is a deterministic volatility function of time $t \in [0, T_i]$, $i = 1, 2, \ldots, n - 1$. The interest rate caplet on $L(T_i, T_i, T_j)$ with strike level κ is priced at time $t \in [0, T_i]$ as*

$$(T_j - T_i)\, \mathbb{E}^* \left[e^{-\int_t^{T_j} r_s ds} (L(T_i, T_i, T_j) - \kappa)^+ \,\middle|\, \mathcal{F}_t \right] \tag{9.4}$$
$$= (P(t, T_i) - P(t, T_j))\Phi(d_+(t, T_i)) - (T_j - T_i)\kappa P(t, T_j)\Phi(d_-(t, T_i)),$$

$0 \leqslant t \leqslant T_i$, *where*

$$d_+(t, T_i) = \frac{\log(L(t, T_i, T_j)/\kappa) + (T_i - t)\sigma_i^2(t, T_i)/2}{\sigma_i(t, T_i)\sqrt{T_i - t}}, \tag{9.5}$$

and

$$d_-(t, T_i) = \frac{\log(L(t, T_i, T_j)/\kappa) - (T_i - t)\sigma_i^2(t, T_i)/2}{\sigma_i(t, T_i)\sqrt{T_i - t}},$$

and

$$\sigma_i^2(t, T_i) := \frac{1}{T_i - t} \int_t^{T_i} |\gamma_i(s)|^2 ds, \qquad 0 \leqslant t < T_i. \tag{9.6}$$

Proof. Taking $P(t, T_j)$ as a numeraire, the forward price

$$\widehat{X}_t := \frac{P(t, T_i)}{P(t, T_j)} = 1 + (T_j - T_i)L(T_i, T_i, T_j)$$

and the forward LIBOR rate process $(L(t, T_i, T_j))_{t \in [0, T_i]}$ are martingales under $\widehat{\mathbb{P}}_j$ by Proposition 8.3, $i = 1, 2, \ldots, n - 1$. More precisely, by (9.3) we have

$$L(T_i, T_i, T_j) = L(t, T_i, T_j) \exp\left(\int_t^{T_i} \gamma_i(s) d\widehat{B}_s^j - \frac{1}{2} \int_t^{T_i} |\gamma_i(s)|^2 ds \right),$$

$0 \leqslant t \leqslant T_i$, *i.e.* $t \mapsto L(t, T_i, T_j)$ is a geometric Brownian motion with volatility $\gamma_i(t)$ under $\widehat{\mathbb{P}}_j$. Hence, by (9.2) we have

$$\mathbb{E}^* \left[e^{-\int_t^{T_j} r_s ds} (L(T_i, T_i, T_j) - \kappa)^+ \,\middle|\, \mathcal{F}_t \right]$$
$$= P(t, T_j)\widehat{\mathbb{E}}_j \left[(L(T_i, T_i, T_j) - \kappa)^+ \,\middle|\, \mathcal{F}_t \right]$$
$$= P(t, T_j) \left(L(t, T_i, T_j)\Phi(d_+(t, T_i)) - \kappa\Phi(d_-(t, T_i)) \right)$$
$$= P(t, T_j)\mathrm{Bl}(L(t, T_i, T_j), \kappa, \sigma_i(t, T_i), 0, T_i - t), \qquad 0 \leqslant t \leqslant T_i,$$

where

$$\mathrm{Bl}(x, \kappa, \sigma, 0, \tau) = x\Phi(d_+(t, T_i)) - \kappa\Phi(d_-(t, T_i))$$

is the zero-interest rate Black-Scholes function, with

$$|\sigma_i(t, T_i)|^2 = \frac{1}{T_i - t} \int_t^{T_i} |\gamma_i(s)|^2 ds, \qquad 0 \leqslant t < T_i. \qquad (9.7)$$

Therefore, we obtain

$$(T_j - T_i) \, \mathbb{E}^* \left[e^{-\int_t^{T_j} r_s ds} (L(T_i, T_i, T_j) - \kappa)^+ \, \Big| \, \mathcal{F}_t \right]$$
$$= P(t, T_j) L(t, T_i, T_j) \Phi(d_+(t, T_i)) - \kappa P(t, T_j) \Phi(d_-(t, T_i))$$
$$= P(t, T_j) \left(\frac{P(t, T_i)}{P(t, T_j)} - 1 \right) \Phi(d_+(t, T_i)) - (T_j - T_i) \kappa P(t, T_j) \Phi(d_-(t, T_i)),$$

$0 \leqslant t < T_i$, which yields (9.4). $\qquad\qquad\qquad\qquad\qquad\qquad\qquad\qquad\square$

In other words, as a consequence of Relations (9.2) and (9.3), the interest rate caplet with payoff

$$(L(T_i, T_i, T_j) - \kappa)^+$$

can be priced as time $t \in [0, T_i]$ using the Black-Scholes formula as

$$\mathbb{E}^* \left[e^{-\int_t^{T_j} r_s ds} (L(T_i, T_i, T_j) - \kappa)^+ \, \Big| \, \mathcal{F}_t \right]$$
$$= P(t, T_j) \widehat{\mathbb{E}}_j \left[(L(T_i, T_i, T_j) - \kappa)^+ \, \big| \, \mathcal{F}_t \right]$$
$$= P(t, T_j) \mathrm{Bl}(L(t, T_i, T_j), \kappa, \sigma_i(t, T_i), 0, T_i - t),$$

where $\mathrm{Bl}(x, \kappa, \sigma, r, \tau)$ is the Black-Scholes function defined in Section 2.3 and $\sigma_i(t, T_i)$ is defined in (9.7). In addition, this yields the self-financing portfolio strategy

$$(\Phi(d_+(t, T_i)), -\Phi(d_+(t, T_i)) - (T_j - T_i) \kappa \Phi(d_-(t, T_i)))$$

in the bonds $(P(t, T_i), P(t, T_j))$ with maturities T_i and T_j, see also Corollary 10.1 in Privault (2014) and Exercise 9.12 for details.

Formula (9.4) is also known as the Black (1976) formula when applied to options on underlying futures or forward contracts on commodities whose prices are modeled according to (9.3). In this case, the bond price $P(t, T_j)$ can be simply modeled as $P(t, T_j) = e^{-(T_j - t)r}$ and (9.4) becomes

$$e^{-(T_j - t)r} L(t, T_i, T_j) \Phi(d_+(t, T_i)) - \kappa e^{-(T_j - t)r} \Phi(d_-(t, T_i)),$$

where $L(t, T_i, T_j)$ is the underlying future price.

By numerical inversion of the Black-Scholes formula, one can compute implied caplet volatilities $\sigma^B_{i,j}(t)$ from market data. Table 9.1 presents such implied volatilities, where the time to maturity $T_i - t$ is in ordinate and the period $T_j - T_i$ is in abscissa.

Vol Cap At the Money

	1M	3M	6M	12M	2Y	3Y	4Y	5Y	7Y	10Y
2D	9,25	9	8,85	18,6	18	16,8	15,7	14,7	13	11,3
1M	15,35	15,1	14,95	17,6	18,03	16,83	15,73	14,73	13,03	11,33
2M	15,75	15,5	15,35	18,1	18,41	17,11	16,01	15,01	13,26	11,56
3M	15,55	15,3	15,15	18,6	18,79	17,39	16,29	15,29	13,49	11,79
6M	17,55	17,3	17,15	18,7	18,28	16,98	15,88	14,98	13,48	11,98
9M	18,35	18,1	17,95	18,3	17,76	16,56	15,51	14,66	13,31	12,01
1Y	19,25	19	18,85	17,9	17,25	16,15	15,15	14,35	13,15	12,05
2Y	17,85	17,6	17,45	16,3	15,96	15,16	14,46	13,86	12,96	12,06
3Y	16,8	16,55	16,4	15,2	15,38	14,58	13,98	13,58	12,88	12,18
4Y	15,6	15,35	15,2	14,4	14,79	14,19	13,69	13,29	12,79	12,29
5Y	14,65	14,4	14,25	13,4	14,5	13,97	13,53	13,2	12,8	12,4
6Y	13,8	13,55	13,45	12,85	14,19	13,66	13,17	12,89	12,54	12,14
7Y	13,35	13,1	13	12,3	13,88	13,35	12,81	12,58	12,28	11,88
8Y	13,1	12,85	12,75	11,97	13,65	13,15	12,65	12,42	12,12	11,75
9Y	12,75	12,5	12,4	11,63	13,43	12,96	12,49	12,26	11,96	11,63
10Y	12,4	12,15	12,05	11,3	13,5	13,02	12,53	12,25	11,89	11,5
12Y	11,85	11,6	11,5	10,8	13,22	12,75	12,28	12,01	11,69	11,3
15Y	11,25	11	10,9	10,2	13	12,55	12,1	11,85	11,57	11,15
20Y	10,45	10,2	10,1	9,5	11,9	11,55	11,2	11,05	11,03	10,8
25Y	9,7	9,45	9,35	8,8	11,68	11,33	10,98	10,83	10,88	10,55
30Y	9,05	8,8	8,7	8,1	11,45	11,1	10,75	10,6	10,72	10,3

Table 9.1: Interest rate caplet volatilities.

Floorlet pricing

The interest rate floorlet on $L(T_i, T_i, T_j)$ with strike level κ is a contract with payoff $(\kappa - L(T_i, T_i, T_j))^+$. Floorlets are analog to put options and can be similarly priced by the call/put parity in the Black-Scholes formula.

Proposition 9.3. *Assume that $L(t, T_i, T_j)$ is modeled in the BGM model as in (9.3). The interest rate floorlet on $L(T_i, T_i, T_j)$ with strike level κ is priced at time $t \in [0, T_i]$ as*

$$
(T_j - T_i)\, \mathbb{E}^* \left[e^{-\int_t^{T_j} r_s ds}(\kappa - L(T_i, T_i, T_j))^+ \,\middle|\, \mathcal{F}_t \right] \tag{9.8}
$$
$$
= (T_j - T_i)\kappa P(t, T_j)\Phi\big(- d_-(t, T_i)\big)
$$
$$
- (P(t, T_i) - P(t, T_j))\Phi\big(- d_+(t, T_i)\big), \qquad 0 \leqslant t \leqslant T_i,
$$

where $d_+(t, T_i)$, $d_-(t, T_i)$ and $|\sigma_i(t, T_i)|^2$ are defined in (9.5)-(9.6).

Proof. We have

$$(T_j - T_i)\, \mathbb{E}^* \left[e^{-\int_t^{T_j} r_s ds} (\kappa - L(T_i, T_i, T_j))^+ \,\Big|\, \mathcal{F}_t \right]$$

$$= (T_j - T_i) P(t, T_j) \widehat{\mathbb{E}}_j \left[(\kappa - L(T_i, T_i, T_j))^+ \mid \mathcal{F}_t \right]$$

$$= (T_j - T_i) P(t, T_j) \big(\kappa \Phi(-d_-(t, T_i)) - (T_j - T_i) L(t, T_i, T_j) \Phi(-d_+(t, T_i)) \big)$$

$$= (T_j - T_i) \kappa P(t, T_j) \Phi(-d_-(t, T_i)) - (P(t, T_i) - P(t, T_j)) \Phi(-d_+(t, T_i)),$$

$0 \leqslant t \leqslant T_i$. $\qquad\qquad\qquad\qquad\qquad\qquad\qquad\qquad\qquad\qquad\qquad\square$

Cap pricing

Given a sequence maturity dates $\{T_i < T_{i+1} < \cdots < T_j\}$ arranged according to a discrete *tenor structure*, the pricing of interest rate caplets extends to caps with discounted payoff

$$\sum_{k=i}^{j-1} (T_{k+1} - T_k)\, e^{-\int_t^{T_{k+1}} r_s ds} (L(T_k, T_k, T_{k+1}) - \kappa)^+.$$

Pricing formulas for interest rate caps are easily deduced from analog formulas for caplets, since the payoff of an interest rate cap can be decomposed into a sum of caplet payoffs. Thus, the price of a cap at time $t \in [0, T_i]$ is given by

$$\mathbb{E}^* \left[\sum_{k=i}^{j-1} (T_{k+1} - T_k)\, e^{-\int_t^{T_{k+1}} r_s ds} (L(T_k, T_k, T_{k+1}) - \kappa)^+ \,\Big|\, \mathcal{F}_t \right]$$

$$= \sum_{k=i}^{j-1} (T_{k+1} - T_k)\, \mathbb{E}^* \left[e^{-\int_t^{T_{k+1}} r_s ds} (L(T_k, T_k, T_{k+1}) - \kappa)^+ \,\Big|\, \mathcal{F}_t \right]$$

$$= \sum_{k=i}^{j-1} (T_{k+1} - T_k) P(t, T_{k+1}) \widehat{\mathbb{E}}_{k+1} \left[(L(T_k, T_k, T_{k+1}) - \kappa)^+ \mid \mathcal{F}_t \right],$$

$$(9.9)$$

where $\widehat{\mathbb{E}}_{k+1}$ denotes the expectation under the forward measure $\widehat{\mathbb{P}}_{k+1}$ defined as

$$\frac{d\widehat{\mathbb{P}}_{k+1}}{d\mathbb{P}^*} = \frac{1}{P(0, T_{k+1})} e^{-\int_0^{T_{k+1}} r_s ds}, \qquad k = i, \dots, j-1,$$

see Definition 8.1. In the BGM model (9.3) the interest rate cap with payoff

$$\sum_{k=i}^{j-1} (T_{k+1} - T_k)(L(T_k, T_k, T_{k+1}) - \kappa)^+$$

can therefore be priced at time $t \in [0, T_1]$ by the Black formula as

$$\sum_{k=i}^{j-1} (T_{k+1} - T_k) P(t, T_{k+1}) \mathrm{Bl}(L(t, T_k, T_{k+1}), \kappa, \sigma_k(t, T_k), 0, T_k - t),$$

where

$$\sigma_k^2(t, T_k) := \frac{1}{T_k - t} \int_t^{T_k} |\gamma_k(s)|^2 ds, \quad 0 \leqslant t < T_k, \quad k = i, \ldots, j-1.$$

9.2 Swaps and Swaptions

Recall, see Section 5.4, that an interest rate swap makes it possible to exchange a sequence of variable forward rates $(f(t, T_k, T_{k+1}))_{i \leqslant k < j}$ against a fixed rate κ. Such an exchange will generate a discounted cash flow

$$\sum_{k=i}^{j-1} (T_{k+1} - T_k) e^{-\int_t^{T_{k+1}} r_s ds} (f(t, T_k, T_{k+1}) - \kappa)$$

valued at time $t \in [0, T_i]$ as

$$\mathbb{E}^* \left[\sum_{k=i}^{j-1} (T_{k+1} - T_k) e^{-\int_t^{T_{k+1}} r_s ds} (f(t, T_k, T_{k+1}) - \kappa) \,\Big|\, \mathcal{F}_t \right]$$

$$= \sum_{k=i}^{j-1} (T_{k+1} - T_k) \mathbb{E}^* \left[e^{-\int_t^{T_{k+1}} r_s ds} \,\Big|\, \mathcal{F}_t \right] (f(t, T_k, T_{k+1}) - \kappa)$$

$$= \sum_{k=i}^{j-1} (T_{k+1} - T_k) P(t, T_{k+1}) (f(t, T_k, T_{k+1}) - \kappa).$$

A swaption is a contract to protect oneself against a risk based on an interest rate swap. A payer (or call) swaption gives the option, but not the obligation, to enter an interest rate swap as payer of a fixed rate κ and as receiver of a sequence of floating LIBOR rates $(L(T_i, T_k, T_{k+1}))_{i \leqslant k < j}$, and has the payoff

$$\left(\mathbb{E}^* \left[\sum_{k=i}^{j-1} (T_{k+1} - T_k) e^{-\int_{T_i}^{T_{k+1}} r_s ds} (L(T_i, T_k, T_{k+1}) - \kappa) \,\Big|\, \mathcal{F}_{T_i} \right] \right)^+$$

$$= \left(\sum_{k=i}^{j-1} (T_{k+1} - T_k) \mathbb{E}^* \left[e^{-\int_{T_i}^{T_{k+1}} r_s ds} \,\Big|\, \mathcal{F}_{T_i} \right] (L(T_i, T_k, T_{k+1}) - \kappa) \right)^+$$

$$= \left(\sum_{k=i}^{j-1} (T_{k+1} - T_k) P(T_i, T_{k+1}) (L(T_i, T_k, T_{k+1}) - \kappa) \right)^+ \qquad (9.10)$$

at time T_i. This swaption can be priced at time $t \in [0, T_i]$ under the risk-neutral probability measure $\mathbb{P}^*$ as

$$
\mathbb{E}^* \left[e^{-\int_t^{T_i} r_s ds} \left(\sum_{k=i}^{j-1} (T_{k+1} - T_k) P(T_i, T_{k+1}) (L(T_i, T_k, T_{k+1}) - \kappa) \right)^+ \Bigg| \mathcal{F}_t \right],
$$
(9.11)

$t \in [0, T_i]$. When $j = i+1$, the swaption price (9.11) coincides with the price at time t of an interest rate caplet on $[T_i, T_{i+1}]$ up to a factor $\delta_i := T_{i+1} - T_i$, since

$$
\mathbb{E}^* \left[e^{-\int_t^{T_i} r_s ds} \left((T_{i+1} - T_i) P(T_i, T_{i+1}) (L(T_i, T_i, T_{i+1}) - \kappa) \right)^+ \Bigg| \mathcal{F}_t \right]
$$

$$
= (T_{i+1} - T_i) \, \mathbb{E}^* \left[e^{-\int_t^{T_i} r_s ds} P(T_i, T_{i+1}) \left(L(T_i, T_i, T_{i+1}) - \kappa \right)^+ \Bigg| \mathcal{F}_t \right]
$$

$$
= (T_{i+1} - T_i)
$$
$$
\times \mathbb{E}^* \left[e^{-\int_t^{T_i} r_s ds} \, \mathbb{E}^* \left[e^{-\int_{T_i}^{T_{i+1}} r_s ds} \Bigg| \mathcal{F}_{T_i} \right] \left(L(T_i, T_i, T_{i+1}) - \kappa \right)^+ \Bigg| \mathcal{F}_t \right]
$$

$$
= (T_{i+1} - T_i)
$$
$$
\times \mathbb{E}^* \left[\mathbb{E}^* \left[e^{-\int_t^{T_i} r_s ds} e^{-\int_{T_i}^{T_{i+1}} r_s ds} \left(L(T_i, T_i, T_{i+1}) - \kappa \right)^+ \Bigg| \mathcal{F}_{T_i} \right] \Bigg| \mathcal{F}_t \right]
$$

$$
= (T_{i+1} - T_i) \, \mathbb{E}^* \left[e^{-\int_t^{T_{i+1}} r_s ds} \left(L(T_i, T_i, T_{i+1}) - \kappa \right)^+ \Bigg| \mathcal{F}_t \right],
$$

$0 \leqslant t \leqslant T_i$, which coincides with the interest rate caplet price (9.2) up to the factor $T_{i+1} - T_i$.

Unlike in the case of interest rate caps, the sum in (9.11) can not be taken out of the positive part. Nevertheless, the price of the swaption can be bounded as in the next proposition.

Proposition 9.4. *The payer swaption price (9.11) can be upper bounded by the cap price (9.9), as*

$$
\mathbb{E}^* \left[e^{-\int_t^{T_i} r_s ds} \left(\sum_{k=i}^{j-1} (T_{k+1} - T_k) P(T_i, T_{k+1}) (L(T_i, T_k, T_{k+1}) - \kappa) \right)^+ \Bigg| \mathcal{F}_t \right]
$$

$$
\leqslant \mathbb{E}^* \left[\sum_{k=i}^{j-1} (T_{k+1} - T_k) e^{-\int_t^{T_{k+1}} r_s ds} \left(L(T_i, T_k, T_{k+1}) - \kappa \right)^+ \Bigg| \mathcal{F}_t \right],
$$

$0 \leqslant t \leqslant T_i$.

Proof. Due to the inequality

$$
(x_1 + x_2 + \cdots + x_m)^+ \leqslant x_1^+ + x_2^+ + \cdots + x_m^+, \quad x_1, x_2, \ldots, x_m \in \mathbb{R},
$$

we have the bound

$$
\mathbb{E}^* \left[e^{-\int_t^{T_i} r_s ds} \left(\sum_{k=i}^{j-1} (T_{k+1} - T_k) P(T_i, T_{k+1})(L(T_i, T_k, T_{k+1}) - \kappa) \right)^+ \bigg| \mathcal{F}_t \right]
$$

$$
\leqslant \mathbb{E}^* \left[e^{-\int_t^{T_i} r_s ds} \sum_{k=i}^{j-1} (T_{k+1} - T_k) P(T_i, T_{k+1}) \left(L(T_i, T_k, T_{k+1}) - \kappa \right)^+ \bigg| \mathcal{F}_t \right]
$$

$$
= \sum_{k=i}^{j-1} (T_{k+1} - T_k) \, \mathbb{E}^* \left[e^{-\int_t^{T_i} r_s ds} P(T_i, T_{k+1}) \left(L(T_i, T_k, T_{k+1}) - \kappa \right)^+ \bigg| \mathcal{F}_t \right]
$$

$$
= \sum_{k=i}^{j-1} (T_{k+1} - T_k)
$$

$$
\times \mathbb{E}^* \left[e^{-\int_t^{T_i} r_s ds} \, \mathbb{E}^* \left[e^{-\int_{T_i}^{T_{k+1}} r_s ds} \bigg| \mathcal{F}_{T_i} \right] \left(L(T_i, T_k, T_{k+1}) - \kappa \right)^+ \bigg| \mathcal{F}_t \right]
$$

$$
= \sum_{k=i}^{j-1} (T_{k+1} - T_k) \, \mathbb{E}^* \left[\mathbb{E}^* \left[e^{-\int_t^{T_{k+1}} r_s ds} \left(L(T_i, T_k, T_{k+1}) - \kappa \right)^+ \bigg| \mathcal{F}_{T_i} \right] \bigg| \mathcal{F}_t \right]
$$

$$
= \sum_{k=i}^{j-1} (T_{k+1} - T_k) \, \mathbb{E}^* \left[e^{-\int_t^{T_{k+1}} r_s ds} \left(L(T_i, T_k, T_{k+1}) - \kappa \right)^+ \bigg| \mathcal{F}_t \right]
$$

$$
= \mathbb{E}^* \left[\sum_{k=i}^{j-1} (T_{k+1} - T_k) e^{-\int_t^{T_{k+1}} r_s ds} \left(L(T_i, T_k, T_{k+1}) - \kappa \right)^+ \bigg| \mathcal{F}_t \right],
$$

$0 \leqslant t \leqslant T_i$. $\qquad\qquad\qquad\qquad\qquad\qquad\qquad\qquad\qquad\qquad\qquad\qquad\qquad\qquad\qquad\square$

The payoff of the payer swaption can be rewritten as in the following lemma which is a direct consequence of the definition (5.12) of the swap rate $S(T_i, T_i, T_j)$.

Lemma 9.1. *The payer swaption payoff* (9.10) *can be rewritten as*

$$
\left(\sum_{k=i}^{j-1} (T_{k+1} - T_k) P(T_i, T_{k+1})(L(T_i, T_k, T_{k+1}) - \kappa) \right)^+
$$

$$
= (P(T_i, T_i) - P(T_i, T_j) - \kappa P(T_i, T_i, T_j))^+ \qquad (9.12)
$$

$$
= P(T_i, T_i, T_j) \left(S(T_i, T_i, T_j) - \kappa \right)^+. \qquad (9.13)
$$

Proof. The relation

$$
\sum_{k=i}^{j-1} (T_{k+1} - T_k) P(t, T_{k+1})(L(t, T_k, T_{k+1}) - S(t, T_i, T_j)) = 0
$$

that defines the forward swap rate $S(t, T_i, T_j)$ shows that

$$\sum_{k=i}^{j-1} (T_{k+1} - T_k) P(t, T_{k+1}) L(t, T_k, T_{k+1})$$

$$= S(t, T_i, T_j) \sum_{k=i}^{j-1} (T_{k+1} - T_k) P(t, T_{k+1})$$

$$= P(t, T_i, T_j) S(t, T_i, T_j)$$

$$= P(t, T_i) - P(t, T_j)$$

as in the proof of Corollary 5.1, hence by the definition (8.24) of $P(t, T_i, T_j)$ we have

$$\sum_{k=i}^{j-1} (T_{k+1} - T_k) P(t, T_{k+1}) (L(t, T_k, T_{k+1}) - \kappa)$$

$$= P(t, T_i) - P(t, T_j) - \kappa P(t, T_i, T_j)$$

$$= P(t, T_i, T_j) \left(S(t, T_i, T_j) - \kappa \right),$$

and for $t = T_i$ we get

$$\left(\sum_{k=i}^{j-1} (T_{k+1} - T_k) P(T_i, T_{k+1}) (L(T_i, T_k, T_{k+1}) - \kappa) \right)^+$$

$$= P(T_i, T_i, T_j) \left(S(T_i, T_i, T_j) - \kappa \right)^+ .$$

$$\square$$

The next proposition states that a payer swaption on the LIBOR rate can be priced as a European call option on the swap rate $S(T_i, T_i, T_j)$ under the forward swap measure $\widehat{\mathbb{P}}_{i,j}$.

Proposition 9.5. *The price (9.11) of the payer swaption with payoff*

$$\left(\sum_{k=i}^{j-1} (T_{k+1} - T_k) P(T_i, T_{k+1}) (L(T_i, T_k, T_{k+1}) - \kappa) \right)^+ \qquad (9.14)$$

on the LIBOR market can be written under the forward swap measure $\widehat{\mathbb{P}}_{i,j}$ as the European call price

$$P(t, T_i, T_j) \widehat{\mathbb{E}}_{i,j} \left[(S(T_i, T_i, T_j) - \kappa)^+ \,\middle|\, \mathcal{F}_t \right], \quad 0 \leqslant t \leqslant T_i,$$

on the swap rate $S(T_i, T_i, T_j)$.

Proof. As a consequence of (8.30) and Lemma 9.1, we find

$$\mathbb{E}^* \left[e^{-\int_t^{T_i} r_s ds} \left(\sum_{k=i}^{j-1} (T_{k+1} - T_k) P(T_i, T_{k+1})(L(T_i, T_k, T_{k+1}) - \kappa) \right)^+ \middle| \mathcal{F}_t \right]$$

$$= \mathbb{E}^* \left[e^{-\int_t^{T_i} r_s ds} \left(P(T_i, T_i) - P(T_i, T_j) - \kappa P(T_i, T_i, T_j) \right)^+ \middle| \mathcal{F}_t \right] \quad (9.15)$$

$$= \mathbb{E}^* \left[e^{-\int_t^{T_i} r_s ds} P(T_i, T_i, T_j) \left(S(T_i, T_i, T_j) - \kappa \right)^+ \middle| \mathcal{F}_t \right]$$

$$= P(t, T_i, T_j) \mathbb{E}^* \left[\frac{d\widehat{\mathbb{P}}_{i,j|\mathcal{F}_t}}{d\mathbb{P}^*_{|\mathcal{F}_t}} \left(S(T_i, T_i, T_j) - \kappa \right)^+ \middle| \mathcal{F}_t \right]$$

$$= P(t, T_i, T_j) \widehat{\mathbb{E}}_{i,j} \left[\left(S(T_i, T_i, T_j) - \kappa \right)^+ \middle| \mathcal{F}_t \right]. \quad (9.16)$$

$\square$

9.3 Black Swaption Pricing

In the next Proposition 9.6 we price the payer swaption with payoff (9.14) or equivalently (9.13), by modeling the swap rate $(S(t, T_i, T_j))_{0 \leqslant t \leqslant T_i}$ using standard Brownian motion $\left(\widehat{B}_t^{i,j} \right)_{0 \leqslant t \leqslant T_i}$ under the forward measure $\widehat{\mathbb{P}}_{i,j}$. See Exercise 9.6 for an example of swaption pricing without the Black-Scholes formula.

Proposition 9.6. *Assume that the LIBOR swap rate (5.18) is modeled as a geometric Brownian motion under $\widehat{\mathbb{P}}_{i,j}$, i.e.*

$$dS(t, T_i, T_j) = \widehat{\sigma}_{i,j}(t) S(t, T_i, T_j) d\widehat{B}_t^{i,j}, \quad (9.17)$$

where $\left(\widehat{\sigma}_{i,j}(t) \right)_{t \in \mathbb{R}_+}$ is a deterministic volatility function of time. Then, the payer swaption with payoff

$$(P(T, T_i) - P(T, T_j) - \kappa P(T_i, T_i, T_j))^+ = P(T_i, T_i, T_j) \left(S(T_i, T_i, T_j) - \kappa \right)^+$$

can be priced using the Black-Scholes call formula as

$$\mathbb{E}^* \left[e^{-\int_t^{T_i} r_s ds} P(T_i, T_i, T_j) \left(S(T_i, T_i, T_j) - \kappa \right)^+ \middle| \mathcal{F}_t \right] \quad (9.18)$$

$$= (P(t, T_i) - P(t, T_j)) \Phi_+(d_+(t, T_i))$$

$$- \kappa \Phi_-(d_-(t, T_i)) \sum_{k=i}^{j-1} (T_{k+1} - T_k) P(t, T_{k+1}),$$

where

$$d_+(t, T_i) = \frac{\log(S(t, T_i, T_j)/\kappa) + (T_i - t)\sigma_{i,j}^2(t, T_i)/2}{\sigma_{i,j}(t, T_i)\sqrt{T_i - t}}, \qquad (9.19)$$

and

$$d_-(t, T_i) = \frac{\log(S(t, T_i, T_j)/\kappa) - (T_i - t)\sigma_{i,j}^2(t, T_i)/2}{\sigma_{i,j}(t, T_i)\sqrt{T_i - t}},$$

and

$$|\sigma_{i,j}(t, T_i)|^2 = \frac{1}{T_i - t}\int_t^{T_i} |\widehat{\sigma}(s)|^2 ds, \qquad 0 \leqslant t < T_i. \qquad (9.20)$$

Proof. Since $S(t, T_i, T_j)$ is a geometric Brownian motion with volatility function $(\widehat{\sigma}(t))_{t \in \mathbb{R}_+}$ under $\widehat{\mathbb{P}}_{i,j}$, by (9.12)-(9.13) in Lemma 9.1 and (9.15)-(9.16), we have

$$\mathbb{E}^*\left[e^{-\int_t^{T_i} r_s ds} P(T_i, T_i, T_j)\left(S(T_i, T_i, T_j) - \kappa\right)^+ \middle| \mathcal{F}_t\right]$$

$$= \mathbb{E}^*\left[e^{-\int_t^{T} r_s ds}(P(T, T_i) - P(T, T_j) - \kappa P(T_i, T_i, T_j))^+ \middle| \mathcal{F}_t\right]$$

$$= P(t, T_i, T_j)\widehat{\mathbb{E}}_{i,j}\left[\left(S(T_i, T_i, T_j) - \kappa\right)^+ \middle| \mathcal{F}_t\right]$$

$$= P(t, T_i, T_j)\mathrm{Bl}(S(t, T_i, T_j), \kappa, \sigma_{i,j}(t, T_i), 0, T_i - t)$$

$$= P(t, T_i, T_j)\left(S(t, T_i, T_j)\Phi_+(t, S(t, T_i, T_j)) - \kappa\Phi_-(t, S(t, T_i, T_j))\right)$$

$$= \left(P(t, T_i) - P(t, T_j)\right)\Phi_+(t, S(t, T_i, T_j)) - \kappa P(t, T_i, T_j)\Phi_-(t, S(t, T_i, T_j))$$

$$= \left(P(t, T_i) - P(t, T_j)\right)\Phi_+(t, S(t, T_i, T_j))$$

$$\quad -\kappa\Phi_-(t, S(t, T_i, T_j))\sum_{k=i}^{j-1}(T_{k+1} - T_k)P(t, T_{k+1}).$$

$\square$

In addition, the hedging strategy

$$(\Phi_+(t, S(t, T_i, T_j)), -\kappa\Phi_-(t, S(t, T_i, T_j))(T_{i+1} - T_i), \dots$$

$$\dots, -\kappa\Phi_-(t, S(t, T_i, T_j))(T_{j-1} - T_{j-2}), -\Phi_+(t, S(t, T_i, T_j)))$$

based on the assets $(P(t, T_i), \dots, P(t, T_j))$ is self-financing by Corollary 10.2 in Privault (2014), Exercise 9.41 and Problem 9.17 for details.

Similarly to the above, a receiver (or put) swaption gives the option, but not the obligation, to enter an interest rate swap as receiver of a fixed rate κ and as payer of a floating LIBOR rate $L(T_i, T_k, T_{k+1})$, and can be priced as in the next proposition.

Proposition 9.7. *Assume that the LIBOR swap rate (5.18) is modeled as the geometric Brownian motion (9.17) under* $\widehat{\mathbb{P}}_{i,j}$. *Then, the receiver swaption with payoff*

$$\left(\kappa P(T_i, T_i, T_j) - \left(P(T, T_i) - P(T, T_j)\right)\right)^+ = P(T_i, T_i, T_j)\left(\kappa - S(T_i, T_i, T_j)\right)^+$$

can be priced using the Black-Scholes put formula as

$$\mathbb{E}^*\left[e^{-\int_t^{T_i} r_s ds} P(T_i, T_i, T_j)\left(\kappa - S(T_i, T_i, T_j)\right)^+ \middle| \mathcal{F}_t \right]$$

$$= \kappa \Phi_-(-d_-(t, T_i)) \sum_{k=i}^{j-1} (T_{k+1} - T_k) P(t, T_{k+1})$$

$$- (P(t, T_i) - P(t, T_j)) \Phi_+(-d_+(t, T_i)),$$

where $d_+(t, T_i)$, *and* $d_-(t, T_i)$ *and* $|\sigma_{i,j}(t, T_i)|^2$ *are defined in (9.19)-(9.20).*

Swaption prices can also be computed by the swaption approximation formula from the exact dynamics of the swap rate $S(t, T_i, T_j)$ under $\widehat{\mathbb{P}}_{i,j}$, based on the bond price dynamics of the form (8.13), see Proposition 9.9 below.

Bermudan swaption pricing in Quantlib

Letting $\delta_k := T_{k+1} - T_k$, $k = i, \ldots, j - 1$. The Bermudan swaption on the tenor structure $\{T_i, \ldots, T_j\}$ is priced as the supremum

$$\sup_{l \in \{i, \ldots, j-1\}} \mathbb{E}^*\left[e^{-\int_t^{T_l} r_s ds} \left(\sum_{k=l}^{j-1} \delta_k P(T_l, T_{k+1})(L(T_l, T_k, T_{k+1}) - \kappa) \right)^+ \middle| \mathcal{F}_t \right]$$

$$= \sup_{l \in \{i, \ldots, j-1\}} \mathbb{E}^*\left[e^{-\int_t^{T_l} r_s ds} \left(P(T_l, T_l) - P(T_l, T_j) - \kappa P(T_l, T_l, T_j) \right)^+ \middle| \mathcal{F}_t \right]$$

$$= \sup_{l \in \{i, \ldots, j-1\}} \mathbb{E}^*\left[e^{-\int_t^{T_l} r_s ds} P(T_l, T_l, T_j)(S(T_l, T_l, T_j) - \kappa)^+ \middle| \mathcal{F}_t \right],$$

where the supremum is over all stopping times taking values in $\{T_i, \ldots, T_j\}$.

Bermudan swaptions can be priced in (R)Quantlib, with the following output:

```
Summary of pricing results for Bermudan Swaption

Price (in bp) of Bermudan swaption is   24.92137
Strike is NULL (ATM strike is   0.05 )
Model used is: Hull-White using analytic formulas
```

```
Calibrated model parameters are:
a =  0.04641
sigma =  0.005869
```

This code can be used in particular on the pricing of ordinary swaptions with the output:

```
Summary of pricing results for Bermudan Swaption


Price (in bp) of Bermudan swaption is  22.45436
Strike is NULL (ATM strike is  0.05 )
Model used is: Hull-White using analytic formulas
Calibrated model parameters are:
a =  0.07107
sigma =  0.006018
```

9.4 Swaption Pricing in the BGM Model

Similarly to the case of LIBOR rates, Relation (5.18) shows that the LIBOR swap rate can be viewed as a forward price with the annuity numeraire $P(t, T_i, T_j)$ and $X_t = P(t, T_i) - P(t, T_j)$. Consequently, the LIBOR swap rate $(S(t, T_i, T_j))_{t \in [T,S]}$ is a martingale under the forward measure $\widehat{\mathbb{P}}_{i,j}$ defined by

$$\frac{d\widehat{\mathbb{P}}_{i,j}}{d\mathbb{P}^*} = \frac{P(T_i, T_i, T_j)}{P(0, T_i, T_j)} \, e^{-\int_0^{T_i} r_t dt},$$

see Proposition 8.10. Letting $\delta_k := T_{k+1} - T_k$, $k = i, \ldots, j-1$, we know from Section 9.3 that the swaption with payoff

$$\left(\sum_{k=i}^{j-1} \delta_k P(T_i, T_{k+1})(L(T_i, T_k, T_{k+1}) - \kappa) \right)^+$$

on the LIBOR market can be priced at time $t \in [0, T_i]$ as

$$P(t, T_i)\widehat{\mathbb{E}}_i \left[P(T_i, T_i, T_j)(S(T_i, T_i, T_j) - \kappa)^+ \mid \mathcal{F}_t \right]$$

$$= P(t, T_i, T_j)\widehat{\mathbb{E}}_{i,j} \left[(S(T_i, T_i, T_j) - \kappa)^+ \mid \mathcal{F}_t \right], \qquad (9.21)$$

where the martingale measure $\widehat{\mathbb{P}}_{i,j}$ has been defined in (8.26) by

$$\frac{d\widehat{\mathbb{P}}_{i,j|\mathcal{F}_t}}{d\mathbb{P}^*_{|\mathcal{F}_t}} = e^{-\int_t^{T_i} r_s ds} \frac{P(T_i, T_i, T_j)}{P(t, T_i, T_j)}, \qquad 0 \leqslant t \leqslant T_i,$$

$1 \leqslant i < j \leqslant n$, see Section 9.2.

Swaption prices can also be computed by the Monte Carlo method using the dynamics of $L(t, T_k, T_{k+1})$ under $\widehat{\mathbb{P}}_i$, $1 \leqslant i \leqslant k \leqslant n$, but the market practice is to use approximation formulas. Recall that by Proposition 5.4, the swap rate $S(t, T_i, T_j)$ satisfies

$$S(t, T_i, T_j) = \frac{1}{P(t, T_i, T_j)} \sum_{k=i}^{j-1} (T_{k+1} - T_k) P(t, T_{k+1}) L(t, T_k, T_{k+1}),$$

where

$$P(t, T_i, T_j) = \sum_{k=i}^{j-1} (T_{k+1} - T_k) P(t, T_{k+1}), \qquad 0 \leqslant t \leqslant T_{i+1},$$

is the annuity numeraire. Moreover, by Proposition 8.9 the process $v_k^{i,j}(t)$ defined by

$$t \mapsto v_k^{i,j}(t) := \frac{P(t, T_k)}{P(t, T_i, T_j)}, \qquad 0 \leqslant t \leqslant \min(T_k, T_{i+1}),$$

is an $(\mathcal{F}_t)_{t \in [0, \min(T_k, T_{i+1})]}$-martingale under $\widehat{\mathbb{P}}_{i,j}$, $1 \leqslant i < j \leqslant n$, and by Proposition 8.11 the process

$$B_t^{i,j} := B_t - \sum_{k=i}^{j-1} \delta_k \int_0^t v_{k+1}^{i,j}(s) \zeta_{k+1}(s) ds, \qquad 0 \leqslant t \leqslant T_{i+1},$$

is a standard Brownian motion under $\widehat{\mathbb{P}}_{i,j}$, for all $1 \leqslant i < j \leqslant n$. In the next proposition, we recompute the dynamics of the swap rate $S(t, T_i, T_j)$ under $\widehat{\mathbb{P}}_{i,j}$, based on Proposition 8.12.

Proposition 9.8. *The dynamics of the swap rate $S(t, T_i, T_j)$ process under $\widehat{\mathbb{P}}_{i,j}$ is given by*

$$dS(t, T_i, T_j) = \sigma_{i,j}(t) S(t, T_i, T_j) d\widehat{B}_t^{i,j}, \qquad 0 \leqslant t \leqslant T_i,$$

where the swap rate volatility is given by

$$\sigma_{i,j}(t) = \frac{1}{S(t, T_i, T_j)} \sum_{k=i}^{j-1} \gamma_k(t) w_k^{i,j}(t) L(t, T_k, T_{k+1}), \qquad 1 \leqslant i < j \leqslant n,$$

with

$$w_k^{i,j}(t) :=$$

$$\frac{\delta_k}{1 + \delta_k L(t, T_k, T_{k+1})} \left(\sum_{l=k}^{j-1} \delta_l v_{l+1}^{i,j}(t) \frac{P(t, T_i) - P(t, T_j)}{P(t, T_i, T_j)} + \frac{P(t, T_j)}{P(t, T_i, T_j)} \right),$$

and $w_i^{i,i+1}(t) := 1$, hence

$$\sigma_{i,j}^2(t) = \tag{9.22}$$

$$\frac{1}{S(t,T_i,T_j)^2} \sum_{l=i}^{j-1} \sum_{k=i}^{j-1} \gamma_l(t)\gamma_k(t)w_l^{i,j}(t)w_k^{i,j}(t)L(t,T_l,T_{l+1})L(t,T_k,T_{k+1}),$$

and $\sigma_{i,i+1}(t) = \gamma_i(t)$, $i = 1, 2, \ldots, n-1$.

Proof. By Proposition 8.12, we have

$$dS(t,T_i,T_j) = \sigma_{i,j}(t)S(t,T_i,T_j)dB_t^{i,j},$$

where, using Relations (7.19)-(7.20), the swap rate volatility $\sigma_{i,j}(t)$ can be computed as

$$\sigma_{i,j}(t) = \frac{P(t,T_j)}{P(t,T_i) - P(t,T_j)}(\zeta_i(t) - \zeta_j(t)) + \sum_{l=i}^{j-1} \delta_l v_{l+1}^{i,j}(t)(\zeta_i(t) - \zeta_{l+1}(t))$$

$$= \frac{P(t,T_j)}{P(t,T_i) - P(t,T_j)} \sum_{k=i}^{j-1} \gamma_k(t)\frac{\delta_k L(t,T_k,T_{k+1})}{1 + \delta_k L(t,T_k,T_{k+1})}$$

$$+ \sum_{l=i}^{j-1} \delta_l v_{l+1}^{i,j}(t) \sum_{k=i}^{l} \gamma_k(t)\frac{\delta_k L(t,T_k,T_{k+1})}{1 + \delta_k L(t,T_k,T_{k+1})}$$

$$= \sum_{k=i}^{j-1} \gamma_k(t)\frac{\delta_k L(t,T_k,T_{k+1})}{1 + \delta_k L(t,T_k,T_{k+1})} \sum_{l=k}^{j-1} \left(\delta_l v_{l+1}^{i,j}(t) + \frac{P(t,T_j)}{P(t,T_i) - P(t,T_j)} \right)$$

$$= \frac{1}{S(t,T_i,T_j)} \sum_{k=i}^{j-1} \gamma_k(t)\frac{\delta_k L(t,T_k,T_{k+1})}{1 + \delta_k L(t,T_k,T_{k+1})}$$

$$\sum_{l=k}^{j-1} \left(\delta_l v_{l+1}^{i,j}(t)\frac{P(t,T_i) - P(t,T_j)}{P(t,T_i,T_j)} + \frac{P(t,T_j)}{P(t,T_i,T_j)} \right)$$

$$= \frac{1}{S(t,T_i,T_j)} \sum_{k=i}^{j-1} \gamma_k(t)w_k^{i,j}(t)L(t,T_k,T_{k+1}).$$

$\square$

When $j = i + 1$, although

$$\frac{d\widehat{\mathbb{P}}_{i,i+1|\mathcal{F}_t}}{d\mathbb{P}_{|\mathcal{F}_t}^*} = e^{-\int_t^{T_i} r_s ds}\frac{P(T_i,T_i,T_{i+1})}{P(t,T_i,T_{i+1})}$$

$$= e^{-\int_t^{T_i} r_s ds}\frac{P(T_i,T_{i+1})}{P(t,T_{i+1})}$$

$$\neq \frac{e^{-\int_t^{T_{i+1}} r_s ds}}{P(t, T_{i+1})}$$

$$= \frac{d\widehat{\mathbb{P}}_{i+1|\mathcal{F}_t}}{d\mathbb{P}^*_{|\mathcal{F}_t}}, \qquad 0 \leqslant t \leqslant T_i,$$

we know by (8.34), (8.35) that $\left(B_t^{i,i+1}\right)_{t \in [0,T_i]}$ and $\left(B_t^{i+1}\right)_{t \in [0,T_i]}$ coincide up to time T_i, and by (5.17) the swaption price (9.21) can be computed as

$$P(t, T_i, T_{i+1})\widehat{\mathbb{E}}_{i,i+1}\left[(S(T_i, T_i, T_{i+1}) - \kappa)^+ \mid \mathcal{F}_t\right]$$
$$= (T_{i+1} - T_i)P(t, T_{i+1})\widehat{\mathbb{E}}_{i+1}\left[(S(T_i, T_i, T_{i+1}) - \kappa)^+ \mid \mathcal{F}_t\right]$$
$$= (T_{i+1} - T_i)P(t, T_{i+1})\widehat{\mathbb{E}}_{i+1}\left[(L(T_i, T_i, T_{i+1}) - \kappa)^+ \mid \mathcal{F}_t\right],$$

which is equal to

$$(T_{i+1} - T_i)P(t, T_{i+1})\mathrm{Bl}(L(t, T_i, T_{i+1}), \kappa, \sigma_i(t), 0, T_i - t), \qquad (9.23)$$

where $|\sigma_i(t)|$ is defined in (9.7). The next proposition extends this relation to general indices $1 \leqslant i < j \leqslant n$ as an approximation, known as the swaption approximation formula, see page 17 of Schoenmakers (2005).

Proposition 9.9. *The payer swaption price*

$$P(t, T_i, T_j)\widehat{\mathbb{E}}_{i,j}\left[(S(T_i, T_i, T_j) - \kappa)^+ \mid \mathcal{F}_t\right]$$

can be approximated by

$$P(t, T_i, T_j)\mathrm{Bl}\big(S(t, T_i, T_j), \kappa, \tilde{\sigma}_{i,j}(t), 0, T_i - t\big), \qquad (9.24)$$

where

$$|\tilde{\sigma}_{i,j}(t)|^2 \qquad (9.25)$$

$$= \sum_{k,l=i}^{j-1} \frac{\delta_k \delta_l v_{l+1}^{i,j}(t) v_{k+1}^{i,j}(t) L(t, T_l, T_{l+1}) L(t, T_k, T_{k+1})}{(T_i - t)|S(t, T_i, T_j)|^2} \int_t^{T_i} \gamma_l(s)\gamma_k(s)ds.$$

Proof. We refer to Chapter 1 of Schoenmakers (2005) for a more complete treatment. Here, we simply note that this approximation can be derived as follows:

$$dS(t, T_i, T_j) = d\left(\frac{1}{P(t, T_i, T_j)} \sum_{k=i}^{j-1} (T_{k+1} - T_k)P(t, T_{k+1})L(t, T_k, T_{k+1})\right)$$

$$\simeq \frac{1}{P(t, T_i, T_j)} \sum_{k=i}^{j-1} (T_{k+1} - T_k)P(t, T_{k+1})dL(t, T_k, T_{k+1})$$

$$= \frac{1}{P(t,T_i,T_j)} \sum_{k=i}^{j-1} \delta_k P(t,T_{k+1}) L(t,T_k,T_{k+1}) \gamma_k(t) dB_t^{(k+1)}$$

$$= S(t,T_i,T_j) \sum_{k=i}^{j-1} \frac{\delta_k P(t,T_{k+1}) \gamma_k(t)}{S(t,T_i,T_j) P(t,T_i,T_j)} L(t,T_k,T_{k+1}) dB_t^{(k+1)}$$

$$= S(t,T_i,T_j) \sum_{k=i}^{j-1} \delta_k v_{k+1}^{i,j}(t) \gamma_k(t) \frac{L(t,T_k,T_{k+1})}{S(t,T_i,T_j)} dB_t^{(k+1)},$$

hence

$$\sigma_{i,j}^2(t) dt \simeq \left| \frac{dS(t,T_i,T_j)}{S(t,T_i,T_j)} \right|^2$$

$$\simeq \sum_{k=i}^{j-1} \sum_{l=i}^{j-1} \delta_k \delta_l v_{k+1}^{i,j}(t) v_{l+1}^{i,j}(t) \gamma_k(t) \gamma_l(t) \frac{L(t,T_l,T_{l+1}) L(t,T_l,T_{l+1})}{S(t,T_i,T_j)^2} dt,$$

which, in view of Relation (9.22), supports the claim that $w_k^{i,j}$ can be approximated by $\delta_k v_{k+1}^{i,j}$, see Chapter 1 of Schoenmakers (2005) for details.

The Black volatility

$$|\sigma_{i,j}(t)|^2 \qquad\qquad\qquad\qquad\qquad\qquad\qquad (9.26)$$

$$= \sum_{k,l=i}^{j-1} \int_t^{T_i} \frac{\delta_l \delta_k v_{l+1}^{i,j}(s) v_{k+1}^{i,j}(s) L(s,T_l,T_{l+1}) L(s,T_k,T_{k+1})}{(T_i - t)|S(t,T_i,T_j)|^2} \gamma_l(s) \gamma_k(s) ds,$$

$0 \leqslant t \leqslant T_i$, is approximated by

$$|\tilde{\sigma}_{i,j}(t)|^2$$

$$\simeq \sum_{k,l=i}^{j-1} \frac{\delta_k \delta_l v_{l+1}^{i,j}(t) v_{k+1}^{i,j}(t) L(t,T_l,T_{l+1}) L(t,T_k,T_{k+1})}{(T_i - t)|S(t,T_i,T_j)|^2} \int_t^{T_i} \gamma_l(s) \gamma_k(s) ds,$$

by "freezing" the random coefficients $v_{l+1}^{i,j}(t)$, $v_{k+1}^{i,j}(t)$, $L(t,T_l,T_{l+1})$, $L(t,T_k,T_{k+1})$ and $S(t,T_i,T_j)$ at time t. $\qquad\square$

The approximation of Proposition 9.9 amounts to saying that $(S(t,T_i,T_j))_{t \in [0,T_i]}$ is an exponential martingale with volatility coefficient $\tilde{\sigma}_{i,j}(t)$ under $\widehat{\mathbb{P}}_{i,j}$. Note also that we have $\tilde{\sigma}_{i,i+1}(t) = \sigma_i(t)$, $t \in [0,T_i]$, hence (9.24) is indeed an extension of (9.23).

9.5 Calibration of the BGM Model

Swaption volatilities can be estimated from swaption prices as implied volatilities from the Black swaption pricing formula (9.18). Table 9.2 shows an example of market data expressed in terms of swaption volatilities $\sigma_{i,j}^{B}(t)$ by inversion of the swaption approximation formula (9.24), which can then be used to calibrate the BGM model, see Schoenmakers (2005), Privault and Wei (2009). Here, the time to maturity $T_i - t$ is in ordinate and the period $T_j - T_i$ is in abscissa.

Vol Swaption At The Money

	1Y	2Y	3Y	4Y	5Y	6Y	7Y	8Y	9Y	10Y	25Y
2D	18,6	18	16,8	15,7	14,7	13,8	13	12,3	11,8	11,3	9,3
1M	17,6	18	16,8	15,7	14,7	13,8	13	12,3	11,8	11,3	9,3
2M	18,1	18,35	17,05	15,95	14,95	14	13,2	12,55	12	11,5	9,45
3M	18,6	18,7	17,3	16,2	15,2	14,2	13,4	12,8	12,2	11,7	9,6
6M	18,7	18,1	16,8	15,7	14,8	13,9	13,3	12,7	12,2	11,8	9,7
9M	18,3	17,5	16,3	15,25	14,4	13,6	13,05	12,55	12,1	11,75	9,7
1Y	17,9	16,9	15,8	14,8	14	13,3	12,8	12,4	12	11,7	9,7
2Y	16,3	15,2	14,4	13,7	13,1	12,6	12,2	11,9	11,6	11,3	9,3
3Y	15,2	14,2	13,4	12,8	12,4	12	11,7	11,5	11,2	11	9,2
4Y	14,4	13,2	12,6	12,1	11,7	11,5	11,2	11	10,8	10,7	8,8
5Y	13,4	12,4	11,9	11,5	11,2	11	10,8	10,7	10,5	10,4	8,6
6Y	12,85	11,95	11,45	11	10,75	10,55	10,4	10,25	10,1	10	8,3
7Y	12,3	11,5	11	10,5	10,3	10,1	10	9,8	9,7	9,6	8
8Y	11,97	11,13	10,67	10,2	10	9,8	9,7	9,53	9,43	9,33	7,83
9Y	11,63	10,77	10,33	9,9	9,7	9,5	9,4	9,27	9,17	9,07	7,67
10Y	11,3	10,4	10	9,6	9,4	9,2	9,1	9	8,9	8,8	7,5
12Y	10,8	10,04	9,58	9,28	9,02	8,92	8,76	8,66	8,56	8,46	7,38
15Y	10,2	9,5	9,1	8,8	8,6	8,5	8,4	8,3	8,2	8,1	7,2
20Y	9,5	8,8	8,5	8,2	8	8	8	8	7,9	7,9	6,9
25Y	8,8	8,1	7,9	7,6	7,4	7,5	7,6	7,7	7,6	7,7	6,6
30Y	8,1	7,4	7,3	7	6,8	7	7,2	7,4	7,3	7,5	6,3

Table 9.2: Swaption volatilities.

This type of data can be also expressed in the form of a graph where the index i refers to the time to maturity $T_i - t$ and the index j refers to the period $T_j - T_i$ as in Figure 9.1.

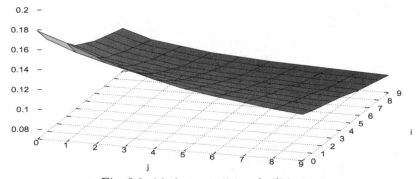

Fig. 9.1: Market swaption volatilities.

The goal of calibration is to estimate the volatility functions $\gamma_i(t)$ appearing in the BGM model (7.16), $i = 1, 2, \ldots, n$, from the data of interest rate caps and swaptions prices observed on the market. This involves several computational and stability issues. Let

$$g_i(t) = |\gamma_i(t)|, \qquad 0 \leqslant t \leqslant T_i, \quad i = 1, 2, \ldots, n.$$

Using the Rebonato (1996) parametrization

$$g_i(t) = g_\infty + (1 + a(T_i - t) - g_\infty) e^{-(T_i - t)b},$$

$a, b, g_\infty > 0$, $i = 1, 2, \ldots, n$, and equating

$$|\sigma_i^B(t)|^2 = \frac{1}{T_i - t} \int_t^{T_i} |\gamma_i(s)|^2 ds, \qquad 0 \leqslant t < T_i,$$

as in (9.7), one obtains from (9.26) an expression $\sigma_{i,j}(t, b, g_\infty)$ of $\sigma_{i,j}(t)$ as a function of b, g_∞, where a has been set equal to 0. Following Schoenmakers (2002), we minimize the mean square distance

$$\mathrm{RMS}(b, g_\infty) := \sqrt{\frac{2}{(n-1)(n-2)} \sum_{i=1}^{k} \sum_{j=i+1}^{n} \left(\frac{\sigma_{i,j}^B(t) - \sigma_{i,j}(t)}{\sigma_{i,j}^B(t)} \right)^2},$$

where n is the number of tenor dates (in multiples of one year) and k is the maximum number of swaption maturities used in the calibration, with non-available data treated as zero in the sum. The data of discount factors and swap rates are interpolated with a fixed tenor $\delta = $ half year.

The volatilities computed in this way are given by the following graph, where the index i refers to $T_i - t$ and j refers to $T_j - T_i$:

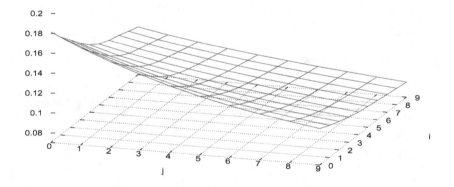

Fig. 9.2: Computed swaption volatilities.

The graph of Figure 9.3 allows us to compare the estimated and computed volatilities.

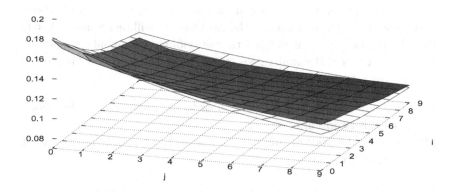

Fig. 9.3: Comparison graphs of swaption volatilities.

A sample of joint numerical estimation of the parameters (b, g_∞) is given in Table 9.3, where the maximum number k of swaption maturities used in each calibration is denoted by Nmat, see Privault and Wei (2009) for details. The total number of swaptions used is bounded by $(n - (k+1)/2)k$.

Nmat	#swaptions	b	g_∞	RMS
1	10	5.03	0.85	0.008
2	20	5.03	0.71	0.010
3	30	5.04	0.73	0.010
4	40	5.03	0.72	0.010
5	50	5.04	0.70	0.011
6	60	5.03	0.65	0.011
7	70	5.02	0.60	0.012
8	80	5.02	0.60	0.012
9	90	5.02	0.72	0.013
10	100	5.04	0.63	0.012
12	110	5.03	0.65	0.012
15	120	5.03	1.00	0.014

Table 9.3: BGM calibration results.

Exercises

Exercise 9.1. Consider a payer swaption giving its holder the right, but not the obligation, to enter into a 3-year annual pay swap in four years, where a fixed rate of 5% will be paid and the LIBOR rate will be received. Assume that the yield curve is flat at 5% with continuous-time annual compounding and the volatility of the swap rate is 20%. The notional principal is \$100,000 per interest rate percentage point.

(1) What are the key assumptions in order to apply Black's formula to value this swaption?
(2) Compute the price of this swaption using Black's formula as an application of Proposition 9.6.

Exercise 9.2.   Consider a floorlet on a three-month LIBOR rate in nine month's time, with a notional principal amount of \$10, 000 per interest rate percentage point. The term structure is flat at 3.95% per year with discrete compounding, the volatility of the forward LIBOR rate in nine months is 10%, and the floor rate is 4.5%.

(1) What are the key assumptions on the LIBOR rate in nine month in order to apply Black's formula to price this floorlet?
(2) Compute the price of this floorlet using Black's formula as an application of Proposition 9.3 and (9.8), using the functions $\Phi(d_+)$ and $\Phi(d_-)$.

Exercise 9.3. Consider a market with three zero-coupon bonds with prices $P(t,T_1)$, $P(t,T_2)$ and $P(t,T_3)$ with maturities $T_1 = \delta$, $T_2 = 2\delta$ and $T_3 = 3\delta$ respectively, and the forward LIBOR $L(t,T_1,T_2)$ and $L(t,T_2,T_3)$ defined by

$$L(t,T_i,T_{i+1}) = \frac{1}{\delta}\left(\frac{P(t,T_i)}{P(t,T_{i+1})} - 1\right), \qquad i = 1,2.$$

Assume that $L(t,T_1,T_2)$ and $L(t,T_2,T_3)$ are modeled as

$$\frac{dL(t,T_1,T_2)}{L(t,T_1,T_2)} = \gamma_1(t)dB_t^{(2)}, \qquad 0 \leqslant t \leqslant T_1, \qquad (9.27)$$

and $L(t,T_2,T_3) = b$, $0 \leqslant t \leqslant T_2$, for some constant $b > 0$ and function $\gamma_1(t)$, where $B_t^{(2)}$ is a standard Brownian motion under the forward measure $\widehat{\mathbb{P}}_2$ defined by

$$\frac{d\widehat{\mathbb{P}}_2}{d\mathbb{P}*} = \frac{e^{-\int_0^{T_2} r_s ds}}{P(0,T_2)}.$$

(1) Compute $L(t, T_1, T_2)$, $0 \leqslant t \leqslant T_2$ by solving Equation (9.27).

(2) Compute the interest rate caplet prices

$$\mathbb{E}^* \left[e^{-\int_t^{T_{i+1}} r_s ds} (L(T_i, T_i, T_{i+1}) - \kappa)^+ \,\middle|\, \mathcal{F}_t \right]$$
$$= P(t, T_i) \widehat{\mathbb{E}}_{i+1} \left[(L(T_i, T_i, T_{i+1}) - \kappa)^+ \,\middle|\, \mathcal{F}_t \right], \qquad 0 \leqslant t \leqslant T_i,$$

at time t, where $\widehat{\mathbb{E}}_{i+1}$ denotes the expectation under the forward measure $\widehat{\mathbb{P}}_{i+1}$, $i = 1, 2$.

(3) Compute

$$\frac{P(t, T_1)}{P(t, T_1, T_3)}, \quad 0 \leqslant t \leqslant T_1, \quad \text{and} \quad \frac{P(t, T_3)}{P(t, T_1, T_3)}, \qquad 0 \leqslant t \leqslant T_2,,$$

in terms of b and $L(t, T_1, T_2)$, where $P(t, T_1, T_3)$ is the annuity numeraire

$$P(t, T_1, T_3) = \delta P(t, T_2) + \delta P(t, T_3), \qquad 0 \leqslant t \leqslant T_2.$$

(4) Compute the dynamics of the swap rate

$$t \mapsto S(t, T_1, T_3) = \frac{P(t, T_1) - P(t, T_3)}{P(t, T_1, T_3)}, \qquad 0 \leqslant t \leqslant T_1,$$

i.e. show that we have

$$dS(t, T_1, T_3) = \sigma_{1,3}(t) S(t, T_1, T_3) dB_t^{(2)},$$

where $(\sigma_{1,3}(t))_{t \in [0,T_1]}$ is a process to be determined from $S(t, T_1, T_3)$ and $L(t, T_1, T_2)$.

Exercise 9.4. Swaption hedging (1). Consider a bond market with tenor structure $\{T_i, \ldots, T_j\}$ and $j - i + 1$ bonds with maturities $T_i, \ldots, T_j$, whose prices $P(t, T_i), \ldots, P(t, T_j)$ at time t are given by

$$\frac{dP(t, T_k)}{P(t, T_k)} = r_t dt + \zeta_k(t) dB_t, \qquad k = i, \ldots, j,$$

where $(r_t)_{t \in \mathbb{R}_+}$ is a short-term interest rate process, $(B_t)_{t \in \mathbb{R}_+}$ is a standard Brownian motion generating the filtration $(\mathcal{F}_t)_{t \in \mathbb{R}_+}$, and $\zeta_i(t), \ldots, \zeta_j(t)$ are volatility processes. The swap rate $S(t, T_i, T_j)$ is defined as in (5.18) by $S(t, T_i, T_j) = (P(t, T_i) - P(t, T_j))/P(t, T_i, T_j)$, where $P(t, T_i, T_j)$ is the annuity numeraire (8.24).

(1) Assume that the swap rate is modeled as a geometric Brownian motion

$$dS(t, T_i, T_j) = S(t, T_i, T_j)\sigma_{i,j}(t)dB_t^{i,j}, \qquad 0 \leqslant t \leqslant T_i,$$

where the swap rate volatility is a deterministic function $\sigma_{i,j}(t)$ and $B_t^{i,j}$ is a standard Brownian motion under the swap measure $\widehat{\mathbb{P}}_{i,j}$, cf. Proposition 8.11. Show that the price at time $t \in [0, T_i]$ of a payer swaption on the LIBOR market can be written as

$$\mathbb{E}^* \left[e^{-\int_t^{T_i} r_s ds} \left(\sum_{k=i}^{j-1} (T_{k+1} - T_k) P(T_i, T_{k+1})(L(T_i, T_k, T_{k+1}) - \kappa) \right)^+ \bigg| \mathcal{F}_t \right]$$

$$= P(t, T_i, T_j) \widehat{\mathbb{E}}_{i,j}[(S(T_i, T_i, T_j) - \kappa)^+ \mid \mathcal{F}_t]$$

$$= P(t, T_i, T_j) C(S(t, T_i, T_j), \kappa, v(t, T_i)),$$

under the forward swap measure $\widehat{\mathbb{P}}_{i,j}$ of Definition 8.2,

$$v^2(t, T_i) = \int_t^{T_i} |\sigma_{i,j}(s)|^2 ds, \qquad 0 \leqslant t \leqslant T_i,$$

and $C(x, \kappa, v)$ is a function to be specified using the Black-Scholes formula of Lemma 2.3.

(2) Consider a portfolio $(\xi_t^i, \dots, \xi_t^j)_{t \in [0, T_i]}$ made of bonds with maturities $T_i, \dots, T_j$, and value

$$V_t = \sum_{k=i}^j \xi_t^k P(t, T_k),$$

at time $t \in [0, T_i]$. We assume that the portfolio is self-financing, *i.e.*

$$dV_t = \sum_{k=i}^j \xi_t^k dP(t, T_k), \qquad 0 \leqslant t \leqslant T_i, \qquad (9.28)$$

and that it *hedges* the claim $(S(T_i, T_i, T_j) - \kappa)^+$, so that

$$V_t = P(t, T_i, T_j) \widehat{\mathbb{E}}_{i,j} \left[(S(T_i, T_i, T_j) - \kappa)^+ \mid \mathcal{F}_t \right], \qquad 0 \leqslant t \leqslant T_i.$$

Show that the forward (or deflated) portfolio value process $\widehat{V}_t = V_t / P(t, T_i, T_j)$ satisfies

$$d\widehat{V}_t = \frac{\partial C}{\partial x}(S_t, \kappa, v(t, T_i))dS_t.$$

(3) Show that we have

$$dV_t = \frac{\partial C}{\partial x}(S_t, \kappa, v(t, T_i))d(P(t, T_i) - P(t, T_j))$$

$$+ \left(\widehat{V}_t - S_t \frac{\partial C}{\partial x}(S_t, \kappa, v(t, T_i)) \right) dP(t, T_i, T_j).$$

(4) Compute the hedging strategy $(\xi_t^i, \ldots, \xi_t^j)$ of the swaption.

Exercise 9.5. Floorlet pricing. Given two bonds with maturities T, S and prices $P(t,T)$, $P(t,S)$, consider the LIBOR rate

$$L(t,T,S) := \frac{P(t,T) - P(t,S)}{(S-T)P(t,S)}$$

at time $t \in [0,T]$, modeled as

$$dL(t,T,S) = \mu_t L(t,T,S)dt + \sigma L(t,T,S)dB_t, \quad 0 \leqslant t \leqslant T, \qquad (9.29)$$

where $(B_t)_{t\in[0,T]}$ is a standard Brownian motion under the risk-neutral measure $\mathbb{P}^*$, $\sigma > 0$ is a constant, and $(\mu_t)_{t\in[0,T]}$ is an $(\mathcal{F}_t)_{t\in\mathbb{R}_+}$-adapted process. Let

$$F_t = \mathbb{E}^* \left[e^{-\int_t^S r_s ds} (\kappa - L(T,T,S))^+ \,\middle|\, \mathcal{F}_t \right]$$

denote the price at time t of an interest rate floorlet with strike level κ, maturity T, and payment date S.

(1) Rewrite the value of F_t using the forward measure $\widehat{\mathbb{P}}_S$ with maturity S.
(2) What is the dynamics of $L(t,T,S)$ under the forward measure $\widehat{\mathbb{P}}_S$?
(3) Write down the value of F_t using the Black-Scholes formula.

Hint: Given X a centered Gaussian random variable with variance $v^2 > 0$, we have

$$\mathbb{E}^* \left[(\kappa - e^{m+X})^+ \right] = \kappa \Phi \left(-\frac{m - \log \kappa}{v} \right) - e^{m+v^2/2} \Phi \left(-v - \frac{m - \log \kappa}{v} \right),$$

where Φ denotes the standard normal cumulative distribution function.

Exercise 9.6.
Jamshidian's trick (Jamshidian (1989)). Consider a family $(P(t,T_k))_{k=i,\ldots,j}$ of bond prices defined from a short rate process $(r_t)_{t\in\mathbb{R}_+}$. We assume that the bond prices are functions $P(T_i,T_{k+1}) = F_{k+1}(T_i, r_{T_i})$ of r_{T_i} that are *increasing* in the variable r_{T_i}, for all $k = i, \ldots, j-1$.

(1) Compute the price $P(t,T_i,T_j)$ of the annuity numeraire paying coupons $c_{i+1}, \ldots, c_j$ at times $T_{i+1}, \ldots, T_j$ in terms of the bond prices

$$P(t,T_{i+1}), \ldots, P(t,T_j).$$

(2) Show that the payer swaption payoff

$$\left(P(T_i, T_i) - P(T_i, T_j) - \kappa P(T_i, T_i, T_j)\right)^+$$

can be rewritten as

$$\left(1 - \kappa \sum_{k=i}^{j-1} \tilde{c}_{k+1} P(T_i, T_{k+1})\right)^+,$$

by writing $\tilde{c}_k$ in terms of c_k, $k = i+1, \ldots, j$.

(3) Assuming that the bond prices are functions $P(T_i, T_{k+1}) = F_k(T_i, r_{T_i})$ of r_{T_i} that are *increasing* in the variable r_{T_i}, for all $k = i, \ldots, j-1$, show, choosing γ_κ such that

$$\kappa \sum_{k=i}^{j-1} c_{k+1} F_{k+1}(T_i, \gamma_\kappa) = 1,$$

that the payer swaption with payoff

$$\left(P(T_i, T_i) - P(T_i, T_j) - \kappa P(T_i, T_i, T_j)\right)^+ = \left(1 - \kappa \sum_{k=i}^{j-1} c_{k+1} P(T_i, T_{k+1})\right)^+,$$

where c_j contains the final coupon payment, can be priced as a weighted sum of bond put options under the forward measure $\widehat{\mathbb{P}}_i$ with numeraire $P(t, T_i)$.

Exercise 9.7. Vasicek caplet pricing. We work in the Vasicek short rate model

$$dr_t = -b r_t dt + \sigma dB_t,$$

where $(B_t)_{t \in \mathbb{R}_+}$ is a standard Brownian motion under $\mathbb{P}^*$, and consider two bond prices $P(t, T_1)$ and $P(t, T_2)$ with maturities T_1 and T_2.

(1) Compute the dynamics

$$\frac{dP(t, T_i)}{P(t, T_i)} = r_t dt + \zeta_i(t) dB_t, \qquad i = 1, 2,$$

of the bond prices $P(t, T_1)$, $P(t, T_2)$ with maturities T_1, T_2 by stating the expressions of $\zeta_1(t)$ and $\zeta_2(t)$ in the Vasicek model

(2) Compute the dynamics of the deflated bond price $P(t, T_1)/P(t, T_2)$ under $\widehat{\mathbb{P}}_2$.

(3) Recall the expression of the forward rate $f(t, T_1, T_2)$ in the Vasicek model.

(4) Compute the dynamics of $f(t, T_1, T_2)$ under the forward measure $\widehat{\mathbb{P}}_2$ defined as

$$\frac{d\widehat{\mathbb{P}}_2}{d\mathbb{P}^*} = \frac{1}{P(0, T_2)} e^{-\int_0^{T_2} r_s ds}.$$

(5) Compute the price

$$(T_2 - T_1) \, \mathbb{E}^* \left[e^{-\int_t^{T_2} r_s ds} (f(T_1, T_1, T_2) - \kappa)^+ \, \Big| \, \mathcal{F}_t \right]$$

at time $t \in [0, T_1]$ of an interest rate caplet with strike level κ using the expectation under the forward measure $\widehat{\mathbb{P}}_2$.
(6) State the expression of the forward rate $f(t, T_1, T_2)$ in the Ho-Lee short rate model $dr_t = \sigma dB_t$, see Exercise 4.1.

Exercise 9.8. Caplet pricing on the LIBOR. (Exercise 9.7 continued).

(1) Compute the dynamics of the forward LIBOR rate process

$$L(t, T_1, T_2) = \frac{P(t, T_1) - P(t, T_2)}{(T_2 - T_1) P(t, T_2)}, \qquad 0 \leqslant t \leqslant T_1,$$

under $\widehat{\mathbb{P}}_2$.
(2) Compute the interest rate caplet price

$$(T_2 - T_1) \, \mathbb{E}^* \left[e^{-\int_t^{T_1} r_s ds} P(T_1, T_2)(L(T_1, T_1, T_2) - \kappa)^+ \, \Big| \, \mathcal{F}_t \right]$$

on $L(T_1, T_1, T_2)$ using the expectation under the forward swap measure $\widehat{\mathbb{P}}_{1,2}$.

Exercise 9.9. Swaption pricing. Consider three zero-coupon bonds $P(t, T_1)$, $P(t, T_2)$ and $P(t, T_3)$ with maturities $T_1 = \delta$, $T_2 = 2\delta$ and $T_3 = 3\delta$ respectively, and the forward LIBOR $L(t, T_1, T_2)$ and $L(t, T_2, T_3)$ defined by

$$L(t, T_i, T_{i+1}) = \frac{1}{\delta} \left(\frac{P(t, T_i)}{P(t, T_{i+1})} - 1 \right), \qquad i = 1, 2.$$

Assume that $L(t, T_1, T_2)$ and $L(t, T_2, T_3)$ are modeled in the BGM model by

$$\frac{dL(t, T_1, T_2)}{L(t, T_1, T_2)} = e^{-at} d\widehat{B}_t^{(2)}, \qquad 0 \leqslant t \leqslant T_1, \tag{9.30}$$

and $L(t, T_2, T_3) = b$, $0 \leqslant t \leqslant T_2$, for some constants $a, b > 0$, where $\widehat{B}_t^{(2)}$ is a standard Brownian motion under the forward rate measure $\widehat{\mathbb{P}}_2$ defined by

$$\frac{d\widehat{\mathbb{P}}_2}{d\mathbb{P}^*} = \frac{e^{-\int_0^{T_2} r_s ds}}{P(0, T_2)}.$$

(1) Compute $L(t, T_1, T_2)$, $0 \leqslant t \leqslant T_2$ by solving Equation (9.30).

(2) Show that the price at time t of the interest rate caplet with strike level κ can be written as

$$\mathbb{E}^* \left[e^{-\int_t^{T_2} r_s ds} (L(T_1, T_1, T_2) - \kappa)^+ \mid \mathcal{F}_t \right]$$
$$= P(t, T_2) \widehat{\mathbb{E}}_2 \left[(L(T_1, T_1, T_2) - \kappa)^+ \mid \mathcal{F}_t \right],$$

where $\widehat{\mathbb{E}}_2$ denotes the expectation under the forward measure $\widehat{\mathbb{P}}_2$.

(3) Using the hint below, compute the price at time t of the interest rate caplet with strike level κ on $L(T_1, T_1, T_2)$.

(4) Compute

$$\frac{P(t, T_1)}{P(t, T_1, T_3)}, \quad 0 \leqslant t \leqslant T_1, \quad \text{and} \quad \frac{P(t, T_3)}{P(t, T_1, T_3)}, \quad 0 \leqslant t \leqslant T_2,$$

in terms of b and $L(t, T_1, T_2)$, where $P(t, T_1, T_3)$ is the annuity numeraire

$$P(t, T_1, T_3) = \delta P(t, T_2) + \delta P(t, T_3), \quad 0 \leqslant t \leqslant T_2.$$

(5) Compute the dynamics of the swap rate

$$t \mapsto S(t, T_1, T_3) = \frac{P(t, T_1) - P(t, T_3)}{P(t, T_1, T_3)}, \quad 0 \leqslant t \leqslant T_1,$$

i.e. show that we have

$$dS(t, T_1, T_3) = \sigma_{1.3}(t) S(t, T_1, T_3) d\widehat{B}_t^{(2)},$$

where $\sigma_{1,3}(t)$ is a process to be determined.

(6) Using the Black-Scholes formula, compute an approximation of the swaption price

$$\mathbb{E}^* \left[e^{-\int_t^{T_1} r_s ds} P(T_1, T_1, T_3)(S(T_1, T_1, T_3) - \kappa)^+ \mid \mathcal{F}_t \right]$$
$$= P(t, T_1, T_3) \widehat{\mathbb{E}}_2 \left[(S(T_1, T_1, T_3) - \kappa)^+ \mid \mathcal{F}_t \right],$$

at time $t \in [0, T_1]$. You will need to approximate $\sigma_{1,3}(s)$, $s \geqslant t$, by "freezing" all random terms at time t.

Hint: Given X a centered Gaussian random variable with variance $v^2 > 0$, we have

$$\mathbb{E}^* \left[(e^{m+X} - \kappa)^+ \right] = e^{m+v^2/2} \Phi \left(v + \frac{m - \log \kappa}{v} \right) - \kappa \Phi \left(\frac{m - \log \kappa}{v} \right),$$

where Φ denotes the standard normal cumulative distribution function.

Exercise 9.10. Caplet pricing. Consider a LIBOR rate $L(t,T,S)$ process, $t \in [0,T]$, modeled as $dL(t,T,S) = \mu_t L(t,T,S)dt + \sigma(t)L(t,T,S)dB_t$, $0 \leqslant t \leqslant T$, where $(B_t)_{t \in [0,T]}$ is a standard Brownian motion under the risk-neutral measure $\mathbb{P}^*$, $(\mu_t)_{t \in [0,T]}$ is an $(\mathcal{F}_t)_{t \in \mathbb{R}_+}$-adapted process, and $\sigma(t) > 0$ is a deterministic function.

(1) What is the dynamics of $L(t,T,S)$ under the forward measure $\widehat{\mathbb{P}}$?

(2) Rewrite the price

$$\mathbb{E}^* \left[e^{-\int_t^S r_s ds} \phi(L(T,T,S)) \, \middle| \, \mathcal{F}_t \right] \tag{9.31}$$

at time $t \in [0,T]$ of an option with payoff function ϕ using the forward measure $\widehat{\mathbb{P}}$.

(3) Write down the above option price (9.31) using an integral.

Exercise 9.11. Receiver swaption pricing. Given n bonds with maturities $T_1, T_2, \ldots, T_n$, consider the annuity numeraire

$$P(t,T_i,T_j) = \sum_{k=i}^{j-1}(T_{k+1} - T_k)P(t,T_{k+1}), \qquad 0 \leqslant t \leqslant T_{i+1},$$

and the swap rate

$$S(t,T_i,T_j) = \frac{P(t,T_i) - P(t,T_j)}{P(t,T_i,T_j)}$$

at time $t \in [0,T_i]$, modeled as

$$dS(t,T_i,T_j) = \mu_t S(t,T_i,T_j)dt + \sigma S(t,T_i,T_j)dB_t, \qquad 0 \leqslant t \leqslant T_i, \tag{9.32}$$

where $(B_t)_{t \in [0,T_i]}$ is a standard Brownian motion under the risk-neutral measure $\mathbb{P}^*$, $(\mu_t)_{t \in [0,T]}$ is an $(\mathcal{F}_t)_{t \in \mathbb{R}_+}$-adapted process and $\sigma > 0$ is a constant. Let

$$\mathbb{E}^* \left[e^{-\int_t^{T_i} r_s ds} P(T_i,T_i,T_j)\phi(S(T_i,T_i,T_j)) \, \middle| \, \mathcal{F}_t \right] \tag{9.33}$$

at time $t \in [0,T_i]$ of an option with payoff function ϕ.

(1) Rewrite the option price (9.33) at time $t \in [0,T_i]$ using the forward swap measure $\widehat{\mathbb{P}}_{i,j}$ defined from the annuity numeraire $P(t,T_i,T_j)$.

(2) What is the dynamics of $S(t,T_i,T_j)$ under the forward swap measure $\widehat{\mathbb{P}}_{i,j}$?

(3) Write down the above option price (9.31) using a Gaussian integral.

(4) Apply the above to the computation at time $t \in [0, T_i]$ of the put (or receiver) swaption price

$$\mathbb{E}^* \left[e^{-\int_t^{T_i} r_s ds} P(T_i, T_i, T_j)(\kappa - S(T_i, T_i, T_j))^+ \,\middle|\, \mathcal{F}_t \right]$$

with strike level κ, using the Black-Scholes formula.

Hint: Given X a centered Gaussian random variable with variance $v^2 > 0$, we have

$$\mathbb{E}\left[(\kappa - e^{m+X})^+ \right] = \kappa \Phi\left(-\frac{m - \log \kappa}{v} \right) - e^{m+v^2/2} \Phi\left(-v - \frac{m - \log \kappa}{v} \right),$$

where Φ denotes the standard normal cumulative distribution function.

Exercise 9.12. Caplet hedging. Consider a bond market with two bonds of maturities T_1, T_2, whose prices $P(t, T_1), P(t, T_2)$ at time t are given by

$$\frac{dP(t, T_1)}{P(t, T_1)} = r_t dt + \zeta_1(t) dB_t, \qquad \frac{dP(t, T_2)}{P(t, T_2)} = r_t dt + \zeta_2(t) dB_t,$$

where $(r_t)_{t \in \mathbb{R}_+}$ is a short-term interest rate process, $(B_t)_{t \in \mathbb{R}_+}$ is a standard Brownian motion generating a filtration $(\mathcal{F}_t)_{t \in \mathbb{R}_+}$, and $\zeta_1(t), \zeta_2(t)$ are volatility processes. The LIBOR rate $L(t, T_1, T_2)$ is defined by

$$L(t, T_1, T_2) := \frac{P(t, T_1) - P(t, T_2)}{P(t, T_2)}, \qquad 0 \leqslant t \leqslant T_1.$$

Recall that an interest rate caplet on the LIBOR market can be priced at time $t \in [0, T_1]$ as

$$\mathbb{E}^* \left[e^{-\int_t^{T_2} r_s ds} (L(T_1, T_1, T_2) - \kappa)^+ \,\middle|\, \mathcal{F}_t \right]$$
$$= P(t, T_2) \widehat{\mathbb{E}}\left[(L(T_1, T_1, T_2) - \kappa)^+ \,\middle|\, \mathcal{F}_t \right],$$

under the forward measure $\widehat{\mathbb{P}}$ defined by

$$\frac{d\widehat{\mathbb{P}}}{d\mathbb{P}^*} = e^{-\int_0^{T_1} r_s ds} \frac{P(T_1, T_2)}{P(0, T_2)},$$

under which

$$\widehat{B}_t := B_t - \int_0^t \zeta_2(s) ds, \qquad 0 \leqslant t \leqslant T_1, \tag{9.34}$$

is a standard Brownian motion.

(1) Using Itô calculus, show that the LIBOR rate satisfies

$$dL(t, T_1, T_2) = L(t, T_1, T_2)\sigma(t)d\widehat{B}_t, \qquad 0 \leqslant t \leqslant T_1, \qquad (9.35)$$

where the LIBOR rate volatility is given by

$$\sigma(t) = \frac{P(t, T_1)(\zeta_1(t) - \zeta_2(t))}{P(t, T_1) - P(t, T_2)}.$$

(2) Solve the equation (9.35) on the interval $[t, T_1]$, and compute $L(T_1, T_1, T_2)$ from the initial condition $L(t, T_1, T_2)$.

(3) Assuming that $\sigma(t)$ in (9.35) is a deterministic function, show that the price

$$P(t, T_2)\widehat{\mathbb{E}}\big[(L(T_1, T_1, T_2) - \kappa)^+ \,\big|\, \mathcal{F}_t\big]$$

of the caplet can be written as $P(t, T_2)C(L(t, T_1, T_2), v(t, T_1))$, where $v^2(t, T_1) = \int_t^{T_1} |\sigma(s)|^2 ds$, and $C(t, v(t, T_1))$ is a function of $L(t, T_1, T_2)$ and $v(t, T_1)$.

(4) Consider a portfolio $(\xi_t^1, \xi_t^2)_{t \in [0, T_1]}$ made of bonds with maturities T_1, T_2 and value

$$V_t = \xi_t^1 P(t, T_1) + \xi_t^2 P(t, T_2),$$

at time $t \in [0, T_1]$. We assume that the portfolio is self-financing, *i.e.*

$$dV_t = \xi_t^1 dP(t, T_1) + \xi_t^2 dP(t, T_2), \qquad 0 \leqslant t \leqslant T_1, \qquad (9.36)$$

and that it *hedges* the claim $(L(T_1, T_1, T_2) - \kappa)^+$, so that

$$V_t = \mathbb{E}^* \left[e^{-\int_t^{T_1} r_s ds} (P(T_1, T_2)(L(T_1, T_1, T_2) - \kappa))^+ \,\Big|\, \mathcal{F}_t \right]$$

$$= P(t, T_2)\widehat{\mathbb{E}}\big[(L(T_1, T_1, T_2) - \kappa)^+ \,\big|\, \mathcal{F}_t\big],$$

$0 \leqslant t \leqslant T_1$. Show that we have

$$\mathbb{E}^* \left[e^{-\int_t^{T_1} r_s ds} (P(T_1, T_2)(L(T_1, T_1, T_2) - \kappa)^+ \,\Big|\, \mathcal{F}_t \right]$$

$$= P(0, T_2)\widehat{\mathbb{E}}\big[(L(T_1, T_1, T_2) - \kappa)^+\big] + \int_0^t \xi_s^1 dP(s, T_1) + \int_0^t \xi_s^2 dP(s, T_1),$$

$0 \leqslant t \leqslant T_1$.

(5) Show that under the self-financing condition (9.36), the discounted portfolio value process $\widetilde{V}_t = e^{-\int_0^t r_s ds} V_t$ satisfies

$$d\widetilde{V}_t = \xi_t^1 d\widetilde{P}(t, T_1) + \xi_t^2 d\widetilde{P}(t, T_2),$$

where $\widetilde{P}(t, T_1) := e^{-\int_0^t r_s ds} P(t, T_1)$ and $\widetilde{P}(t, T_2) := e^{-\int_0^t r_s ds} P(t, T_2)$ denote the discounted bond prices.

(6) Show that
$$\widehat{\mathbb{E}}\big[(L(T_1, T_1, T_2) - \kappa)^+ \,\big|\, \mathcal{F}_t\big] = \widehat{\mathbb{E}}\big[(L(T_1, T_1, T_2) - \kappa)^+\big]$$
$$+ \int_0^t \frac{\partial C}{\partial x}(L(u, T_1, T_2), v(u, T_1))dL(t, T_1, T_2),$$
and that the deflated portfolio value process $\widehat{V}_t = V_t/P(t, T_2)$ satisfies
$$d\widehat{V}_t = \frac{\partial C}{\partial x}(L(t, T_1, T_2), v(t, T_1))dL(t, T_1, T_2)$$
$$= \sigma(t)L(t, T_1, T_2)\frac{\partial C}{\partial x}(L(t, T_1, T_2), v(t, T_1))d\widehat{B}_t, \qquad t \geqslant 0.$$
Hint: Use the martingale property and the Itô formula.

(7) Show that
$$dV_t = (P(t, T_1) - P(t, T_2))\frac{\partial C}{\partial x}(L(t, T_1, T_2), v(t, T_1))\sigma(t)dB_t + \widehat{V}_t dP(t, T_2).$$

(8) Show that
$$d\widetilde{V}_t = \frac{\partial C}{\partial x}(L(t, T_1, T_2), v(t, T_1))d\big(\widetilde{P}(t, T_1) - \widetilde{P}(t, T_2)\big)$$
$$+ \bigg(\widehat{V}_t - L(t, T_1, T_2)\frac{\partial C}{\partial x}(L(t, T_1, T_2), v(t, T_1))\bigg)d\widetilde{P}(t, T_2),$$
and deduce the values of the self-financing portfolio $(\xi_t^1, \xi_t^2)_{t \in [0, T_1]}$ hedging the interest rate caplet with payoff $(L(T_1, T_1, T_2) - \kappa)^+$.

Exercise 9.13. Consider a market with short-term interest rate $(r_t)_{t \in \mathbb{R}_+}$ and two zero-coupon bonds $P(t, T_1)$, $P(t, T_2)$ with maturities $T_1 = \delta$ and $T_2 = 2\delta$, where $P(t, T_i)$ is modeled according to
$$\frac{dP(t, T_i)}{P(t, T_i)} = r_t dt + \zeta_i(t)dB_t, \qquad i = 1, 2.$$
Consider also the forward LIBOR $L(t, T_1, T_2)$ defined by
$$L(t, T_1, T_2) = \frac{1}{\delta}\left(\frac{P(t, T_1)}{P(t, T_2)} - 1\right), \qquad 0 \leqslant t \leqslant T_1,$$
and assume that $L(t, T_1, T_2)$ is modeled in the BGM model as
$$\frac{dL(t, T_1, T_2)}{L(t, T_1, T_2)} = \gamma dB_t^{(2)}, \qquad 0 \leqslant t \leqslant T_1, \tag{9.37}$$
where γ is a deterministic constant, and
$$B_t^{(2)} = B_t - \int_0^t \zeta_2(s)ds$$
is a standard Brownian motion under the forward measure $\widehat{\mathbb{P}}_2$ defined by
$$\frac{d\widehat{\mathbb{P}}_2}{d\mathbb{P}^*} = \exp\left(\int_0^{T_2} \zeta_2(s)dB_s - \frac{1}{2}\int_0^{T_2} |\zeta_2(s)|^2 ds\right).$$

(1) Compute $L(t, T_1, T_2)$ by solving Equation (9.37).
(2) Compute the price at time t:

$$P(t, T_2)\widehat{\mathbb{E}}_2 \left[(L(T_1, T_1, T_2) - \kappa)^+ \mid \mathcal{F}_t \right], \qquad 0 \leqslant t \leqslant T_1,$$

of an interest rate caplet with strike level κ, where $\widehat{\mathbb{E}}_2$ denotes the expectation under the forward measure $\widehat{\mathbb{P}}_2$.

Exercise 9.14. (Exercise 9.3 continued). Compute the price at time t:

$$\mathbb{E}^* \left[e^{-\int_t^{T_1} r_s ds} P(T_1, T_1, T_3)(S(T_1, T_1, T_3) - \kappa)^+ \mid \mathcal{F}_t \right]$$
$$= P(t, T_1, T_3)\widehat{\mathbb{E}}_{1,3} \left[(S(T_1, T_1, T_3) - \kappa)^+ \mid \mathcal{F}_t \right],$$

of the swaption on $S(t, T_1, T_3)$ with strike level κ, where $\widehat{\mathbb{E}}_{1,3}$ denotes the expectation under the forward swap measure $\widehat{\mathbb{P}}_{1,3}$ defined by

$$\frac{d\widehat{\mathbb{P}}_{1,3}}{d\mathbb{P}^*} = e^{-\int_0^{T_1} r_s ds} \frac{P(T_1, T_1, T_3)}{P(0, T_1, T_3)}.$$

You will need to use an approximation of $\sigma_{1,3}(s)$, for this it can be useful to "freeze" at time t all the random terms appearing in $\sigma_{1,3}(s)$, $s \geqslant t$.

Exercise 9.15. (Exercise 9.13 continued).

(1) Derive the stochastic differential equation satisfied by $P(t, T_1)$, and determine the process $\zeta_1(t)$ from the problem data.
(2) Show that $L(t, T_1, T_2)$ satisfies the stochastic differential equation

$$\frac{dL(t, T_1, T_2)}{L(t, T_1, T_2)} = \gamma dB_t - \gamma \zeta_2(t)dt, \qquad 0 \leqslant t \leqslant T_1. \tag{9.38}$$

(3) Assume that $r_t = r > 0$ is a deterministic constant and that $\zeta_1(t) = 0$, $t \in \mathbb{R}_+$. Compute an approximation for the bond call option price

$$P(t, T_1) \mathbb{E}^* \left[(P(T_1, T_2) - K)^+ \mid \mathcal{F}_t \right]$$

as a function of $L(t, T_1, T_2)$. In order to derive an approximated price you may "freeze" the drift $\gamma \zeta_2(t)$ of $L(s, T_1, T_2)$ under $\mathbb{P}$ by assuming that (9.38) is written as

$$\frac{dL(s, T_1, T_2)}{L(s, T_1, T_2)} = \gamma dB_s - \gamma \zeta_2(t)ds, \qquad t \leqslant s \leqslant T_1.$$

The final result may be expressed as an integral over the real line $\mathbb{R}$, whose explicit computation is not required.

Exercise 9.16. Path freezing. Consider n bonds with prices $P(t, T_i)$, $i = 1, 2, \ldots, n$, and the bond option with payoff

$$\left(\sum_{i=2}^{n} c_i P(T_0, T_i) - \kappa P(T_0, T_1) \right)^+ = P(T_0, T_1) (X_{T_0} - \kappa)^+ ,$$

where

$$X_t := \frac{1}{P(t, T_1)} \sum_{i=2}^{n} c_i P(t, T_i) = \sum_{i=2}^{n} c_i \widehat{P}(t, T_i), \quad 0 \leqslant t \leqslant T_1,$$

with $\widehat{P}(t, T_i) := P(t, T_i)/P(t, T_1)$, $i = 2, 3, \ldots, n$.

(1) Assuming that the deflated bond price $(\widehat{P}(t, T_i))_{t \in [0, T_i]}$ has the (martingale) dynamics $d\widehat{P}(t, T_i) = \sigma_i(t)\widehat{P}(t, T_i)d\widehat{B}_t$ under the forward measure $\widehat{\mathbb{P}}_1$, write down the dynamics of X_t as $dX_t = \sigma_t X_t d\widehat{B}_t$, where σ_t is to be computed explicitly.

(2) Approximating $(\widehat{P}(t, T_i))_{t \in [0, T_i]}$ by $\widehat{P}(0, T_i)$ and $(P(t, T_2, T_n))_{t \in [0, T_2]}$ by $P(0, T_2, T_n)$, find a deterministic approximation $\widehat{\sigma}(t)$ of σ_t, and deduce an expression of the option price

$$\mathbb{E}^* \left[e^{-\int_0^{T_1} r_s ds} \left(\sum_{i=2}^{n} c_i P(T_0, T_i) - \kappa P(T_0, T_1) \right)^+ \right]$$

$$= P(0, T_1) \widehat{\mathbb{E}} \left[(X_{T_0} - \kappa)^+ \right]$$

using the Black-Scholes formula.

Hint: Given X a centered Gaussian random variable with variance $v^2 > 0$, we have:

$$\mathbb{E}^* \left[(x e^{X - v^2/2} - \kappa)^+ \right] = x\Phi \left(\frac{v}{2} + \frac{1}{v} \log \frac{x}{\kappa} \right) - \kappa\Phi \left(-\frac{v}{2} + \frac{1}{v} \log \frac{x}{\kappa} \right).$$

Problem 9.17. Swaption hedging (2). Consider a bond market with tenor structure $\{T_i, \ldots, T_j\}$ and bonds with maturities $T_i, \ldots, T_j$, whose prices $P(t, T_i), \ldots, P(t, T_j)$ at time t are given by

$$\frac{dP(t, T_k)}{P(t, T_k)} = r_t dt + \zeta_k(t)dB_t, \qquad k = i, \ldots, j,$$

where $(r_t)_{t \in \mathbb{R}_+}$ is a short-term interest rate process and $(B_t)_{t \in \mathbb{R}_+}$ denotes a standard Brownian motion generating a filtration $(\mathcal{F}_t)_{t \in \mathbb{R}_+}$, and

$\zeta_i(t), \ldots, \zeta_j(t)$ are volatility processes. The swap rate $S(t, T_i, T_j)$ is defined by

$$S(t, T_i, T_j) = \frac{P(t, T_i) - P(t, T_j)}{P(t, T_i, T_j)},$$

where

$$P(t, T_i, T_j) = \sum_{k=i}^{j-1} (T_{k+1} - T_k) P(t, T_{k+1})$$

is the annuity numeraire. Recall that a payer swaption on the LIBOR market can be priced at time $t \in [0, T_i]$ as

$$\mathbb{E}^* \left[e^{-\int_t^{T_i} r_s ds} \left(\sum_{k=i}^{j-1} (T_{k+1} - T_k) P(T_i, T_{k+1})(S(T_i, T_k, T_{k+1}) - \kappa) \right)^+ \Bigg| \mathcal{F}_t \right]$$

$$= P(t, T_i, T_j) \widehat{\mathbb{E}}_{i,j} \left[(S(T_i, T_i, T_j) - \kappa)^+ \,\big|\, \mathcal{F}_t \right],$$

under the forward swap measure $\widehat{\mathbb{P}}_{i,j}$ defined by

$$\frac{d\widehat{\mathbb{P}}_{i,j}}{d\mathbb{P}^*} = e^{-\int_0^{T_i} r_s ds} \frac{P(T_i, T_i, T_j)}{P(0, T_i, T_j)}, \qquad 1 \leqslant i < j \leqslant n,$$

under which

$$\widehat{B}_t^{i,j} := B_t - \sum_{k=i}^{j-1} (T_{k+1} - T_k) \frac{P(t, T_{k+1})}{P(t, T_i, T_j)} \zeta_{k+1}(t) dt \tag{9.39}$$

is a standard Brownian motion. Recall that the swap rate can be modeled as

$$dS(t, T_i, T_j) = S(t, T_i, T_j)\sigma_{i,j}(t) d\widehat{B}_t^{i,j}, \qquad 0 \leqslant t \leqslant T_i, \tag{9.40}$$

where the swap rate volatilities are given by

$$\sigma_{i,j}(t) = \sum_{l=i}^{j-1} (T_{l+1} - T_l) \frac{P(t, T_{l+1})}{P(t, T_i, T_j)} (\zeta_i(t) - \zeta_{l+1}(t)) \tag{9.41}$$

$$+ \frac{P(t, T_j)}{P(t, T_i) - P(t, T_j)} (\zeta_i(t) - \zeta_j(t))$$

$1 \leqslant i, j \leqslant n$, cf. Proposition 8.12. In the sequel, we denote $S_t := S(t, T_i, T_j)$ for simplicity of notation.

(1) Solve the equation (9.40) on the interval $[t, T_i]$, and compute $S(T_i, T_i, T_j)$ from the initial condition $S(t, T_i, T_j)$.

(2) Assuming that $\sigma_{i,j}(t)$ is a deterministic function of t for $1 \leqslant i, j \leqslant n$, show that the price (9.16) of the swaption can be written as

$$P(t, T_i, T_j)C(S_t, v(t, T_i)),$$

where

$$v^2(t, T_i) = \int_t^{T_i} |\sigma_{i,j}(s)|^2 ds, \qquad 0 \leqslant t \leqslant T_i,$$

and $C(x, v)$ is a function to be specified using the Black-Scholes formula $\text{Bl}(x, K, \sigma, r, \tau)$, with

$$\mathbb{E}[(x e^{m+X} - K)^+] = \Phi(v + (m + \log(x/K))/v) - K\Phi((m + \log(x/K))/v),$$

where $m = r\tau - v^2/2$ and X is a centered Gaussian random variable with variance $v^2 > 0$.

(3) Consider a portfolio $(\xi_t^i, \ldots, \xi_t^j)_{t \in [0, T_i]}$ made of bonds with maturities $T_i, \ldots, T_j$ and value

$$V_t = \sum_{k=i}^{j} \xi_t^k P(t, T_k),$$

at time $t \in [0, T_i]$. We assume that the portfolio is self-financing, *i.e.*

$$dV_t = \sum_{k=i}^{j} \xi_t^k dP(t, T_k), \qquad 0 \leqslant t \leqslant T_i, \tag{9.42}$$

and that it *hedges* the claim $(S(T_i, T_i, T_j) - \kappa)^+$, so that

$$V_t =$$

$$\mathbb{E}^* \left[e^{-\int_t^{T_i} r_s ds} \left(\sum_{k=i}^{j-1} (T_{k+1} - T_k) P(T_i, T_{k+1}) (L(T_i, T_k, T_{k+1}) - \kappa) \right)^+ \bigg| \mathcal{F}_t \right]$$

$$= P(t, T_i, T_j) \widehat{\mathbb{E}}_{i,j} [(S(T_i, T_i, T_j) - \kappa)^+ | \mathcal{F}_t], \qquad 0 \leqslant t \leqslant T_i.$$

Show that

$$\mathbb{E}^* \left[e^{-\int_t^{T_i} r_s ds} \left(\sum_{k=i}^{j-1} (T_{k+1} - T_k) P(T_i, T_{k+1}) (L(T_i, T_k, T_{k+1}) - \kappa) \right)^+ \bigg| \mathcal{F}_t \right]$$

$$= P(0, T_i, T_j) \widehat{\mathbb{E}}_{i,j} [(S(T_i, T_i, T_j) - \kappa)^+] + \sum_{k=i}^{j} \int_0^t \xi_s^k dP(s, T_i),$$

$$0 \leqslant t \leqslant T_i.$$

(4) Show that under the self-financing condition (9.42), the discounted portfolio value process $\widetilde{V}_t = e^{-\int_0^t r_s ds} V_t$ satisfies

$$d\widetilde{V}_t = \sum_{k=i}^{j} \xi_t^k d\widetilde{P}(t, T_k),$$

where $\widetilde{P}(t, T_k) = e^{-\int_0^t r_s ds} P(t, T_k)$, $k = i, \ldots, j$, denote the discounted bond prices.

(5) Show that

$$\widehat{\mathbb{E}}_{i,j} \left[(S(T_i, T_i, T_j) - \kappa)^+ \mid \mathcal{F}_t \right]$$

$$= \widehat{\mathbb{E}}_{i,j} \left[(S(T_i, T_i, T_j) - \kappa)^+ \right] + \int_0^t \frac{\partial C}{\partial x} (S_u, v(u, T_i)) dS_u.$$

Hint: Use the martingale property and the Itô formula.

(6) Show that the deflated portfolio value process $\widehat{V}_t = V_t / P(t, T_i, T_j)$ satisfies

$$d\widehat{V}_t = \frac{\partial C}{\partial x} (S_t, v(t, T_i)) dS_t = S_t \frac{\partial C}{\partial x} (S_t, v(t, T_i)) \sigma_t^{i,j} d\widehat{B}_t^{i,j}.$$

(7) Show that

$$dV_t = (P(t, T_i) - P(t, T_j)) \frac{\partial C}{\partial x} (S_t, v(t, T_i)) \sigma_t^{i,j} dB_t + \widehat{V}_t dP(t, T_i, T_j).$$

(8) Show that

$$dV_t = S_t \zeta_i(t) \frac{\partial C}{\partial x} (S_t, v(t, T_i)) \sum_{k=i}^{j-1} (T_{k+1} - T_k) P(t, T_{k+1}) dB_t$$

$$+ \left(\widehat{V}_t - S_t \frac{\partial C}{\partial x} (S_t, v(t, T_i)) \right) \sum_{k=i}^{j-1} (T_{k+1} - T_k) P(t, T_{k+1}) \zeta_{k+1}(t) dB_t$$

$$+ \frac{\partial C}{\partial x} (S_t, v(t, T_i)) P(t, T_j) (\zeta_i(t) - \zeta_j(t)) dB_t.$$

(9) Show that

$$d\widetilde{V}_t = \frac{\partial C}{\partial x} (S_t, v(t, T_i)) d(\widetilde{P}(t, T_i) - \widetilde{P}(t, T_j))$$

$$+ \left(\widehat{V}_t - S_t \frac{\partial C}{\partial x} (S_t, v(t, T_i)) \right) d\widetilde{P}(t, T_i, T_j).$$

(10) Show that

$$\frac{\partial C}{\partial x} (x, v(t, T_i)) = \Phi \left(\frac{\log(x/K)}{v(t, T_i)} + \frac{v(t, T_i)}{2} \right).$$

(11) Show that we have

$$d\widetilde{V}_t = \Phi\left(\frac{\log(S_t/K)}{v(t,T_i)} + \frac{v(t,T_i)}{2}\right) d\big(\widetilde{P}(t,T_i) - \widetilde{P}(t,T_j)\big)$$
$$-\kappa\Phi\left(\frac{\log(S_t/K)}{v(t,T_i)} - \frac{v(t,T_i)}{2}\right) d\widetilde{P}(t,T_i,T_j).$$

(12) Show that the hedging strategy of the swaption is given by

$$\xi_t^i = \Phi\left(\frac{\log(S_t/K)}{v(t,T_i)} + \frac{v(t,T_i)}{2}\right),$$
$$\xi_t^j = -\Phi\left(\frac{\log(S_t/K)}{v(t,T_i)} + \frac{v(t,T_i)}{2}\right)$$
$$-(T_j - T_{j-1})\kappa\Phi\left(\frac{\log(S_t/K)}{v(t,T_i)} - \frac{v(t,T_i)}{2}\right),$$

and

$$\xi_t^k = -(T_{k+1} - T_k)\kappa\Phi\left(\frac{\log(S_t/K)}{v(t,T_i)} - \frac{v(t,T_i)}{2}\right), \quad i+1 \leqslant k \leqslant j-1.$$

Chapter 10

Default Bond Pricing

The bond pricing models considered in the previous chapters rely on the assumption that the bond principal payment at maturity is always equal to $1, therefore excluding the possibility of default. In this chapter we study the pricing of bonds that may default at a random time τ, in the reduced-form approach to credit modeling. We also consider the associated options (credit default swaps) that are designed as a protection against default.

10.1 Survival Probabilities

Given $t > 0$, let $\mathbb{P}(\tau > t)$ denote the probability that a random system with lifetime τ survives at least t years. Assuming that survival probabilities $\mathbb{P}(\tau > t)$ are strictly positive for all $t > 0$, we can compute the conditional probability for that system to survive up to time T, given that it was still functioning at time $t \in [0, T]$, as

$$\mathbb{P}(\tau > T \mid \tau > t) = \frac{\mathbb{P}(\tau > T \text{ and } \tau > t)}{\mathbb{P}(\tau > t)} = \frac{\mathbb{P}(\tau > T)}{\mathbb{P}(\tau > t)}, \quad 0 \leqslant t \leqslant T,$$

with

$$\begin{aligned}
\mathbb{P}(\tau \leqslant T \mid \tau > t) &= 1 - \mathbb{P}(\tau > T \mid \tau > t) \\
&= \frac{\mathbb{P}(\tau > t) - \mathbb{P}(\tau > T)}{\mathbb{P}(\tau > t)} \\
&= \frac{\mathbb{P}(\tau \leqslant T) - \mathbb{P}(\tau \leqslant t)}{\mathbb{P}(\tau > t)} \\
&= \frac{\mathbb{P}(t < \tau \leqslant T)}{\mathbb{P}(\tau > t)}, \quad 0 \leqslant t \leqslant T,
\end{aligned}$$

and the conditional survival probability distribution

$$\mathbb{P}(\tau \in dx \mid \tau > t) = \mathbb{P}(x < \tau \leqslant x + dx \mid \tau > t)$$

$$= \mathbb{P}(\tau \leqslant x + dx \mid \tau > t) - \mathbb{P}(\tau \leqslant x \mid \tau > t)$$

$$= \frac{\mathbb{P}(\tau \leqslant x + dx) - \mathbb{P}(\tau \leqslant x)}{\mathbb{P}(\tau > t)}$$

$$= \frac{1}{\mathbb{P}(\tau > t)} d\mathbb{P}(\tau \leqslant x)$$

$$= -\frac{1}{\mathbb{P}(\tau > t)} d\mathbb{P}(\tau > x), \qquad x > t.$$

Such survival probabilities are typically found in life (or mortality) tables:

Age t	$\mathbb{P}(\tau \leqslant t + 1 \mid \tau > t)$
20	0.0894%
30	0.1008%
40	0.2038%
50	0.4458%
60	0.9827%

Table 10.1: Mortality table.

Proposition 10.1. *The* failure rate *function, defined as*

$$\lambda(t) := \frac{\mathbb{P}(\tau \leqslant t + dt \mid \tau > t)}{dt},$$

satisfies

$$\mathbb{P}(\tau > t) = e^{-\int_0^t \lambda(u) du}, \qquad t \geqslant 0. \tag{10.1}$$

Proof. We have

$$\lambda(t) := \frac{\mathbb{P}(\tau \leqslant t + dt \mid \tau > t)}{dt}$$

$$= \frac{1}{\mathbb{P}(\tau > t)} \frac{\mathbb{P}(t < \tau \leqslant t + dt)}{dt}$$

$$= \frac{1}{\mathbb{P}(\tau > t)} \frac{\mathbb{P}(\tau > t) - \mathbb{P}(\tau > t + dt)}{dt}$$

$$= -\frac{d}{dt} \log \mathbb{P}(\tau > t)$$

$$= -\frac{1}{\mathbb{P}(\tau > t)} \frac{d}{dt} \mathbb{P}(\tau > t), \qquad t > 0,$$

and the differential equation

$$\frac{d}{dt} \mathbb{P}(\tau > t) = -\lambda(t)\mathbb{P}(\tau > t),$$

which can be solved as in (10.1) under the initial condition $\mathbb{P}(\tau > 0) = 1$. $\square$

Proposition 10.1 allows us to rewrite the (conditional) survival probability as

$$\mathbb{P}(\tau > T \mid \tau > t) = \frac{\mathbb{P}(\tau > T)}{\mathbb{P}(\tau > t)} = e^{-\int_t^T \lambda(u)du}, \qquad 0 \leqslant t \leqslant T,$$

with

$$\mathbb{P}(\tau > t + h \mid \tau > t) = e^{-\lambda(t)h} \simeq 1 - \lambda(t)h, \qquad [h \searrow 0],$$

and

$$\mathbb{P}(\tau \leqslant t + h \mid \tau > t) = 1 - e^{-\lambda(t)h} \simeq \lambda(t)h, \qquad [h \searrow 0],$$

as h tends to 0. When the failure rate $\lambda(t) = \lambda > 0$ is a constant function of time, Relation (10.1) shows that

$$\mathbb{P}(\tau > T) = e^{-\lambda T}, \qquad T \geqslant 0,$$

i.e. τ has the exponential distribution with parameter λ. Note that given $(\tau_n)_{n \geqslant 1}$ a sequence of *i.i.d.* exponentially distributed random variables, letting

$$T_n = \tau_1 + \tau_2 + \cdots + \tau_n, \qquad n \geqslant 1,$$

defines the sequence of jump times of a standard Poisson process with intensity $\lambda > 0$.

10.2 Stochastic Default

When the random time τ is a *stopping time* with respect to $(\mathcal{F}_t)_{t \in \mathbb{R}_+}$ we have

$$\{\tau > t\} \in \mathcal{F}_t, \qquad t \geqslant 0,$$

i.e. the knowledge of whether default or bankruptcy has already occurred at time t is contained in $\mathcal{F}_t$, $t \in \mathbb{R}_+$. As a consequence, we can write

$$\mathbb{P}(\tau > t \mid \mathcal{F}_t) = \mathbb{E}\left[\mathbf{1}_{\{\tau > t\}} \mid \mathcal{F}_t\right] = \mathbf{1}_{\{\tau > t\}}, \qquad t \geqslant 0.$$

In the sequel we do not assume that τ is an $\mathcal{F}_t$-stopping time, and by analogy with (10.1) we write $\mathbb{P}(\tau > t \mid \mathcal{F}_t)$ as

$$\mathbb{P}(\tau > t \mid \mathcal{F}_t) = e^{-\int_0^t \lambda_u du}, \qquad t \geqslant 0, \qquad (10.2)$$

where the failure rate function $(\lambda_t)_{t \in \mathbb{R}_+}$ is modeled as a random process adapted to a filtration $(\mathcal{F}_t)_{t \in \mathbb{R}_+}$.

The process $(\lambda_t)_{t \in \mathbb{R}_+}$ can also be chosen among the classical mean-reverting diffusion processes, including jump-diffusion processes. In Lando (1998), the process $(\lambda_t)_{t \in \mathbb{R}_+}$ is constructed as $\lambda_t := h(X_t)$, $t \in \mathbb{R}_+$, where h is a nonnegative function and $(X_t)_{t \in \mathbb{R}_+}$ is a stochastic process generating the filtration $(\mathcal{F}_t)_{t \in \mathbb{R}_+}$. The default time τ is then defined as

$$\tau := \inf \left\{ t \geqslant 0 : \int_0^t h(X_u) du \geqslant L \right\},$$

where L is an exponentially distributed random variable independent of $(\mathcal{F}_t)_{t \in \mathbb{R}_+}$. In this case, as τ is not an $(\mathcal{F}_t)_{t \in \mathbb{R}_+}$-stopping time, we have

$$\mathbb{P}(\tau > t \mid \mathcal{F}_t) = \mathbb{P}\left(\int_0^t h(X_u) du < L \;\middle|\; \mathcal{F}_t \right)$$

$$= e^{-\int_0^t h(X_u) du}$$

$$= e^{-\int_0^t \lambda_u du}, \qquad t \geqslant 0.$$

Definition 10.1. *Let $(\mathcal{G}_t)_{t \in \mathbb{R}_+}$ be the filtration defined by $\mathcal{G}_\infty := \mathcal{F}_\infty \vee \sigma(\tau)$ and*

$$\mathcal{G}_t := \left\{ B \in \mathcal{G}_\infty : \exists A \in \mathcal{F}_t \text{ such that } A \cap \{\tau > t\} = B \cap \{\tau > t\} \right\}, \quad (10.3)$$

with $\mathcal{F}_t \subset \mathcal{G}_t$, $t \geqslant 0$.

In other words, $\mathcal{G}_t$ contains insider information on whether default at time τ has occurred or not before time t, and τ is a $(\mathcal{G}_t)_{t \in \mathbb{R}_+}$-stopping time. Note that this information on τ may not be available to a generic user who has only access to the smaller filtration $(\mathcal{F}_t)_{t \in \mathbb{R}_+}$. The next key Lemma 10.1 allows us to price a contingent claim given the information in the larger filtration $(\mathcal{G}_t)_{t \in \mathbb{R}_+}$, by only using information in $(\mathcal{F}_t)_{t \in \mathbb{R}_+}$ and factoring in the default rate factor $e^{-\int_t^T \lambda_u du}$.

Lemma 10.1. *(Guo et al. (2007)) For any $\mathcal{F}_T$-measurable integrable random variable F, we have*

$$\mathbb{E}\left[F \mathbb{1}_{\{\tau > T\}} \mid \mathcal{G}_t \right] = \mathbb{1}_{\{\tau > t\}} \mathbb{E}\left[F e^{-\int_t^T \lambda_u du} \;\middle|\; \mathcal{F}_t \right].$$

Proof. By (10.2) we have

$$\frac{\mathbb{P}(\tau > T \mid \mathcal{F}_T)}{\mathbb{P}(\tau > t \mid \mathcal{F}_t)} = \frac{e^{-\int_0^T \lambda_u du}}{e^{-\int_0^t \lambda_u du}} = e^{-\int_t^T \lambda_u du}, \qquad 0 \leqslant t \leqslant T,$$

hence, since F is $\mathcal{F}_T$-measurable,

$$\mathbb{1}_{\{\tau > t\}} \mathbb{E}\left[F e^{-\int_t^T \lambda_u du} \;\middle|\; \mathcal{F}_t \right] = \mathbb{1}_{\{\tau > t\}} \mathbb{E}\left[F \frac{\mathbb{P}(\tau > T \mid \mathcal{F}_T)}{\mathbb{P}(\tau > t \mid \mathcal{F}_t)} \;\middle|\; \mathcal{F}_t \right]$$

$$= \frac{\mathbb{1}_{\{\tau>t\}}}{\mathbb{P}(\tau > t \mid \mathcal{F}_t)} \mathbb{E}\left[F \, \mathbb{E}\left[\mathbb{1}_{\{\tau>T\}} \mid \mathcal{F}_T\right] \mid \mathcal{F}_t\right]$$

$$= \frac{\mathbb{1}_{\{\tau>t\}}}{\mathbb{P}(\tau > t \mid \mathcal{F}_t)} \mathbb{E}\left[\mathbb{E}\left[F\mathbb{1}_{\{\tau>T\}} \mid \mathcal{F}_T\right] \mid \mathcal{F}_t\right]$$

$$= \mathbb{1}_{\{\tau>t\}} \frac{\mathbb{E}\left[F\mathbb{1}_{\{\tau>T\}} \mid \mathcal{F}_t\right]}{\mathbb{P}(\tau > t \mid \mathcal{F}_t)}$$

$$= \mathbb{1}_{\{\tau>t\}} \mathbb{E}\left[F\mathbb{1}_{\{\tau>T\}} \mid \mathcal{G}_t\right]$$

$$= \mathbb{E}\left[F\mathbb{1}_{\{\tau>T\}} \mid \mathcal{G}_t\right], \qquad 0 \leqslant t \leqslant T.$$

In the last step of the above argument we used the key relation

$$\mathbb{1}_{\{\tau>t\}} \mathbb{E}\left[F\mathbb{1}_{\{\tau>T\}} \,\Big|\, \mathcal{G}_t\right] = \frac{\mathbb{1}_{\{\tau>t\}}}{\mathbb{P}(\tau > t \mid \mathcal{F}_t)} \mathbb{E}\left[F\mathbb{1}_{\{\tau>T\}} \mid \mathcal{F}_t\right],$$

see Relation (75.2) in § XX-75 page 186 of Dellacherie *et al.* (1992), Theorem VI-3-14 page 371 of Protter (2004), and Lemma 3.1 of Elliott *et al.* (2000), under the conditional probability measure $\mathbb{P}_{|\mathcal{F}_t}$, $0 \leqslant t \leqslant T$. Indeed, according to (10.3), for any $B \in \mathcal{G}_t$ we have, for some event $A \in \mathcal{F}_t$,

$$\mathbb{E}\left[\mathbb{1}_B\mathbb{1}_{\{\tau>t\}}F\mathbb{1}_{\{\tau>T\}}\right] = \mathbb{E}\left[\mathbb{1}_{B\cap\{\tau>t\}}F\mathbb{1}_{\{\tau>T\}}\right]$$

$$= \mathbb{E}\left[\mathbb{1}_{A\cap\{\tau>t\}}F\mathbb{1}_{\{\tau>T\}}\right]$$

$$= \mathbb{E}\left[\mathbb{1}_A\mathbb{1}_{\{\tau>t\}}F\mathbb{1}_{\{\tau>T\}}\right]$$

$$= \mathbb{E}\left[\mathbb{1}_A\mathbb{1}_{\{\tau>t\}}\frac{\mathbb{E}[\mathbb{1}_{\{\tau>t\}} \mid \mathcal{F}_t]}{\mathbb{P}(\tau > t \mid \mathcal{F}_t)}F\mathbb{1}_{\{\tau>T\}}\right]$$

$$= \mathbb{E}\left[\frac{\mathbb{E}[\mathbb{1}_A\mathbb{1}_{\{\tau>t\}} \mid \mathcal{F}_t]}{\mathbb{P}(\tau > t \mid \mathcal{F}_t)}F\mathbb{1}_{\{\tau>T\}}\right]$$

$$= \mathbb{E}\left[\frac{\mathbb{E}[\mathbb{1}_A\mathbb{1}_{\{\tau>t\}} \mid \mathcal{F}_t]}{\mathbb{P}(\tau > t \mid \mathcal{F}_t)} \mathbb{E}\left[F\mathbb{1}_{\{\tau>T\}} \mid \mathcal{F}_t\right]\right]$$

$$= \mathbb{E}\left[\frac{\mathbb{1}_A\mathbb{1}_{\{\tau>t\}}}{\mathbb{P}(\tau > t \mid \mathcal{F}_t)} \mathbb{E}\left[F\mathbb{1}_{\{\tau>T\}} \mid \mathcal{F}_t\right]\right]$$

$$= \mathbb{E}\left[\frac{\mathbb{E}[\mathbb{1}_A\mathbb{1}_{\{\tau>t\}} \mid \mathcal{F}_t]}{\mathbb{P}(\tau > t \mid \mathcal{F}_t)} \mathbb{E}\left[F\mathbb{1}_{\{\tau>T\}} \mid \mathcal{F}_t\right]\right]$$

$$= \mathbb{E}\left[\frac{\mathbb{1}_A\mathbb{1}_{\{\tau>t\}}}{\mathbb{P}(\tau > t \mid \mathcal{F}_t)} \mathbb{E}\left[F\mathbb{1}_{\{\tau>T\}} \mid \mathcal{F}_t\right]\right]$$

$$= \mathbb{E}\left[\frac{\mathbb{1}_B\mathbb{1}_{\{\tau>t\}}}{\mathbb{P}(\tau > t \mid \mathcal{F}_t)} \mathbb{E}\left[F\mathbb{1}_{\{\tau>T\}} \mid \mathcal{F}_t\right]\right],$$

hence by a standard characterization of conditional expectations, see *e.g.* Relation (11.4), we have

$$\mathbb{E}\left[\mathbb{1}_{\{\tau>t\}}F\mathbb{1}_{\{\tau>T\}} \mid \mathcal{G}_t\right] = \frac{\mathbb{1}_{\{\tau>t\}}}{\mathbb{P}(\tau > t \mid \mathcal{F}_t)} \mathbb{E}\left[F\mathbb{1}_{\{\tau>T\}} \mid \mathcal{F}_t\right].$$

$\square$

Taking $F = 1$ in Lemma 10.1 allows one to write the survival probability up to time T, given the information known up to time t, as

$$\mathbb{P}(\tau > T \mid \mathcal{G}_t) = \mathbb{E}\left[\mathbb{1}_{\{\tau > T\}} \mid \mathcal{G}_t\right] \tag{10.4}$$

$$= \mathbb{1}_{\{\tau > t\}}\, \mathbb{E}\left[e^{-\int_t^T \lambda_u du} \,\middle|\, \mathcal{F}_t\right], \quad 0 \leqslant t \leqslant T.$$

In particular, applying Lemma 10.1 for $t = T$ and $F = 1$ shows that

$$\mathbb{E}\left[\mathbb{1}_{\{\tau > t\}} \mid \mathcal{G}_t\right] = \mathbb{1}_{\{\tau > t\}},$$

which shows that $\{\tau > t\} \in \mathcal{G}_t$ for all $t > 0$, and recovers the fact that τ is a $(\mathcal{G}_t)_{t \in \mathbb{R}_+}$-stopping time, while in general, τ is not an $(\mathcal{F}_t)_{t \in \mathbb{R}_+}$-stopping time.

The computation of $\mathbb{P}(\tau > T \mid \mathcal{G}_t)$ according to (10.4) is then similar to that of a bond price, by considering the failure rate $\lambda(t)$ as a "virtual" short-term interest rate. In particular, the failure rate $\lambda(t, T)$ can be modeled in the HJM framework, see Chapter 7, and

$$\mathbb{P}(\tau > T \mid \mathcal{G}_t) = \mathbb{E}\left[e^{-\int_t^T \lambda(t,u)du} \,\middle|\, \mathcal{F}_t\right]$$

can then be computed by applying HJM bond pricing techniques.

The computation of expectations given $\mathcal{G}_t$ as in Lemma 10.1 can be useful for pricing under insider trading, in which the insider has access to the augmented filtration $\mathcal{G}_t$ while the ordinary trader has only access to $\mathcal{F}_t$, therefore generating two different prices $\mathbb{E}^*[F \mid \mathcal{F}_t]$ and $\mathbb{E}^*[F \mid \mathcal{G}_t]$ for the same claim payoff F under the same risk-neutral probability measure $\mathbb{P}^*$.

This leads to the problem of computing the dynamics of the underlying asset price by decomposing it using a $(\mathcal{G}_t)_{t \in \mathbb{R}_+}$-martingale *vs* an $(\mathcal{F}_t)_{t \in \mathbb{R}_+}$-martingale. This decomposition can be obtained by the technique of enlargement of filtration, see Jeulin (1980), Jacod (1985), Yor (1985), Elliott and Jeanblanc (1999).

10.3   Defaultable Bonds

Bond pricing models are generally based on the terminal condition $P(T, T) = \$1$ according to which the bond payoff at maturity is always equal to $\$1$, and default does not occurs. In this chapter we allow for the possibility of default at a random time τ, in which case the terminal payoff of a bond is allowed to vanish at maturity.

The price $P_d(t, T)$ at time t of a default bond with maturity T, (random) default time τ and (possibly random) recovery rate $\xi \in [0, 1]$ is given by

$$P_d(t, T) = \mathbb{E}^* \left[\mathbb{1}_{\{\tau > T\}} e^{-\int_t^T r_u du} \, \Big| \, \mathcal{G}_t \right]$$
$$+ \mathbb{E}^* \left[\xi \mathbb{1}_{\{\tau \leqslant T\}} e^{-\int_t^T r_u du} \, \Big| \, \mathcal{G}_t \right], \qquad 0 \leqslant t \leqslant T.$$

Proposition 10.2. *The default bond with maturity T and default time τ can be priced at time $t \in [0, T]$ as*

$$P_d(t, T) = \mathbb{1}_{\{\tau > t\}} \mathbb{E}^* \left[e^{-\int_t^T (r_u + \lambda_u) du} \, \Big| \, \mathcal{F}_t \right]$$
$$+ \mathbb{E}^* \left[\xi \mathbb{1}_{\{\tau \leqslant T\}} e^{-\int_t^T r_u du} \, \Big| \, \mathcal{G}_t \right], \quad 0 \leqslant t \leqslant T.$$

Proof. We take $F = e^{-\int_t^T r_u du}$ in Lemma 10.1, which shows that

$$\mathbb{E}^* \left[\mathbb{1}_{\{\tau > T\}} e^{-\int_t^T r_u du} \, \Big| \, \mathcal{G}_t \right] = \mathbb{1}_{\{\tau > t\}} \mathbb{E}^* \left[e^{-\int_t^T (r_u + \lambda_u) du} \, \Big| \, \mathcal{F}_t \right],$$

cf. *e.g.* Lando (1998), Duffie and Singleton (2003), Guo *et al.* (2007). □

In the case of complete default (zero-recovery), we have $\xi = 0$ and

$$P_d(t, T) = \mathbb{1}_{\{\tau > t\}} \mathbb{E}^* \left[e^{-\int_t^T (r_s + \lambda_s) ds} \, \Big| \, \mathcal{F}_t \right], \quad 0 \leqslant t \leqslant T. \tag{10.5}$$

From the above expression (10.5) we note that the effect of the presence of a default time τ is to decrease the bond price, which can be viewed as an increase of the short rate by the amount λ_u. In a simple setting where the interest rate $r > 0$ and failure rate $\lambda > 0$ are constant, the default bond price becomes

$$P_d(t, T) = \mathbb{1}_{\{\tau > t\}} e^{-(r+\lambda)(T-t)}, \qquad 0 \leqslant t \leqslant T.$$

Finally, from *e.g.* Proposition 12.1 of Privault (2014) the bond price (10.5) can also be expressed under the forward measure $\widehat{\mathbb{P}}$ with maturity T, as

$$P_d(t, T) = \mathbb{1}_{\{\tau > t\}} \mathbb{E}^* \left[e^{-\int_t^T (r_s + \lambda_s) ds} \, \Big| \, \mathcal{F}_t \right]$$
$$= \mathbb{1}_{\{\tau > t\}} \mathbb{E}^* \left[e^{-\int_t^T r_s ds} \, \Big| \, \mathcal{F}_t \right] \widehat{\mathbb{E}} \left[e^{-\int_t^T \lambda_s ds} \, \Big| \, \mathcal{F}_t \right]$$
$$= \mathbb{1}_{\{\tau > t\}} P(t, T) \widehat{\mathbb{P}}(\tau > T \mid \mathcal{G}_t),$$

where

$$P(t, T) = \mathbb{E}^* \left[e^{-\int_t^T r_s ds} \, \Big| \, \mathcal{F}_t \right], \qquad 0 \leqslant t \leqslant T,$$

and by (10.4),

$$\widehat{\mathbb{P}}(\tau > T \mid \mathcal{G}_t) = \mathbb{1}_{\{\tau > t\}} \widehat{\mathbb{E}}\left[e^{-\int_t^T \lambda_s ds} \,\middle|\, \mathcal{F}_t \right]$$

is the survival probability under the forward measure $\widehat{\mathbb{P}}$ defined as

$$\frac{d\widehat{\mathbb{P}}}{d\mathbb{P}} := \frac{1}{P(0,T)} e^{-\int_0^T r_t dt},$$

see Chen and Huang (2001), Chen *et al.* (2008).

10.4   Estimating Default Rates

Recall that the price of a default bond with maturity T, (random) default time τ and (possibly random) recovery rate $\xi \in [0,1]$ is given by

$$P_d(t,T) = \mathbb{1}_{\{\tau > t\}} \mathbb{E}^*\left[e^{-\int_t^T (r_u + \lambda_u) du} \,\middle|\, \mathcal{F}_t \right]$$
$$+ \mathbb{E}^*\left[\xi \mathbb{1}_{\{\tau \leqslant T\}} e^{-\int_t^T r_u du} \,\middle|\, \mathcal{G}_t \right], \qquad 0 \leqslant t \leqslant T,$$

where ξ denotes the recovery rate. We consider a simplified deterministic step function model with zero recovery rate and tenor structure

$$\{t = T_0 < T_1 < \cdots < T_n = T\},$$

where

$$r(t) = \sum_{l=0}^{n-1} r_l \mathbb{1}_{(T_l, T_{l+1}]}(t) \quad \text{and} \quad \lambda(t) = \sum_{l=0}^{n-1} \lambda_l \mathbb{1}_{(T_l, T_{l+1}]}(t), \quad t \geqslant 0. \quad (10.6)$$

(1) Estimating the default rates from default bond prices.

We have

$$P_d(t,T_k) = \mathbb{1}_{\{\tau > t\}} e^{-\int_t^{T_k} (r(u) + \lambda(u)) du}$$
$$= \mathbb{1}_{\{\tau > t\}} \exp\left(-\sum_{l=0}^{k-1} (r_l + \lambda_l)(T_{l+1} - T_l) \right),$$

$k = 1, 2, \ldots, n$, from which we can infer

$$\lambda_k = -r_k - \frac{1}{T_{k+1} - T_k} \log \frac{P_d(t, T_{k+1})}{P_d(t, T_k)} > 0, \quad k = 0, 1, \ldots, n-1.$$

(2) Estimating (implied) default probabilities $\mathbb{P}^*(\tau < T \mid \mathcal{G}_t)$ from default rates.

Based on the expression

$$\mathbb{P}^*(\tau > T \mid \mathcal{G}_t) = \mathbb{E}^* \left[\mathbb{1}_{\{\tau > T\}} \mid \mathcal{G}_t \right]$$
$$= \mathbb{1}_{\{\tau > t\}} \, \mathbb{E}^* \left[e^{- \int_t^T \lambda_u du} \,\Big|\, \mathcal{F}_t \right], \qquad 0 \leqslant t \leqslant T,$$

of the survival probability up to time T, and given the information known up to time t, in terms of the hazard rate process $(\lambda_u)_{u \in \mathbb{R}_+}$ adapted to a filtration $(\mathcal{F}_t)_{t \in \mathbb{R}_+}$, we find

$$\mathbb{P}(\tau > T \mid \mathcal{G}_{T_k}) = \mathbb{1}_{\{\tau > T_k\}} \, e^{- \int_{T_k}^T \lambda_u du}$$
$$= \mathbb{1}_{\{\tau > t\}} \exp \left(- \sum_{l=k}^{n-1} \lambda_l (T_{l+1} - T_l) \right), \qquad k = 0, 1, \ldots, n-1,$$

where

$$\mathcal{G}_t = \mathcal{F}_t \vee \sigma(\{\tau \leqslant u\} : 0 \leqslant u \leqslant t), \quad t \geqslant 0,$$

i.e. $\mathcal{G}_t$ contains the additional information on whether default at time τ has occurred or not before time t.

In Table 10.2, bond ratings are determined according to hazard (or failure) rate thresholds.

Bond Credit Ratings	Moody's		S & P	
	Municipal	Corporate	Municipal	Corporate
Aaa/AAAs	0.00	0.52	0.00	0.60
Aa/AA	0.06	0.52	0.00	1.50
A/A	0.03	1.29	0.23	2.91
Baa/BBB	0.13	4.64	0.32	10.29
Ba/BB	2.65	19.12	1.74	29.93
B/B	11.86	43.34	8.48	53.72
Caa-C/CCC-C	16.58	69.18	44.81	69.19
Investment Grade	0.07	2.09	0.20	4.14
Non-Invest. Grade	4.29	31.37	7.37	42.35
All	0.10	9.70	0.29	12.98

Table 10.2: Cumulative historic default rates (in percentage).

10.5   Credit Default Swaps

In this section we consider the credit default options (swaps) that can act as a protection against default. According to the Bank for International Settlements, the outstanding notional amounts of Credit Default Swap (CDS) contracts has decreased from \$61.2 trillion at year-end 2007 to \$7.6 trillion at year-end 2019.

We work with a tenor structure $\{t = T_i < \cdots < T_j = T\}$. Here, τ is a default time and the filtration $\mathcal{G}_t := \mathcal{F}_t \vee \sigma(\tau)$ contains the additional information on τ, as defined in (10.3).

A Credit Default Swap (CDS) is a contract consisting in

- *A premium leg:* the buyer is purchasing protection at time t against default at time T_k, $k = i+1, \ldots, j$, and has to make a fixed spread payment $S_t^{i,j}$ at times $T_{i+1}, \ldots, T_j$ between t and T in compensation.

  The discounted value at time t of the premium leg is

$$V^p(t,T) = \mathbb{E}\left[\sum_{k=i}^{j-1} S_t^{i,j} \delta_k \mathbb{1}_{\{\tau > T_{k+1}\}} e^{-\int_t^{T_{k+1}} r_s ds} \,\middle|\, \mathcal{G}_t\right]$$

$$= \sum_{k=i}^{j-1} S_t^{i,j} \delta_k \, \mathbb{E}\left[\mathbb{1}_{\{\tau > T_{k+1}\}} e^{-\int_t^{T_{k+1}} r_s ds} \,\middle|\, \mathcal{G}_t\right]$$

$$= S_t^{i,j} \sum_{k=i}^{j-1} \delta_k P(t, T_{k+1})$$

$$= S_t^{i,j} P(t, T_i, T_j), \tag{10.7}$$

where $\delta_k = T_{k+1} - T_k$,

$$P(t, T_k) = \mathbb{1}_{\{\tau > t\}} \mathbb{E}\left[e^{-\int_t^{T_k} (r_s + \lambda_s) ds} \,\middle|\, \mathcal{F}_t\right], \quad 0 \leqslant t \leqslant T_k,$$

is the defaultable bond price with maturity T_k, $k = i, \ldots, j-1$, see Lemma 10.1, and

$$P(t, T_i, T_j) = \sum_{k=i}^{j-1} \delta_k P(t, T_{k+1})$$

is the (default) annuity numeraire, cf. *e.g.* Relation (12.10) in Privault (2014).

For simplicity, we have ignored a possible accrual interest term over the time interval $[T_k, \tau]$ when $\tau \in [T_k, T_{k+1}]$ in the above value (10.7) of the premium leg.

- **A protection leg:** the seller or issuer of the contract makes a compensation payment $1 - \xi_{k+1}$ to the buyer in case default occurs at time T_{k+1}, $k = i, \ldots, j-1$, where ξ_{k+1} is the *recovery rate*.

The value at time t of the protection leg is

$$V^d(t, T) := \mathbb{E}\left[\sum_{k=i}^{j-1} \mathbb{1}_{(T_k, T_{k+1}]}(\tau)(1 - \xi_{k+1}) e^{-\int_t^{T_{k+1}} r_s ds} \,\middle|\, \mathcal{G}_t \right], \quad (10.8)$$

where ξ_{k+1} is the recovery rate associated with the maturity T_{k+1}, $k = i, \ldots, j-1$.

In the case of a non-random recovery rate ξ_k, the value of the protection leg becomes

$$\sum_{k=i}^{j-1} (1 - \xi_{k+1}) \, \mathbb{E}\left[\mathbb{1}_{(T_k, T_{k+1}]}(\tau) e^{-\int_t^{T_{k+1}} r_s ds} \,\middle|\, \mathcal{G}_t \right].$$

The spread $S_t^{i,j}$ is computed by equating the values of the premium (10.7) and protection (10.8) legs, *i.e.* from the relation

$$V^p(t, T) = S_t^{i,j} P(t, T_i, T_j)$$

$$= \mathbb{E}\left[\sum_{k=i}^{j-1} \mathbb{1}_{(T_k, T_{k+1}]}(\tau)(1 - \xi_{k+1}) e^{-\int_t^{T_{k+1}} r_s ds} \,\middle|\, \mathcal{G}_t \right],$$

which yields

$$S_t^{i,j} = \frac{1}{P(t, T_i, T_j)} \sum_{k=i}^{j-1} \mathbb{E}\left[\mathbb{1}_{(T_k, T_{k+1}]}(\tau)(1 - \xi_{k+1}) e^{-\int_t^{T_{k+1}} r_s ds} \,\middle|\, \mathcal{G}_t \right].$$

$$(10.9)$$

The spread $S_t^{i,j}$, which is quoted in basis points per year and paid at regular time intervals, gives protection against defaults on payments of \$1. For a notional amount N the premium payment will become $N \times S_t^{i,j}$.

In the case of a constant recovery rate $\xi \in [0, 1]$, we find

$$S_t^{i,j} = \frac{1 - \xi}{P(t, T_i, T_j)} \sum_{k=i}^{j-1} \mathbb{E}\left[\mathbb{1}_{(T_k, T_{k+1}]}(\tau) e^{-\int_t^{T_{k+1}} r_s ds} \,\middle|\, \mathcal{G}_t \right],$$

and if τ is constrained to take values in the tenor structure $\{T_i, \ldots, T_j\}$, we get

$$S_t^{i,j} = \frac{1-\xi}{P(t,T_i,T_j)} \, \mathbb{E}\left[\mathbb{1}_{(t,T]}(\tau) \, e^{-\int_t^\tau r_s ds} \,\Big|\, \mathcal{G}_t \right].$$

The buyer of a Credit Default Swap (CDS) is purchasing protection at time t against default at time T_k, $k = i+1, \ldots, j$, by making a fixed payment $S_t^{i,j}$ (the premium leg) at times $T_{i+1}, \ldots, T_j$. On the other hand, the issuer of the contract makes a payment $1 - \xi_{k+1}$ to the buyer in case default occurs at time T_{k+1}, $k = i, \ldots, j-1$.

The contract is priced in terms of the swap rate $S_t^{i,j}$ (or spread) computed by equating the values $V^d(t,T)$ and $V^p(t,T)$ of the protection and premium legs, and acts as a compensation that makes the deal fair to both parties. Recall that from (10.9) and Lemma 10.1, we have

$$S_t^{i,j} = \frac{1}{P(t,T_i,T_j)} \sum_{k=i}^{j-1} \mathbb{E}\left[\mathbb{1}_{(T_k,T_{k+1}]}(\tau)(1-\xi_{k+1}) \, e^{-\int_t^{T_{k+1}} r_s ds} \,\Big|\, \mathcal{G}_t \right]$$

$$= \frac{1}{P(t,T_i,T_j)} \sum_{k=i}^{j-1} \mathbb{E}\left[(\mathbb{1}_{\{T_k<\tau\}} - \mathbb{1}_{\{T_{k+1}<\tau\}})(1-\xi_{k+1}) \, e^{-\int_t^{T_{k+1}} r_s ds} \,\Big|\, \mathcal{G}_t \right]$$

$$= \frac{\mathbb{1}_{\{\tau>t\}}}{P(t,T_i,T_j)}$$

$$\times \sum_{k=i}^{j-1} \mathbb{E}\left[(1-\xi_{k+1}) \left(e^{-\int_t^{T_k} \lambda_s ds} - e^{-\int_t^{T_{k+1}} \lambda_s ds} \right) e^{-\int_t^{T_{k+1}} r_s ds} \,\Big|\, \mathcal{F}_t \right].$$

Estimating a deterministic failure rate

In case the rates $r(s)$, $\lambda(s)$ and the recovery rate ξ_{k+1} are deterministic, the above spread can be written as

$$S_t^{i,j} P(t,T_i,T_j)$$

$$= \mathbb{1}_{\{\tau>t\}} \sum_{k=i}^{j-1} (1-\xi_k) \, e^{-\int_t^{T_{k+1}} r(s)ds} \left(e^{-\int_t^{T_k} \lambda_s ds} - e^{-\int_t^{T_{k+1}} \lambda_s ds} \right).$$

Given that

$$P(t,T_i,T_j) = \sum_{k=i}^{j-1} (T_{k+1} - T_k) P(t,T_{k+1}), \qquad 0 \leqslant t \leqslant T_{i+1},$$

we can write

$$S_t^{i,j} \sum_{k=i}^{j-1} (T_{k+1} - T_k) \, e^{- \int_t^{T_{k+1}} (r(s)+\lambda(s))ds}$$

$$= \mathbb{1}_{\{\tau>t\}} \sum_{k=i}^{j-1} (1 - \xi_k) \, e^{- \int_t^{T_{k+1}} r(s)ds} \left(e^{- \int_t^{T_k} \lambda(s)ds} - e^{- \int_t^{T_{k+1}} \lambda(s)ds} \right).$$

In particular, when $r(t)$ and $\lambda(t)$ are written as in (10.6) and assuming that $\xi_k = \xi$, $k = i, \ldots, j$, we get, with $t = T_i$ and writing $\delta_k = T_{k+1} - T_k$, $k = i, \ldots, j-1$,

$$S_{T_i}^{i,j} \sum_{k=i}^{j-1} \delta_k \exp\left(- \sum_{p=i}^{k} \delta_p(r_p + \lambda_p) \right)$$

$$= \mathbb{1}_{\{\tau>t\}} (1 - \xi) \sum_{k=i}^{j-1} \exp\left(- \sum_{p=i}^{k} \delta_p(r_p + \lambda_p) \right) \left(e^{\delta_k \lambda_k} - 1 \right).$$

Assuming further that $\lambda_k = \lambda$, $k = i, \ldots, j$, we have

$$S_{T_i}^{i,j} \sum_{k=i}^{j-1} \delta_k \exp\left(- \sum_{p=i}^{k} \delta_p r_p - \lambda \sum_{p=i}^{k} \delta_p \right) \tag{10.10}$$

$$= (1 - \xi) \sum_{k=i}^{j-1} \left(e^{\delta_k \lambda_k} - 1 \right) \exp\left(- \sum_{p=i}^{k} \delta_p r_p - \lambda \sum_{p=i}^{k} \delta_p \right),$$

which can be solved numerically for λ, cf. Sections 4 and 5 of Castellacci (2008) and Exercise 10.3. Note that, as λ tends to ∞, the ratio

$$\frac{S_{T_i}^{i,j} \sum_{k=i}^{j-1} \delta_k \exp\left(- \sum_{p=i}^{k} \delta_p r_p - \lambda \sum_{p=i}^{k} \delta_p \right)}{(1 - \xi) \sum_{k=i}^{j-1} \left(e^{\delta_k \lambda_k} - 1 \right) \exp\left(- \sum_{p=i}^{k} \delta_p r_p - \lambda \sum_{p=i}^{k} \delta_p \right)}$$

converges to 0, while it tends to $+\infty$ as λ goes to 0. Therefore, the equation (10.10) admits a numerical solution.

Exercises

Exercise 10.1. Consider a standard zero coupon bond with constant yield $r > 0$ and a defaultable (risky) bond with constant yield r_d and default probability $\alpha \in (0, 1)$. Find a relation between r, r_d, α and the bond maturity T.

Exercise 10.2. A standard zero coupon bond with constant yield $r > 0$ and maturity T is priced $P(t, T) = e^{-(T-t)r}$ at time $t \in [0, T]$. Assume that the company can get bankrupt at a random time $t + \tau$, and default on its final \$1 payment if $\tau < T - t$.

(1) Explain why the defaultable bond price $P_d(t, T)$ can be expressed as

$$P_d(t, T) = e^{-(T-t)r} \, \mathbb{E} \left[\mathbb{1}_{\{\tau > T-t\}} \right]. \tag{10.11}$$

(2) Assuming that the default time τ is exponentially distributed with parameter $\lambda > 0$, compute the default bond price $P_d(t, T)$ using (10.11).
(3) Find a formula that can estimate the parameter λ from the risk-free rate r and the market data $P_M(t, T)$ of the defaultable bond price at time $t \in [0, T]$.

Exercise 10.3. Credit default swaps. Estimate the first default rate λ_1 and the associated default probability in the framework of (10.10), based on the following CDS market data from the McDonald's Corp, with spread $S_{T_i} = 0.10790\%$, $t =$ Apr 12, 2015, $T_i =$ Mar 20, 2015, and the recovery rate $\rho = 40\%$.

Exercise 10.4. Credit default swaps. We work with a tenor structure $\{T_i < \cdots < T_j = T\}$. Let

$$\sum_{k=i}^{j-1} \mathbb{E} \left[\mathbb{1}_{(T_k, T_{k+1}]}(\tau)(1 - \xi_{k+1}) e^{-\int_t^{T_{k+1}} r_s \, ds} \, \Big| \, \mathcal{G}_t \right]$$

$$= \sum_{k=i}^{j-1} \mathbb{E} \left[(\mathbb{1}_{\{T_k < \tau\}} - \mathbb{1}_{\{T_{k+1} < \tau\}})(1 - \xi_{k+1}) e^{-\int_t^{T_{k+1}} r_s \, ds} \, \Big| \, \mathcal{G}_t \right]$$

$$= \mathbb{1}_{\{\tau > t\}} \sum_{k=i}^{j-1} \mathbb{E} \left[(1 - \xi_{k+1}) \left(e^{-\int_t^{T_k} \lambda_s \, ds} - e^{-\int_t^{T_{k+1}} \lambda_s \, ds} \right) e^{-\int_t^{T_{k+1}} r_s \, ds} \, \Big| \, \mathcal{F}_t \right]$$

$$= \mathbb{1}_{\{\tau > t\}} (1 - \xi) \sum_{k=i}^{j-1} e^{-\int_t^{T_{k+1}} r_s \, ds} \, \mathbb{E} \left[e^{-\int_t^{T_k} \lambda_s \, ds} - e^{-\int_t^{T_{k+1}} \lambda_s \, ds} \, \Big| \, \mathcal{F}_t \right]$$

$$= \mathbb{1}_{\{\tau > t\}} (1 - \xi) \sum_{k=i}^{j-1} P(t, T_{k+1}) \left(Q(t, T_k) - Q(t, T_{k+1}) \right),$$

denote the discounted value at time t of the protection leg, where

$$P(t, T_k) = e^{-\int_t^{T_k} r_s ds} = e^{-r_k(T_k - t)}, \qquad k = i, \ldots, j,$$

is a deterministic discount factor, and

$$Q(t, T_k) = \mathbb{E}\left[e^{-\int_t^{T_k} \lambda_s ds} \,\middle|\, \mathcal{F}_t\right]$$

is the survival probability. Let

$$V(t, T) = S_t^{i,j} \sum_{k=i}^{j-1} \delta_k \, \mathbb{E}\left[\mathbb{1}_{\{\tau > T_{k+1}\}} e^{-\int_t^{T_{k+1}} r_s ds} \,\middle|\, \mathcal{G}_t\right]$$

$$= S_t^{i,j} \sum_{k=i}^{j-1} \delta_k \, \mathbb{E}\left[\mathbb{1}_{\{T_{k+1} < \tau\}} e^{-\int_t^{T_{k+1}} r_s ds} \,\middle|\, \mathcal{G}_t\right]$$

$$= \mathbb{1}_{\{\tau > t\}} S_t^{i,j} \sum_{k=i}^{j-1} \delta_k \, \mathbb{E}\left[e^{-\int_t^{T_{k+1}} \lambda_s ds} e^{-\int_t^{T_{k+1}} r_s ds} \,\middle|\, \mathcal{F}_t\right]$$

$$= S_t^{i,j} \mathbb{1}_{\{\tau > t\}} \sum_{k=i}^{j-1} \delta_k e^{-\int_t^{T_{k+1}} r_s ds} \, \mathbb{E}\left[e^{-\int_t^{T_{k+1}} \lambda_s ds} \,\middle|\, \mathcal{F}_t\right]$$

$$= \mathbb{1}_{\{\tau > t\}} S_t^{i,j} \sum_{k=i}^{j-1} \delta_k P(t, T_{k+1}) Q(t, T_{k+1}),$$

denote the discounted value at time t of the premium leg, where $\delta_k :=$ $T_{k+1} - T_k$, $k = i, \ldots, j - 1$.

(1) By equating the protection and premium legs, find the value of $Q(t, T_{i+1})$ with $Q(t, T_i) = 1$, and derive a recurrence relation between $Q(t, T_{j+1})$ and $Q(t, T_i), \ldots, Q(t, T_j)$.

(2) From the following data on the Coca-Cola Company compute the discount factors $P(t, T_i), \ldots, P(t, T_n)$, and estimate the corresponding survival probabilities $Q(t, T_i), \ldots, Q(t, T_n)$.

k	Maturity	T_k	$S_t^{1,k}$ (bp)
1	6M	0.5	10.97
2	1Y	1	12.25
3	2Y	2	14.32
4	3Y	3	19.91
5	4Y	4	26.48
6	5Y	5	33.29
7	7Y	7	52.91
8	10Y	10	71.91

Table 10.3: Spread market data.

Problem 10.5. Defaultable bonds. Consider a (random) default time τ with cumulative distribution function

$$\mathbb{P}(\tau > t \mid \mathcal{F}_t) = e^{-\int_0^t \lambda_u du}, \qquad t \geqslant 0,$$

where λ_t is a (random) default rate process which is adapted to the filtration $(\mathcal{F}_t)_{t \in \mathbb{R}_+}$. Recall that the probability of survival up to time T, given the information known up to time t, is given by

$$\mathbb{P}(\tau > T \mid \mathcal{G}_t) = \mathbb{1}_{\{\tau > t\}} \mathbb{E}\left[e^{-\int_t^T \lambda_u du} \,\middle|\, \mathcal{F}_t \right],$$

where $\mathcal{G}_t = \mathcal{F}_t \vee \sigma(\{\tau < u\} : 0 \leqslant u \leqslant t)$, $t \in \mathbb{R}_+$, is the filtration defined by adding the default time information to the history $(\mathcal{F}_t)_{t \in \mathbb{R}_+}$. In this framework, the price $P(t, T)$ of defaultable bond with maturity T, short-term interest rate r_t and (random) default time τ is given by

$$P(t, T) = \mathbb{E}\left[\mathbb{1}_{\{\tau > T\}} e^{-\int_t^T r_u du} \,\middle|\, \mathcal{G}_t \right] \tag{10.12}$$

$$= \mathbb{1}_{\{\tau > t\}} \mathbb{E}\left[e^{-\int_t^T (r_u + \lambda_u) du} \,\middle|\, \mathcal{F}_t \right].$$

In the sequel we assume that the processes $(r_t)_{t \in \mathbb{R}_+}$ and $(\lambda_t)_{t \in \mathbb{R}_+}$ are modeled according to the Vasicek processes

$$\begin{cases} dr_t = -ar_t dt + \sigma dB_t^{(1)}, \\[2mm] d\lambda_t = -b\lambda_t dt + \eta dB_t^{(2)}, \end{cases}$$

where $\left(B_t^{(1)}\right)_{t \in \mathbb{R}_+}$ and $\left(B_t^{(2)}\right)_{t \in \mathbb{R}_+}$ are two standard $\mathcal{F}_t$-Brownian motions with correlation $\rho \in [-1, 1]$, and $dB_t^{(1)} \cdot dB_t^{(2)} = \rho dt$.

(1) Give a justification for the fact that
$$\mathbb{E}\left[e^{-\int_t^T (r_u + \lambda_u)du} \,\Big|\, \mathcal{F}_t\right]$$
can be written as a function $F(t, r_t, \lambda_t)$ of t, r_t and λ_t, $t \in [0, T]$.

(2) Show that
$$t \mapsto e^{-\int_0^t (r_s + \lambda_s)ds}\, \mathbb{E}\left[e^{-\int_t^T (r_u + \lambda_u)du} \,\Big|\, \mathcal{F}_t\right]$$
is an $(\mathcal{F}_t)_{t\in[0,T]}$-martingale under $\mathbb{P}$.

(3) Use the Itô formula with two variables to derive a PDE on $\mathbb{R}^2$ for the function $F(t, x, y)$.

(4) Show that we have
$$\int_t^T r_s ds = C(a, t, T)r_t + \sigma \int_t^T C(a, s, T)dB_s^{(1)},$$
and
$$\int_t^T \lambda_s ds = C(b, t, T)\lambda_t + \eta \int_t^T C(b, s, T)dB_s^{(2)},$$
where
$$C(a, t, T) = -\frac{1}{a}(e^{-a(T-t)} - 1).$$

(5) Show that the random variable
$$\int_t^T r_s ds + \int_t^T \lambda_s ds$$
has a Gaussian distribution, and compute its conditional mean
$$\mathbb{E}\left[\int_t^T r_s ds + \int_t^T \lambda_s ds \,\Big|\, \mathcal{F}_t\right]$$
and conditional variance
$$\mathrm{Var}\left[\int_t^T r_s ds + \int_t^T \lambda_s ds \,\Big|\, \mathcal{F}_t\right],$$
conditionally to $\mathcal{F}_t$.

(6) Compute $P(t, T)$ from its expression (10.12) as a conditional expectation.

(7) Show that the solution $F(t, x, y)$ to the 2-dimensional PDE of Question (3) is
$$F(t, x, y) = e^{-C(a,t,T)x - C(b,t,T)y}$$
$$\times \exp\left(\frac{\sigma^2}{2}\int_t^T C^2(a, s, T)ds + \frac{\eta^2}{2}\int_t^T C^2(b, s, T)ds\right)$$
$$\times \exp\left(\rho\sigma\eta\int_t^T C(a, s, T)C(b, s, T)ds\right).$$

(8) Show that the defaultable bond price $P(t,T)$ can also be written as

$$P(t,T) = e^{U(t,T)} \mathbb{P}(\tau > T \mid \mathcal{G}_t) \, \mathbb{E}\left[e^{-\int_t^T r_s ds} \,\Big|\, \mathcal{F}_t \right],$$

where

$$U(t,T) = \rho \frac{\sigma \eta}{ab} \left(T - t - C(a,t,T) - C(b,t,T) + C(a+b,t,T) \right).$$

(9) By partial differentiation of $\log P(t,T)$ with respect to T, compute the corresponding instantaneous short rate

$$f(t,T) = -\frac{\partial}{\partial T} \log P(t,T).$$

(10) Show that $\mathbb{P}(\tau > T \mid \mathcal{G}_t)$ can be written using an HJM-type default rate as

$$\mathbb{P}(\tau > T \mid \mathcal{G}_t) = \mathbb{1}_{\{\tau > t\}} \, e^{-\int_t^T f_2(t,u) du},$$

where

$$f_2(t,u) = \lambda_t \, e^{-b(u-t)} - \frac{\eta^2}{2} C^2(b,t,u).$$

(11) Show how the result of Question (8) can be simplified when $\left(B_t^{(1)}\right)_{t \in \mathbb{R}_+}$ and $\left(B_t^{(2)}\right)_{t \in \mathbb{R}_+}$ are independent.

Chapter 11

Appendix: Mathematical Tools

This appendix surveys some basic results in probability and measure theory used in the lectures. It does not aim for completeness and the reader is referred to standard texts in probability such as Jacod and Protter (2000), Protter (2004) for more details.

In the sequel we work on a probability space $(\Omega, \mathcal{F}, \mathbb{P})$, and we denote by $\mathbb{E}$ the expectation under $\mathbb{P}$.

Measurability

Given a sequence $(Y_n)_{n \geqslant 1}$ of random variables, a random variable F is said to be $\mathcal{F}_n$-measurable if it can be written as a function

$$F = f_n(Y_1, \ldots, Y_n)$$

of $Y_1, \ldots, Y_n$, where $f_n : \mathbb{R}^n \to \mathbb{R}$ is (Borel) measurable. This defines the natural filtration $(\mathcal{F}_n)_{n \geqslant 0}$ generated by $(Y_k)_{k \geqslant 1}$, as

$$\mathcal{F}_n = \sigma(Y_1, \ldots, Y_n), \qquad n \geqslant 1,$$

and $\mathcal{F}_0 = \{\emptyset, \Omega\}$, where $\sigma(Y_1, \ldots, Y_n)$ is the smallest σ-algebra making $Y_1, \ldots, Y_n$ measurable.

A random variable X is said to be *integrable* if $\mathbb{E}[|X|] < \infty$.

Covariance and Correlation

The *covariance* of two random variables X and Y is defined as

$$\mathrm{Cov}(X, Y) = \mathbb{E}[(X - \mathbb{E}[X])(Y - \mathbb{E}[Y])],$$

with

$$\begin{aligned}
\text{Cov}(X,X) &= \mathbb{E}\left[(X - \mathbb{E}[X])^2\right] \\
&= \mathbb{E}[X^2] - (\mathbb{E}[X])^2 \\
&= \text{Var}[X].
\end{aligned}$$

Moreover, for all $\alpha \in \mathbb{R}$ the variance satisfies the relation

$$\begin{aligned}
\text{Var}[\alpha X] &= \mathbb{E}\left[(\alpha X - \mathbb{E}[\alpha X])^2\right] \\
&= \mathbb{E}\left[(\alpha X - \alpha\,\mathbb{E}[X])^2\right] \\
&= \mathbb{E}\left[\alpha(X - \mathbb{E}[X])^2\right] \\
&= \alpha^2\,\mathbb{E}\left[(X - \mathbb{F}[X])^2\right] \\
&= \alpha^2\,\text{Var}[X]. \tag{11.1}
\end{aligned}$$

The *correlation* of X and Y is the coefficient

$$c(X,Y) := \frac{\text{Cov}(X,Y)}{\sqrt{\text{Var}[X]}\sqrt{\text{Var}[Y]}}.$$

Clearly we have $c(X,Y) = 0$ when X is independent of Y, and $c(X,Y) = 1$ when $X = Y$. The conditional variance and covariance given a σ-algebra $\mathcal{G}$ can be defined similarly, with

$$\text{Cov}(X,Y \mid \mathcal{G}) = \mathbb{E}[(X - \mathbb{E}[X \mid \mathcal{G}])(Y - \mathbb{E}[Y \mid \mathcal{G}]) \mid \mathcal{G}],$$

and

$$\text{Var}[X \mid \mathcal{G}] = \text{Cov}(X,X \mid \mathcal{G}) = \mathbb{E}\left[X^2 \mid \mathcal{G}\right] - (\mathbb{E}[X \mid \mathcal{G}])^2.$$

Note that if Y is square-integrable and $\mathcal{G}$-measurable we have the relation

$$\begin{aligned}
\text{Var}[X + Y \mid \mathcal{G}] &= \mathbb{E}\left[(X + Y - \mathbb{E}[X + Y \mid \mathcal{G}])^2 \mid \mathcal{G}\right] \\
&= \mathbb{E}\left[(X - \mathbb{E}[X \mid \mathcal{G}])^2 \mid \mathcal{G}\right] \\
&= \text{Var}[X \mid \mathcal{G}].
\end{aligned}$$

An integrable random variable X is said to be *centered* if $\mathbb{E}[X] = 0$.

Gaussian Random Variables

A random variable X has a Gaussian distribution with mean μ and variance σ^2 if and only if its characteristic function satisfies

$$\mathbb{E}\left[e^{i\alpha X}\right] = e^{i\alpha\mu - \alpha^2\sigma^2/2}, \qquad \alpha \in \mathbb{R},$$

i.e., in terms of Laplace transform,

$$\mathbb{E}\left[e^{\alpha X}\right] = e^{\alpha\mu + \alpha^2\sigma^2/2}, \qquad \alpha \in \mathbb{R}. \tag{11.2}$$

More generally, a family $(X_1, X_2, \ldots, X_d)$ of $d \geqslant 2$ random variables is said to be jointly Gaussian (or multivariate normal) with mean $\mu \in \mathbb{R}^d$ and covariance matrix Σ if its characteristic function satisfies

$$\mathbb{E}\left[e^{i\langle X, v\rangle_{\mathbb{R}^d}}\right] = \exp\left(i\langle \mu, v\rangle_{\mathbb{R}^d} - \frac{1}{2}\langle \Sigma v, v\rangle_{\mathbb{R}^d}\right), \quad v \in \mathbb{R}^d. \qquad (11.3)$$

From *e.g.* Corollary 16.1 of Jacod and Protter (2000), we have the following result.

Proposition 11.1. *Let $(X_1, \ldots, X_d)$ be a jointly Gaussian vector of centered Gaussian variables which are assumed to be orthogonal to each other, i.e.*

$$\mathbb{E}[X_i X_j] = 0, \qquad 1 \leqslant i \neq j \leqslant d.$$

Then the random variables $X_1, \ldots, X_d$ are independent.

Finally, if $X_1, \ldots, X_n$ are independent Gaussian random variables with probability distributions $\mathcal{N}(m_1, \sigma_1^2), \ldots, \mathcal{N}(m_n, \sigma_n^2)$ then the sum $X_1 + \cdots + X_n$ is a Gaussian random variable with probability distribution

$$\mathcal{N}(m_1 + \cdots + m_n, \sigma_1^2 + \cdots + \sigma_n^2).$$

Conditional Expectation

Consider $\mathcal{G}$ a sub σ-algebra of $\mathcal{F}$. The conditional expectation $\mathbb{E}[F \mid \mathcal{G}]$ of $F \in L^2(\Omega, \mathcal{F}, \mathbb{P})$ given $\mathcal{G}$ can be defined as the orthogonal projection of F on $L^2(\Omega, \mathcal{G}, \mathbb{P})$ for the inner product $\langle F, G \rangle := \mathbb{E}[FG]$. In other words, $\mathbb{E}[F \mid \mathcal{G}]$ is the only $\mathcal{G}$-measurable random variable that satisfies

$$\mathbb{E}[G(F - \mathbb{E}[F \mid \mathcal{G}])] = 0,$$

or

$$\mathbb{E}[GF] = \mathbb{E}[G\,\mathbb{E}[F \mid \mathcal{G}]], \qquad (11.4)$$

for all bounded and $\mathcal{G}$-measurable random variables G.

The conditional expectation has the following properties

a) $\mathbb{E}[F \mid \mathcal{G}] = \mathbb{E}[F]$ if F is independent of $\mathcal{G}$.

b) $\mathbb{E}[\mathbb{E}[F \mid \mathcal{F}] \mid \mathcal{G}] = \mathbb{E}[F \mid \mathcal{G}]$ if $\mathcal{G} \subset \mathcal{F}$ (Tower Property).

c) $\mathbb{E}[GF \mid \mathcal{G}] = G\,\mathbb{E}[F \mid \mathcal{G}]$ if G is $\mathcal{G}$-measurable and sufficiently integrable.

d) We have

$$\mathbb{E}[f(X,Y) \mid \mathcal{F}] = \mathbb{E}[f(X,y)]_{y=Y} \tag{11.5}$$

if X, Y are independent and Y is $\mathcal{F}$-measurable.

Martingales in Discrete Time

Consider $(\mathcal{F}_n)_{n \geqslant 0}$ an increasing family of sub σ-algebra of $\mathcal{F}$. A discrete-time martingale in L^2 with respect to $(\mathcal{F}_n)_{n \geqslant 0}$ is a family $(M_n)_{n \geqslant 0}$ of random variables such that

i) $M_n \in L^2(\Omega, \mathcal{F}_n, \mathbb{P})$, $n \in \mathbb{N}$,

ii) $\mathbb{E}[M_{n+1} \mid \mathcal{F}_n] = M_n$, $n \in \mathbb{N}$.

As an example, the process $(Z_n)_{n \geqslant 0}$ defined by $Z_0 := 0$ and

$$Z_n = Y_1 + \cdots + Y_n, \qquad n \geqslant 1,$$

whose sequence $(Y_n)_{n \geqslant 1}$ of increments satisfies

$$\mathbb{E}[Y_n \mid \mathcal{F}_{n-1}] = 0, \qquad n \geqslant 1, \tag{11.6}$$

is a martingale with respect to its own filtration defined as

$$\mathcal{F}_0 = \{\emptyset, \Omega\}$$

and

$$\mathcal{F}_n = \sigma(Y_1, \ldots, Y_n), \qquad n \geqslant 1.$$

In particular, Condition (11.6) is satisfied when the increments $(Y_n)_{n \geqslant 1}$ are independent centered random variables.

Martingales in Continuous Time

Let $(\mathcal{F}_t)_{t \in \mathbb{R}_+}$ denote a continuous-time filtration, *i.e.* an increasing family of sub σ-algebras of $\mathcal{F}$. We assume that $(\mathcal{F}_t)_{t \in \mathbb{R}_+}$ is continuous on the right, *i.e.*

$$\mathcal{F}_t = \bigcap_{s>t} \mathcal{F}_s, \qquad t \in \mathbb{R}_+.$$

By a stochastic process we mean a family of random variables indexed by a time interval.

Definition 11.1. *A stochastic process* $(M_t)_{t \in \mathbb{R}_+}$ *such that* $\mathbb{E}[|M_t|] < \infty$, $t \in \mathbb{R}_+$, *is called an* $(\mathcal{F}_t)_{t \in \mathbb{R}_+}$-*martingale if*

$$\mathbb{E}[M_t \mid \mathcal{F}_s] = M_s, \qquad 0 \leqslant s \leqslant t.$$

A process $(X_t)_{t \in \mathbb{R}_+}$ is said to have independent increments if $X_t - X_s$ is independent of $\sigma(X_u \; : \; 0 \leqslant u \leqslant s)$, $0 \leqslant s < t$.

Proposition 11.2. *Every integrable process* $(X_t)_{t \in \mathbb{R}_+}$ *with centered independent increments is a martingale with respect to the filtration*

$$\mathcal{F}_t := \sigma(X_u \; : \; u \leqslant t), \qquad t \geqslant 0,$$

it generates, called the natural filtration.

Note that for all square-integrable random variable F the process $(\mathbb{E}[F \mid \mathcal{F}_t])_{t \in \mathbb{R}_+}$ is a martingale, due to the relation

$$\mathbb{E}[\mathbb{E}[F \mid \mathcal{F}_t] \mid \mathcal{F}_s] = \mathbb{E}[F \mid \mathcal{F}_s], \qquad 0 \leqslant s \leqslant t, \tag{11.7}$$

that follows from the Tower Property (*b*) of conditional expectation.

Markov Processes

Let $\mathcal{C}_0(\mathbb{R}^n)$ denote the class of continuous functions tending to 0 at infinity. Recall that f is said to tend to 0 at infinity if for all $\varepsilon > 0$ there exists a compact subset K of $\mathbb{R}^n$ such that $|f(x)| \leqslant \varepsilon$ for all $x \in \mathbb{R}^n \setminus K$.

Definition 11.2. *An* $\mathbb{R}^n$-*valued stochastic process, i.e. a family* $(X_t)_{t \in \mathbb{R}_+}$ *of* $\mathbb{R}^n$-*valued random variables on* $(\Omega, \mathcal{F}, \mathbb{P})$, *is a Markov process if for all* $t \in \mathbb{R}_+$ *the* σ-*fields*

$$\mathcal{F}_t^+ := \sigma(X_s \; : \; s \geqslant t) \quad and \quad \mathcal{F}_t := \sigma(X_s \; : \; 0 \leqslant s \leqslant t)$$

are conditionally independent given X_t, $t \in \mathbb{R}_+$.

This condition can be restated by saying that for all $A \in \mathcal{F}_t^+$ and $B \in \mathcal{F}_t$ we have

$$\mathbb{P}(A \cap B \mid X_t) = \mathbb{P}(A \mid X_t)\mathbb{P}(B \mid X_t),$$

cf. Chung (2002). This definition naturally entails that:

i) $(X_t)_{t \in \mathbb{R}_+}$ is adapted with respect to $(\mathcal{F}_t)_{t \in \mathbb{R}_+}$, *i.e.* X_t is $\mathcal{F}_t$-measurable, $t \in \mathbb{R}_+$, and

ii) X_u is conditionally independent of $\mathcal{F}_t$ given X_t, for all $u \geqslant t$, *i.e.*

$$\mathbb{E}[f(X_u) \mid \mathcal{F}_t] = \mathbb{E}[f(X_u) \mid X_t], \qquad 0 \leqslant t \leqslant u,$$

for any bounded measurable function f on $\mathbb{R}^n$.

In particular,

$$\mathbb{P}(X_u \in A \mid \mathcal{F}_t) = \mathbb{E}[\mathbf{1}_A(X_u) \mid \mathcal{F}_t] = \mathbb{E}[\mathbf{1}_A(X_u) \mid X_t] = \mathbb{P}(X_u \in A \mid X_t),$$

for A in the Borel σ-algebra $\mathcal{B}(\mathbb{R}^n)$ on $\mathbb{R}^n$. Processes with independent increments provide simple examples of Markov processes. Indeed, for all bounded measurable functions f, g we have

$$
\begin{aligned}
&\mathbb{E}[f(X_{t_1}, \ldots, X_{t_n}) g(X_{s_1}, \ldots, X_{s_n}) \mid X_t] \\
&= \mathbb{E}[f(X_{t_1} - X_t + x, \ldots, X_{t_n} - X_t + x) \\
&\qquad g(X_{s_1} - X_t + x, \ldots, X_{s_n} - X_t + x)]_{x=X_t} \\
&= \mathbb{E}[f(X_{t_1} - X_t + x, \ldots, X_{t_n} - X_t + x)]_{x=X_t} \\
&\qquad \mathbb{E}[g(X_{s_1} - X_t + x, \ldots, X_{s_n} - X_t + x)]_{x=X_t} \\
&= \mathbb{E}[f(X_{t_1}, \ldots, X_{t_n}) \mid X_t] \, \mathbb{E}[g(X_{s_1}, \ldots, X_{s_n}) \mid X_t],
\end{aligned}
$$

$0 \leqslant s_1 < \cdots < s_n < t < t_1 < \cdots < t_n$.

A transition kernel is a mapping $\mathbb{P}(x, dy)$ such that

i) for every $x \in \mathbb{R}^n$, $A \mapsto \mathbb{P}(x, A)$ is a probability measure, and

ii) for every $A \in \mathcal{B}(\mathbb{R}^n)$, the mapping $x \mapsto \mathbb{P}(x, A)$ is a measurable function.

The transition kernel $\mu_{s,t}$ associated to $(X_t)_{t \in \mathbb{R}_+}$ is defined as

$$\mu_{s,t}(x, A) = \mathbb{P}(X_t \in A \mid X_s = x) \qquad 0 \leqslant s \leqslant t,$$

and we have

$$\mu_{s,t}(X_s, A) = \mathbb{P}(X_t \in A \mid X_s) = \mathbb{P}(X_t \in A \mid \mathcal{F}_s), \qquad 0 \leqslant s \leqslant t.$$

The transition operator $(P_{s,t})_{0 \leqslant s \leqslant t}$ associated to $(X_t)_{t \in \mathbb{R}_+}$ is defined as

$$P_{s,t} f(x) = \mathbb{E}[f(X_t) \mid X_s = x] = \int_{\mathbb{R}^n} f(y) \mu_{s,t}(x, dy), \qquad x \in \mathbb{R}^n.$$

Letting $p_{s,t}(x)$ denote the probability density of $X_t - X_s$ we have

$$\mu_{s,t}(x, A) = \int_A p_{s,t}(y - x)dy, \qquad A \in \mathcal{B}(\mathbb{R}^n),$$

and

$$P_{s,t}f(x) = \int_{\mathbb{R}^n} f(y)p_{s,t}(y - x)dy.$$

In the sequel we will assume that $(X_t)_{t\in\mathbb{R}_+}$ is time homogeneous, *i.e.* $\mu_{s,t}$ depends only on the difference $t - s$, and we will denote it by μ_{t-s}. In this case the family $(P_{0,t})_{t\in\mathbb{R}_+}$ is denoted by $(P_t)_{t\in\mathbb{R}_+}$ and defines a transition semigroup associated to $(X_t)_{t\in\mathbb{R}_+}$, with

$$P_t f(x) = \mathbb{E}[f(X_t) \mid X_0 = x] = \int_{\mathbb{R}^n} f(y)\mu_t(x, dy), \qquad x \in \mathbb{R}^n.$$

It satisfies the semigroup property

$$\begin{aligned}
P_t P_s f(x) &= \mathbb{E}[P_s f(X_t) \mid X_0 = x]\\
&= \mathbb{E}[\mathbb{E}[f(X_{t+s}) \mid X_s] \mid X_0 = x]]\\
&= \mathbb{E}[\mathbb{E}[f(X_{t+s}) \mid \mathcal{F}_s] \mid X_0 = x]]\\
&= \mathbb{E}[f(X_{t+s}) \mid X_0 = x]\\
&= P_{t+s}f(x),
\end{aligned}$$

which leads to the Chapman-Kolmogorov equation

$$\mu_{s+t}(x, A) = \mu_s * \mu_t(x, A) = \int_{\mathbb{R}^n} \mu_s(x, dy)\mu_t(y, A).$$

By induction we obtain

$$\mathbb{P}_x((X_{t_1}, \dots, X_{t_n}) \in B_1 \times \cdots \times B_n)$$
$$= \int_{B_1} \cdots \int_{B_n} \mu_{t_{n-1},t_n}(x_{n-1}, dx_n) \cdots \mu_{0,t_1}(x, dx_1),$$

for $0 < t_1 < \cdots < t_n$ and $B_1, \dots, B_n$ Borel subsets of $\mathbb{R}^n$.

If $(X_t)_{t\in\mathbb{R}_+}$ is a homogeneous Markov processes with independent increments then the probability density $p_t(x)$ of X_t satisfies the convolution property

$$p_{s+t}(x) = \int_{\mathbb{R}^n} p_s(y - x)p_t(y)dy, \qquad x \in \mathbb{R}^n,$$

which is satisfied in particular by processes with stationary and independent increments such as Lévy processes. A typical example of a probability density satisfying such a convolution property is the Gaussian probability density

$$p_t(x) = \frac{1}{(2\pi t)^{n/2}} \exp\left(-\frac{1}{2t}\|x\|_{\mathbb{R}^n}^2\right), \qquad x \in \mathbb{R}^n.$$

Chapter 12

Solutions to the Exercises

Chapter 1

Exercise 1.1. According to Definition 1.1, we need to check the following five properties of Brownian motion:

(i) starts at 0 at time 0,

(ii) independence of increments,

(iii) almost sure continuity of trajectories,

(iv) stationarity of the increments,

(v) Gaussianity of increments.

Checking conditions (i) to (iv) does not pose any particular problem since the time changes $t \mapsto c + t$ and $t \mapsto t/c^2$ are deterministic and continuous. Concerning (v), $B_{c+t} - B_c$ clearly has a centered Gaussian distribution with variance t, and the same property holds for cB_{t/c^2}, since

$$\mathrm{Var}[cB_{t/c^2}] = c^2 \, \mathrm{Var}[B_{t/c^2}] = c^2 t/c^2 = t, \qquad t \geqslant 0.$$

Exercise 1.2. We search for a solution of the form $S_t = f(t, B_t)$, where $f(t, x)$ is a function to be determined. By Itô's formula (1.10) we have

$$dS_t = df(t, B_t) = \frac{\partial f}{\partial t}(t, B_t)dt + \frac{\partial f}{\partial x}(t, B_t)dB_t + \frac{1}{2}\frac{\partial^2 f}{\partial x^2}(t, B_t)dt.$$

Comparing this expression to (1.13) and identifying the terms in dB_t we get

$$\begin{cases} \dfrac{\partial f}{\partial x}(t, B_t) = \sigma S_t, \\[3mm] \dfrac{\partial f}{\partial t}(t, B_t) + \dfrac{1}{2}\dfrac{\partial^2 f}{\partial x^2}(t, B_t) = \mu S_t. \end{cases}$$

Using the relation $S_t = f(t, B_t)$, these two equations will rewrite as

$$\begin{cases} \dfrac{\partial f}{\partial x}(t, B_t) = \sigma f(t, B_t), \\[3mm] \dfrac{\partial f}{\partial t}(t, B_t) + \dfrac{1}{2}\dfrac{\partial^2 f}{\partial x^2}(t, B_t) = \mu f(t, B_t). \end{cases}$$

Since B_t is a Gaussian random variable taking all possible values in $\mathbb{R}$, the equations should hold for all $x \in \mathbb{R}$, as follows:

$$\begin{cases} \dfrac{\partial f}{\partial x}(t, x) = \sigma f(t, x), & \text{(S.1a)} \\[3mm] \dfrac{\partial f}{\partial t}(t, x) + \dfrac{1}{2}\dfrac{\partial^2 f}{\partial x^2}(t, x) = \mu f(t, x). & \text{(S.1b)} \end{cases}$$

To find the solution $f(t, x) = f(t, 0)\, e^{\sigma x}$ of (S.1a) we let $g(t, x) = \log f(t, x)$ and rewrite (S.1a) as

$$\frac{\partial g}{\partial x}(t, x) = \frac{\partial}{\partial x}\log f(t, x) = \frac{1}{f(t, x)}\frac{\partial f}{\partial x}(t, x) = \sigma,$$

i.e.

$$\frac{\partial g}{\partial x}(t, x) = \sigma,$$

which is solved as

$$g(t, x) = g(t, 0) + \sigma x,$$

hence

$$f(t, x) = e^{g(t,0)}\, e^{\sigma x} = f(t, 0)\, e^{\sigma x}.$$

Plugging back this expression into the second equation (S.1b) yields

$$e^{\sigma x}\frac{\partial f}{\partial t}(t, 0) + \frac{\sigma^2}{2} e^{\sigma x} f(t, 0) = \mu f(t, 0)\, e^{\sigma x},$$

i.e.

$$\frac{\partial f}{\partial t}(t, 0) = \left(\mu - \frac{\sigma^2}{2}\right) f(t, 0).$$

In other words, we have $\frac{\partial g}{\partial t}(t, 0) = \mu - \sigma^2/2$, which yields

$$g(t, 0) = g(0, 0) + \left(\mu - \frac{\sigma^2}{2}\right) t,$$

i.e.

$$\begin{aligned}
f(t, x) &= e^{g(t,x)} = e^{g(t,0)+\sigma x} \\
&= e^{g(0,0)+\sigma x+(\mu-\sigma^2/2)t} \\
&= f(0, 0) \, e^{\sigma x+(\mu-\sigma^2/2)t}, \qquad t \geqslant 0.
\end{aligned}$$

We conclude that

$$S_t = f(t, B_t) = f(0, 0) \, e^{\sigma B_t+(\mu-\sigma^2/2)t},$$

and the solution to (1.13) is given by

$$S_t = S_0 \, e^{\sigma B_t+(\mu-\sigma^2/2)t}, \qquad t \geqslant 0.$$

The next R code can be used to generate Figure 1.3.

```
N=1000; t <- 0:N; dt <- 1.0/N; nsim <- 10; sigma=0.6; mu=0.001
Z <- c(rnorm(n = N, sd = sqrt(dt)));
plot(t*dt, exp(mu*t), xlab = "time", ylab = "Geometric Brownian motion", type =
    "l", ylim = c(0, 4), col = 1,lwd=3)
lines(t*dt, exp(sigma*c(0,cumsum(Z))+mu*t-sigma*sigma*t*dt/2),xlab = "time",type
    = "l",ylim = c(0, 4), col = 4)
```

Exercise 1.3. We look for a solution of the form

$$X_t = a(t)Y_t = a(t) \left(X_0 + \int_0^t b(s) dB_s\right)$$

where $a(\cdot)$ and $b(\cdot)$ are deterministic functions. After applying the Itô formula (1.10) to the Itô process $X_0 + \int_0^t b(s) dB_s$ of the form (1.8) with $u_t = b(t)$ and $v(t) = 0$, and to the function $f(t, x) = xa(t)$, we find

$$\begin{aligned}
dX_t &= d(a(t)Y_t) \\
&= Y_t a'(t)dt + a(t)dY_t \\
&= Y_t a'(t)dt + a(t)b(t)dB_t. \qquad\qquad \text{(S.2)}
\end{aligned}$$

By identification of (1.14) with (S.2), we get

$$\begin{cases} a'(t) = -\alpha a(t) \\ a(t)b(t) = \sigma, \end{cases}$$

hence $a(t) = a(0)\,\mathrm{e}^{-\alpha t} = \mathrm{e}^{-\alpha t}$ and $b(t) = \sigma/a(t) = \sigma\,\mathrm{e}^{\alpha t}$, which shows that

$$X_t = X_0\,\mathrm{e}^{-\alpha t} + \sigma \int_0^t \mathrm{e}^{-(t-s)\alpha}dB_s, \qquad t \geqslant 0. \tag{S.3}$$

Using integration by parts on the interval $[0,t]$, we can also write

$$X_t = X_0\,\mathrm{e}^{-\alpha t} + \sigma B_t - \sigma\alpha \int_0^t \mathrm{e}^{-(t-s)\alpha}B_s ds, \qquad t \geqslant 0.$$

Remark: the solution of the equation (1.14) *cannot* be written as a function $f(t, B_t)$ of t and B_t as in Exercise 1.2. The next R code can be used to generate Figure 1.4.

```
[j]=X[j-1]-alpha*X[j-1]*dt+sigma*Z[j]
N=10000; t <- 0:(N-1); dt <- 1.0/N;alpha=5; sigma=0.4;
Z <- c(rnorm(n = N, sd = sqrt(dt)));X <- c(1,N);X[1]=0.5
for (j in 2:N)X
  plot(t, X, xlab = "t", ylab = "", type = "l", ylim = c(-0.5,1), col = "blue")
  abline(h=0)
```

Exercise 1.4. Looking for a solution of the form

$$X_t = a(t)\left(X_0 + \int_0^t b(s)dB_s\right),$$

where $a(\cdot)$ and $b(\cdot)$ are deterministic functions of time, we get $a'(t)/a(t) = t$ and $a(t)b(t) = \mathrm{e}^{t^2/2}$, hence $a(t) = \mathrm{e}^{t^2/2}$ and $b(t) = 1$, which yields

$$X_t = \mathrm{e}^{t^2/2}(X_0 + B_t), \qquad t \geqslant 0.$$

The next R code can be used to generate Figure 1.5.

```
[j]=X[j-1]+j*X[j-1]*dt*dt+exp(j*dt*j*dt/2)*Z[j]
N=10000; T<-2.0; t <- 0:(N-1); dt <- T/N;
Z <- c(rnorm(n = N, sd = sqrt(dt)));X <- c(1,N);X[1]=0.5
for (j in 2:N)X
  plot(t, X, xlab = "t", ylab = "", type = "l", ylim = c(-0.5,10), col = "blue")
  abline(h=0)
```

Exercise 1.5. Letting $X_t = \sqrt{Y_t}$, by the Itô formula we find $dX_t = \alpha X_t dt + \sigma dB_t$, hence by the solution of Exercise 1.3 we have

$$Y_t = (X_t)^2 = \left(e^{\alpha t}\sqrt{Y_0} + \sigma \int_0^t e^{(t-s)\alpha}dB_s \right)^2, \quad t \geq 0.$$

The next R code can be used to generate Figure 1.6.

```
[j]=max(0,Y[j-1]+(2*mu*Y[j-1]+sg*sg)*dt+2*sg*sqrt(Y[j-1])*Z[j])
N=10000; t <- 0:(N-1); dt <- 1.0/N;mu=-5;sg=1;
Z <- c(rnorm(n = N, sd = sqrt(dt)));Y <- c(1,N);Y[1]=0.5
for (j in 2:N)Y
plot(t, Y, xlab = "t", ylab = "", type = "l", ylim = c(-0.1,1), col = "blue")
abline(h=0)
```

Exercise 1.6. For any $f \in L^2([0,T])$, we have

$$E\left[e^{\int_0^T f(s)dB_s} \,\Big|\, \mathcal{F}_t \right] = E\left[e^{\int_0^t f(s)dB_s} e^{\int_t^T f(s)dB_s} \,\Big|\, \mathcal{F}_t \right]$$

$$= e^{\int_0^t f(s)dB_s} E\left[e^{\int_t^T f(s)dB_s} \,\Big|\, \mathcal{F}_t \right]. \qquad (S.4)$$

Next, we use the Gaussian moment generating function identity $\mathbb{E}\left[e^X \right] = e^{\sigma^2/2}$ for $X \simeq \mathcal{N}(0, \sigma^2)$ and the fact that $\int_t^T f(s)dB_s \simeq \mathcal{N}\left(0, \int_t^T f^2(s)ds \right)$ by Proposition 1.2, showing that

$$E\left[e^{\int_t^T f(s)dB_s} \right] = \exp\left(\frac{1}{2}\int_t^T |f(s)|^2 ds \right).$$

We conclude to

$$E\left[e^{\int_0^T f(s)dB_s} \,\Big|\, \mathcal{F}_t \right] = \exp\left(\int_0^t f(s)dB_s + \frac{1}{2}\int_t^T |f(s)|^2 ds \right), \quad 0 \leq t \leq T.$$

Exercise 1.7. For all $\beta < 1/T$, by (1.11) we have

$$\mathbb{E}\left[\exp\left(\beta \int_0^T B_t dB_t \right) \right] = \mathbb{E}\left[\exp\left((B_T^2 - T)\beta/2 \right) \right]$$

$$= e^{-\beta T/2} \mathbb{E}\left[e^{\beta |B_T|^2/2} \right]$$

$$= \frac{e^{-\beta T/2}}{\sqrt{2\pi T}} \int_{-\infty}^{\infty} e^{\beta x^2/2 - x^2/(2T)}dx$$

$$= \frac{e^{-\beta T/2}}{\sqrt{2\pi T}} \int_{-\infty}^{\infty} e^{(\beta - 1/T)x^2/2} dx$$

$$= \frac{e^{-\beta T/2}}{\sqrt{1 - \beta T}}.$$

Exercise 1.8.

(1) We have

$$d\frac{X_t^T}{T - t} = \frac{dX_t^T}{T - t} + \frac{X_t^T}{|T - t|^2} dt = \sigma \frac{dB_t}{T - t},$$

hence, by integration over the time interval $[0, t]$ and using the initial condition $X_0^T = 0$, we find

$$\frac{X_t^T}{T - t} = \frac{X_0^T}{T} + \int_0^t d\frac{X_s^T}{T - s} = \sigma \int_0^t \frac{dB_s}{T - s}, \qquad 0 \leqslant t < T.$$

(2) By (1.5), we have

$$\mathbb{E}[X_t^T] = \sigma(T - t) \mathbb{E}\left[\int_0^t \frac{dB_s}{T - s}\right] = 0, \qquad 0 \leqslant t < T.$$

(3) Using the Itô isometry, we find

$$\mathrm{Var}[X_t^T] = \sigma^2 |T - t|^2 \mathrm{Var}\left[\int_0^t \frac{dB_s}{T - s}\right]$$

$$= \sigma^2 |T - t|^2 \int_0^t \frac{ds}{|T - s|^2}$$

$$= \sigma^2 |T - t|^2 \left(\frac{1}{T - t} - \frac{1}{T}\right)$$

$$= \sigma^2 \left(1 - \frac{t}{T}\right), \qquad 0 \leqslant t < T.$$

(4) We have

$$\lim_{t \to T} \|X_t^T\|_{L^2(\Omega)} = \lim_{t \to T} \mathrm{Var}[X_t^T] = \sigma^2 \lim_{t \to T} \left(1 - \frac{t}{T}\right) = 0.$$

Chapter 2

Exercise 2.1.

(1) By the solution (S.3) of Exercise 1.3, we have

$$S_t = S_0 \, e^{\alpha t} + \sigma \int_0^t e^{(t-s)\alpha} dB_s, \qquad t \geqslant 0.$$

(2) We have $\alpha_M = r$.

(3) After computing the conditional expectation $\mathbb{E}^* \left[|S_T|^2 \, | \, \mathcal{F}_t \right]$, we find

$$C(t, x) = e^{-(T-t)r} \exp \left(x e^{(T-t)r} + \frac{\sigma^2}{4r} \left(e^{2(T-t)r} - 1 \right) \right), \qquad 0 \leqslant t \leqslant T.$$

(4) Here, we note that the usual Black-Scholes argument (2.6) applies, and yields $\eta_t = \partial C(t, S_t)/\partial x$, that is,

$$\eta_t = \exp \left(S_t \, e^{(T-t)r} + \frac{\sigma^2}{4r} \left(e^{2(T-t)r} - 1 \right) \right), \qquad 0 \leqslant t \leqslant T.$$

Exercise 2.2.

(1) By the Girsanov Theorem 2.1, the probability $\mathbb{P}^*$ is given by its Radon-Nikodym density

$$\frac{d\mathbb{P}^*}{d\mathbb{P}} = \exp \left(-\frac{\alpha - r}{\sigma} B_T - \frac{|\alpha - r|^2 T}{2\sigma^2} \right).$$

(2) For all $t \in [0, T_0]$, we have

$$\begin{aligned}
C(t, S_t) &= e^{-(T-t)r} \, \mathbb{E}^* \left[\frac{S_T}{S_{T_0}} \, \middle| \, \mathcal{F}_t \right] \\
&= e^{-(T-t)r} \, \mathbb{E}^* \left[\frac{S_T}{S_{T_0}} \right] \\
&= e^{-(T-t)r} \, \mathbb{E}^* \left[e^{(\hat{B}_T - \hat{B}_{T_0})\sigma - (T-T_0)\sigma^2/2 + (T-T_0)r} \right] \\
&= e^{-(T-t)r + (T-T_0)r} \\
&= e^{-(T_0-t)r}, \qquad 0 \leqslant t \leqslant T_0,
\end{aligned}$$

and, for $t \in (T_0, T]$,

$$\begin{aligned}
C(t, S_t) &= e^{-(T-t)r} \, \mathbb{E}^* \left[\frac{S_T}{S_{T_0}} \, \middle| \, \mathcal{F}_t \right] \\
&= e^{-(T-t)r} \frac{S_t}{S_{T_0}} \, \mathbb{E}^* \left[\frac{S_T}{S_t} \, \middle| \, \mathcal{F}_t \right]
\end{aligned}$$

$$= e^{-(T-t)r} \frac{S_t}{S_{T_0}} \, \mathbb{E}^* \left[\frac{S_T}{S_t} \right]$$

$$= e^{-(T-t)r} \frac{S_t}{S_{T_0}} \, \mathbb{E}^* \left[e^{\sigma(\widehat{B}_T - \widehat{B}_t) - (T-t)\sigma^2/2 + (T-t)r} \right]$$

$$= \frac{S_t}{S_{T_0}}, \qquad T_0 < t \leqslant T.$$

(3) For all $t \in [0, T_0]$ we have $\eta_t = 0$ and $\zeta_t = e^{-rT_0}/A_0$, and for $t \in (T_0, T]$ we have $\zeta_t = 0$ and $\eta_t = 1/S_{T_0}$. We also have $d\eta_t = d\zeta_t = 0$ for $t \in [0, T_0]$ and $t \in (T_0, T]$, hence the portfolio is self-financing on these time intervals. On the other hand, at $t = T_0$ we also have $S_{T_0} d\eta_{T_0} + A_{T_0} d\zeta_{T_0} = S_{T_0} \times 1/S_{T_0} - A_{T_0} e^{-rT_0}/A_0 = 0$, hence the self-financing condition (2.2) is satisfied.

Exercise 2.3. (Exercise 2.2 continued).

(1) For all $t \in [0, T]$, we have

$$C(t, S_t) = e^{-(T-t)r} S_t^2 \, \mathbb{E}^* \left[\frac{S_T^2}{S_t^2} \, \Big| \, \mathcal{F}_t \right]$$

$$= e^{-(T-t)r} S_t^2 \, \mathbb{E}^* \left[e^{2\sigma(\widehat{B}_T - \widehat{B}_t) - (T-t)\sigma^2 + 2(T-t)r} \, \Big| \, \mathcal{F}_t \right]$$

$$= S_t^2 \, e^{(T-t)(r+\sigma^2)},$$

where $\widehat{B}_t = B_t + (\alpha - r)t/\sigma$, $t \in [0, T]$, is a standard Brownian motion under $\mathbb{P}^*$.

(2) For all $t \in [0, T]$ we have

$$\eta_t = \frac{\partial C}{\partial x}(t, x)_{|x = S_t} = 2 S_t \, e^{(T-t)(r+\sigma^2)},$$

and

$$\zeta_t = \frac{C(t, S_t) - \eta_t S_t}{A_t} = \frac{e^{-rt}}{A_0} \left(S_t^2 \, e^{(T-t)(r+\sigma^2)} - 2 S_t^2 \, e^{(T-t)(r+\sigma^2)} \right)$$

$$= -\frac{S_t^2}{A_0} \, e^{(T-t)\sigma^2 + (T-2t)r}.$$

Exercise 2.4.

(1) By the solution (S.3) of Exercise 1.3, we have

$$X_t^{(\alpha)} = e^{\alpha t} X_0^{(\alpha)} + \int_0^t e^{(t-s)\alpha} dB_s,$$

and the Girsanov Theorem 2.1 yields

$$\frac{d\mathbb{Q}}{d\mathbb{P}} = \exp \left(-\alpha \int_0^T X_t^{(\alpha)} dB_t - \frac{\alpha^2}{2} \int_0^T |X_t^{(\alpha)}|^2 dt \right).$$

(2) For all $\beta < 1/T$, we have

$$\mathbb{E}\left[\exp\left((\beta - \alpha)\int_0^T X_t^{(\alpha)}dX_t^{(\alpha)} + \frac{\alpha^2}{2}\int_0^T |X_t^{(\alpha)}|^2 dt\right)\right]$$

$$= \mathbb{E}\left[\exp\left(\beta\int_0^T X_t^{(\alpha)}dX_t^{(\alpha)} - \alpha\int_0^T X_t^{(\alpha)}dB_t\right.\right.$$

$$\left.\left. -\alpha^2\int_0^T |X_t^{(\alpha)}|^2 dt + \frac{\alpha^2}{2}\int_0^T |X_t^{(\alpha)}|^2 dt\right)\right]$$

$$= \mathbb{E}\left[\exp\left(\beta\int_0^T X_t^{(\alpha)}dX_t^{(\alpha)} - \alpha\int_0^T X_t^{(\alpha)}dB_t - \frac{\alpha^2}{2}\int_0^T |X_t^{(\alpha)}|^2 dt\right)\right]$$

$$= \mathbb{E}_{\mathbb{Q}}\left[\exp\left(\beta\int_0^T X_t^{(\alpha)}dX_t^{(\alpha)}\right)\right]$$

$$= \mathbb{E}\left[\exp\left(\beta\int_0^T B_t dB_t\right)\right]$$

$$= \frac{e^{-\beta T/2}}{\sqrt{1 - \beta T}},$$

using the fact that $\int_0^T B_t dB_t = (B_T^2 - T)/2$, where $B_T \simeq \mathcal{N}(0, T)$.

(3) Taking $\beta = \alpha$ in Question (2) yields

$$\mathbb{E}\left[\exp\left(\frac{\alpha^2}{2}\int_0^T |X_t^{(\alpha)}|^2 dt\right)\right] = \frac{e^{-\alpha T/2}}{\sqrt{1 - \alpha T}},$$

for all $\alpha < 1/T$.

Chapter 3

Exercise 3.1. We have

$$dr_t = r_0 d e^{-bt} + \frac{a}{b}d(1 - e^{-bt}) + \sigma d\left(e^{-bt}\int_0^t e^{bs}dB_s\right)$$

$$= -br_0 e^{-bt}dt + a e^{-bt}dt + \sigma e^{-bt}d\int_0^t e^{bs}dB_s + \sigma\int_0^t e^{bs}dB_s d e^{-bt}$$

$$= -br_0 e^{-bt}dt + a e^{-bt}dt + \sigma e^{-bt} e^{bt}dB_t - \sigma b\int_0^t e^{bs}dB_s e^{-bt}dt$$

$$= -br_0\,\mathrm{e}^{-bt}dt + a\,\mathrm{e}^{-bt}dt + \sigma dB_t - \sigma b\int_0^t \mathrm{e}^{bs}dB_s\,\mathrm{e}^{-bt}dt$$

$$= -br_0\,\mathrm{e}^{-bt}dt + a\,\mathrm{e}^{-bt}dt + \sigma dB_t - b\left(r_t - r_0\,\mathrm{e}^{-bt} - \frac{a}{b}(1 - \mathrm{e}^{-bt})\right)dt$$

$$= (a - br_t)dt + \sigma dB_t,$$

which shows that r_t solves (3.11).

Exercise 3.2. We check from (3.13) and the differentiation rule $d\int_0^t f(u)du = f(t)dt$ that

$$dr_t = \alpha\beta d\left(S_t\int_0^t \frac{1}{S_u}du\right) + r_0 dtS_t$$

$$= \alpha\beta S_t d\int_0^t \frac{1}{S_u}du + \alpha\beta\int_0^t \frac{1}{S_u}du dS_t + r_0 dtS_t$$

$$= \alpha\beta\frac{S_t}{S_t}dt + \alpha\beta\int_0^t \frac{S_t}{S_u}du\frac{dS_t}{S_t} + r_0 dS_t$$

$$= \alpha\beta dt + (r_t - r_0 S_t)\frac{dS_t}{S_t} + r_0 dS_t$$

$$= \alpha\beta dt + r_t\frac{dS_t}{S_t}$$

$$= \alpha\beta dt + r_t(-\beta dt + \sigma dB_t)$$

$$= \beta(\alpha - r_t)dt + \sigma dB_t, \qquad t \geqslant 0.$$

Exercise 3.3. Exponential Vasicek model (1). Applying the Itô formula to $X_t = \mathrm{e}^{r_t} = f(r_t)$ with $f(x) = \mathrm{e}^x$, we have

$$dX_t = d\mathrm{e}^{r_t}$$

$$= \mathrm{e}^{r_t}dr_t + \frac{1}{2}\mathrm{e}^{r_t}dr_t \cdot dr_t$$

$$= \mathrm{e}^{r_t}((a - br_t)dt + \sigma dB_t) + \frac{1}{2}\mathrm{e}^{r_t}((a - br_t)dt + \sigma dB_t)^2$$

$$= \mathrm{e}^{r_t}((a - br_t)dt + \sigma dB_t) + \frac{\sigma^2}{2}\mathrm{e}^{r_t}dt$$

$$= X_t\left(a + \frac{\sigma^2}{2} - b\log(X_t)\right)dt + \sigma X_t dB_t$$

$$= X_t\big(\tilde{a} - \tilde{b}f(X_t)\big)dt + \sigma g(X_t)dB_t,$$

hence we have

$$\tilde{a} = a + \frac{\sigma^2}{2} \qquad \text{and} \qquad \tilde{b} = b,$$

and the functions $f(x)$ and $g(x)$ are given by $f(x) = \log x$ and $g(x) = x$. Note that this stochastic differential equation is that of the exponential Vasicek model.

Exercise 3.4. Exponential Vasicek model (2).

(1) We have $Y_t = e^{-at}y_0 + \frac{\theta}{a}(1 - e^{-at}) + \sigma \int_0^t e^{-(t-s)a}dB_s$.

(2) We have $dX_t = X_t\left(\theta + \frac{\sigma^2}{2} - a\log X_t\right)dt + \sigma X_t dB_t$.

(3) We have $r_t = \exp\left(e^{-at}\log r_0 + \frac{\theta}{a}(1 - e^{-at}) + \sigma \int_0^t e^{-(t-s)a}dB_s\right)$.

(4) Using the Gaussian moment generating function identity $\mathbb{E}\left[e^X\right] = e^{\alpha^2/2}$ for $X \simeq \mathcal{N}(0, \alpha^2)$, see (11.2), Property (b) of conditional expectations in the appendix, and Proposition 1.2, we have

$$\mathbb{E}[r_t \mid \mathcal{F}_u]$$

$$= \mathbb{E}\left[\exp\left(e^{-at}\log r_0 + \frac{\theta}{a}(1 - e^{-at}) + \sigma\int_0^t e^{-(t-s)a}dB_s\right)\,\bigg|\,\mathcal{F}_u\right]$$

$$= e^{e^{-at}\log r_0 + \theta(1-e^{-at})/a + \sigma\int_0^u e^{-(t-s)a}dB_s}\,\mathbb{E}\left[e^{\sigma\int_u^t e^{-(t-s)a}dB_s}\,\bigg|\,\mathcal{F}_u\right]$$

$$= e^{e^{-at}\log r_0 + \theta(1-e^{-at})/a + \sigma\int_0^u e^{-(t-s)a}dB_s}\,\mathbb{E}\left[e^{\sigma\int_u^t e^{-(t-s)a}dB_s}\right]$$

$$= e^{e^{-at}\log r_0 + \frac{\theta}{a}\left(1-e^{-at}\right) + \sigma\int_0^u e^{-(t-s)a}dB_s}\exp\left(\frac{\sigma^2}{2}\int_u^t e^{-2(t-s)a}ds\right)$$

$$= e^{e^{-at}\log r_0 + \frac{\theta}{a}\left(1-e^{-at}\right) + \sigma\int_0^u e^{-(t-s)a}dB_s}\exp\left(\frac{\sigma^2}{4a}\left(1 - e^{-2(t-u)a}\right)\right)$$

$$= \exp\left(e^{-(t-u)a}\left(e^{-au}\log r_0 + \frac{\theta}{a}(1 - e^{-au}) + \sigma\int_0^u e^{-(u-s)a}dB_s\right)\right.$$

$$\left. + \frac{\theta}{a}(1 - e^{-(t-u)a}) + \frac{\sigma^2}{4a}(1 - e^{-2(t-u)a})\right)$$

$$= \exp\left(e^{-(t-u)a}\log r_u + \frac{\theta}{a}(1 - e^{-(t-u)a}) + \frac{\sigma^2}{4a}(1 - e^{-2(t-u)a})\right)$$

$$= (r_u)^{e^{-(t-u)a}}\exp\left(\frac{\theta}{a}(1 - e^{-(t-u)a}) + \frac{\sigma^2}{4a}(1 - e^{-2(t-u)a})\right).$$

In particular, for $u = 0$ we find

$$\mathbb{E}[r_t] = (r_0)^{e^{-at}} \exp\left(+\frac{\theta}{a}(1 - e^{-at}) + \frac{\sigma^2}{4a}(1 - e^{-2at})\right).$$

(5) Similarly, we have

$$\mathbb{E}[r_t^2 \mid \mathcal{F}_u]$$

$$= \mathbb{E}\left[\exp\left(2 e^{-at} \log r_0 + \frac{2\theta}{a}(1 - e^{-at}) + 2\sigma \int_0^t e^{-(t-s)a} dB_s\right) \middle| \mathcal{F}_u\right]$$

$$= e^{2 e^{-at} \log r_0 + 2(1 - e^{-at})/a + 2\sigma \int_0^u e^{-(t-s)a} dB_s} \mathbb{E}\left[e^{2\sigma \int_u^t e^{-(t-s)a} dB_s} \middle| \mathcal{F}_u\right]$$

$$= e^{2 e^{-at} \log r_0 + 2\theta(1 - e^{-at})/a + 2\sigma \int_0^u e^{-(t-s)a} dB_s} \mathbb{E}\left[e^{2\sigma \int_u^t e^{-(t-s)a} dB_s}\right]$$

$$= e^{2 e^{-at} \log r_0 + \frac{2\theta}{a}\left(1 - e^{-at}\right) + 2\sigma \int_0^u e^{-(t-s)a} dB_s} \exp\left(2\sigma^2 \int_u^t e^{-2(t-s)a} ds\right)$$

$$= e^{2 e^{-at} \log r_0 + \frac{2\theta}{a}\left(1 - e^{-at}\right) + 2\sigma \int_0^u e^{-(t-s)a} dB_s} \exp\left(\frac{\sigma^2}{a}(1 - e^{-2(t-u)a})\right)$$

$$= \exp\left(\frac{2\theta}{a}(1 - e^{-(t-u)a}) + \frac{\sigma^2}{a}(1 - e^{-2(t-u)a})\right.$$

$$\left. + 2 e^{-(t-u)a}\left(2 e^{-au} \log r_0 + \frac{2\theta}{a}(1 - e^{-au}) + 2\sigma \int_0^u e^{-(u-s)a} dB_s\right)\right)$$

$$= \exp\left(2 e^{-(t-u)a} \log r_u + \frac{2\theta}{a}(1 - e^{-(t-u)a}) + \frac{\sigma^2}{a}(1 - e^{-2(t-u)a})\right)$$

$$= r_u^{2 e^{-(t-u)a}} \exp\left(\frac{2\theta}{a}(1 - e^{-(t-u)a}) + \frac{\sigma^2}{a}(1 - e^{-2(t-u)a})\right),$$

hence $\text{Var}[r_t \mid \mathcal{F}_u]$ can be computed using the above two expressions along with the identity $\text{Var}[r_t \mid \mathcal{F}_u] = \mathbb{E}[r_t^2 \mid \mathcal{F}_u] - (\mathbb{E}[r_t \mid \mathcal{F}_u])^2$, as

$$\text{Var}[r_t \mid \mathcal{F}_u] = \mathbb{E}\left[r_t^2 \mid \mathcal{F}_u\right] - (\mathbb{E}[r_t \mid \mathcal{F}_u])^2$$

$$= r_u^{2 e^{-(t-u)a}} \exp\left(\frac{2\theta}{a}(1 - e^{-(t-u)a}) + \frac{\sigma^2}{a}(1 - e^{-2(t-u)a})\right)$$

$$- r_u^{2 e^{-(t-u)a}} \exp\left(\frac{2\theta}{a}(1 - e^{-(t-u)a}) + \frac{\sigma^2}{2a}(1 - e^{-2(t-u)a})\right)$$

$$= r_u^{2 e^{-(t-u)a}} \exp\left(\frac{2\theta}{a}(1 - e^{-(t-u)a}) + \frac{\sigma^2}{a}(1 - e^{-2(t-u)a})\right)$$

$$\times \left(1 - \exp\left(-\frac{\sigma^2}{2a}(1 - e^{-2(t-u)a})\right)\right).$$

(6) We have $\lim_{t\to\infty} \mathbb{E}[r_t] = \exp\left(\frac{\theta}{a} + \frac{\sigma^2}{4a}\right)$ and

$$\lim_{t\to\infty} \text{Var}[r_t] = \exp\left(\frac{2\theta}{a} + \frac{\sigma^2}{a}\right) - \exp\left(\frac{2\theta}{a} + \frac{\sigma^2}{2a}\right).$$

We find $\lim_{t\to\infty} \mathbb{E}[r_t] = r_0 \exp\left(\dfrac{\theta}{a} + \dfrac{\sigma^2}{4a}\right)$, and

$$\lim_{t\to\infty} \text{Var}[r_t] = \exp\left(\frac{2\theta}{a} + \frac{\sigma^2}{a}\right)\left(1 - e^{-\sigma^2/(2a)}\right)$$
$$= e^{2\theta/a}\left(e^{\sigma^2/a} - 1\right).$$

Exercise 3.5.

(1) We have $r_t = r_0 + \displaystyle\int_0^t (\alpha - \beta r_s)ds + \sigma \int_0^t \sqrt{r_s}dB_s$.

(2) Using the fact that the expectation of the stochastic integral with respect to Brownian motion is zero, after taking expectations and then the derivative with respect to t on both sides of the above integral equation, we get the differential equation $u'(t) = \alpha - \beta u(t)$ with initial condition $u(s) = r_s$. This equation is solved as

$$u(t) = \frac{\alpha}{\beta}\left(1 - e^{-(t-s)\beta}\right) + r_s e^{-(t-s)\beta}, \qquad 0 \leqslant s \leqslant t.$$

(3) Apply the Itô formula to

$$r_t^2 = f\left(r_0 + \int_0^t (\alpha - \beta r_s)ds + \sigma \int_0^t \sqrt{r_s}dB_s\right),$$

with $f(x) = x^2$, to obtain $d(r_t)^2 = r_t(\sigma^2 + 2\alpha - 2\beta r_t)dt + 2r_t\sigma\sqrt{r_t}dB_t$.

(4) Taking again the conditional expectation and then derivatives on both sides of the above equation we get

$$v'(t) = \left(2\alpha + \sigma^2\right)u(t) - 2\beta v(t),$$

which implies

$$v'(t) = \frac{\alpha}{\beta}\left(2\alpha + \sigma^2\right) + \left(2\alpha + \sigma^2\right)\left(r_s - \frac{\alpha}{\beta}\right)e^{-(t-s)\beta} - 2\beta v(t).$$

The solution of this equation can written as the sum $v(t) = w(t) + p(t)$ of a solution $w(t) = ce^{-2(t-s)\beta}$ to the homogeneous equation $w'(t) = -2\beta w(t)$ and a particular solution $p(t)$ to the original equation. Searching for a particular solution of the form $p(t) = \zeta + \xi e^{-(t-s)\beta}$ yields

$$p(t) = \frac{(2\alpha + \sigma^2)\alpha}{2\beta^2} + \frac{2\alpha + \sigma^2}{\beta}\left(r_s - \frac{\alpha}{\beta}\right)e^{-(t-s)\beta},$$

hence

$$v(t) = ce^{-2(t-s)\beta} + \frac{(2\alpha + \sigma^2)\alpha}{2\beta^2} + \frac{2\alpha + \sigma^2}{\beta}\left(r_s - \frac{\alpha}{\beta}\right)e^{-(t-s)\beta},$$

and the initial condition $v(s) = r_s^2$ yields

$$v(t) = r_s^2 e^{-2(t-s)\beta} + \frac{(2\alpha + \sigma^2)\alpha}{2\beta^2}\left(1 - e^{-2(t-s)\beta}\right)$$

$$+ \frac{2\alpha + \sigma^2}{\beta}\left(r_s - \frac{\alpha}{\beta}\right)\left(e^{-(t-s)\beta} - e^{-2(t-s)\beta}\right),$$

$0 \leqslant s \leqslant t$. In addition, we can compute the conditional variance

$$\text{Var}[r_t \mid r_s] = \text{Var}[r_t \mid \mathcal{F}_s] = \mathbb{E}\left[r_t^2 \mid \mathcal{F}_s\right] - (\mathbb{E}[r_t \mid \mathcal{F}_s])^2$$

$$= v(t) - |u(t)|^2$$

$$= r_s^2 e^{-2(t-s)\beta} + \frac{(2\alpha + \sigma^2)\alpha}{2\beta^2}\left(1 - e^{-2(t-s)\beta}\right)$$

$$+ \frac{2\alpha + \sigma^2}{\beta}\left(r_s - \frac{\alpha}{\beta}\right)\left(e^{-(t-s)\beta} - e^{-2(t-s)\beta}\right)$$

$$- \left(\frac{\alpha}{\beta}\left(1 - e^{-(t-s)\beta}\right) + r_s e^{-(t-s)\beta}\right)^2$$

$$= r_s^2 e^{-2(t-s)\beta} + \frac{(2\alpha + \sigma^2)\alpha}{2\beta^2}\left(1 - e^{-2(t-s)\beta}\right)$$

$$+ \frac{2\alpha + \sigma^2}{\beta}\left(r_s - \frac{\alpha}{\beta}\right)\left(e^{-(t-s)\beta} - e^{-2(t-s)\beta}\right)$$

$$- \left(\frac{\alpha}{\beta}\left(1 - e^{-(t-s)\beta}\right)\right)^2 - r_s^2 e^{-2(t-s)\beta}$$

$$- \frac{2\alpha}{\beta}\left(1 - e^{-(t-s)\beta}\right)r_s e^{-(t-s)\beta}$$

$$= \frac{(2\alpha + \sigma^2)\alpha}{2\beta^2}\left(1 - e^{-2(t-s)\beta}\right)$$

$$+ \frac{2\alpha + \sigma^2}{\beta}\left(r_s - \frac{\alpha}{\beta}\right)\left(e^{-(t-s)\beta} - e^{-2(t-s)\beta}\right)$$

$$- \left(\frac{\alpha}{\beta}\left(1 - e^{-(t-s)\beta}\right)\right)^2 - \frac{2\alpha}{\beta}\left(1 - e^{-(t-s)\beta}\right)r_s e^{-(t-s)\beta}$$

$$= r_s \frac{\sigma^2}{\beta}\left(e^{-(t-s)\beta} - e^{-2(t-s)\beta}\right) + \frac{\alpha\sigma^2}{\beta^2}\left|1 - e^{-(t-s)\beta}\right|^2, \quad 0 \leqslant s \leqslant t,$$

which satisfies the initial condition $\text{Var}[r_s \mid r_s] = 0$.

(5) The result follows by a direct application of the Itô formula.

(6) Again, from the Itô formula we have

$$dR_t = 2X_t dX_t + \frac{\sigma^2}{4}dt$$

$$= \left(\frac{\sigma^2}{4} - \beta X_t^2\right) dt + \sigma X_t dB_t$$

$$= \left(\frac{\sigma^2}{4} - \beta R_t\right) dt + \sigma |X_t| dW_t$$

$$= \left(\frac{\sigma^2}{4} - \beta R_t\right) dt + \sigma \sqrt{R_t} dW_t, \qquad t > 0.$$

One could also show that $(W_t)_{t\in\mathbb{R}_+}$ given by (3.17) is a standard Brownian motion, thus providing an explicit solution to (3.15).

Exercise 3.6. An estimator of σ can be obtained from the orthogonality relation

$$\sum_{l=0}^{n-1} \left(\left(\tilde{r}_{t_{l+1}} - a\Delta t - (1 - b\Delta t)\tilde{r}_{t_l}\right)^2 - \sigma^2 \Delta t \left(\tilde{r}_{t_l}\right)^{2\gamma} \right)$$

$$= \sigma^2 \sum_{l=0}^{n-1} \left(\tilde{r}_{t_l}\right)^{2\gamma} \left((Z_l)^2 - \Delta t\right)$$

$$\simeq 0,$$

which is due to the independence of t_{t_l} and Z_l, $l = 0, \ldots, n-1$, and yields

$$\hat{\sigma}^2 = \frac{\displaystyle\sum_{l=0}^{n-1} \left(\tilde{r}_{t_{l+1}} - a\Delta t - (1 - b\Delta t)\tilde{r}_{t_l}\right)^2}{\Delta t \displaystyle\sum_{l=0}^{n-1} \left(\tilde{r}_{t_l}\right)^{2\gamma}}.$$

One may also attempt to minimize the residual

$$\sum_{l=0}^{n-1} \left(\left(\tilde{r}_{t_{l+1}} - a\Delta t - (1 - b\Delta t)\tilde{r}_{t_l}\right)^2 - \sigma^2 \Delta t \left(\tilde{r}_{t_l}\right)^{2\gamma} \right)^2$$

by equating the following derivatives to zero, as

$$\frac{\partial}{\partial \sigma} \sum_{l=0}^{n-1} \left(\left(\tilde{r}_{t_{l+1}} - a\Delta t - (1 - b\Delta t)\tilde{r}_{t_l}\right)^2 - \sigma^2 \Delta t \left(\tilde{r}_{t_l}\right)^{2\gamma} \right)^2$$

$$= -4\sigma \sum_{l=0}^{n-1} \left(\tilde{r}_{t_l}\right)^{2\gamma} \left(\left(\tilde{r}_{t_{l+1}} - a\Delta t - (1 - b\Delta t)\tilde{r}_{t_l}\right)^2 - \sigma^2 \Delta t \left(\tilde{r}_{t_l}\right)^{2\gamma} \right)$$

$$= 0,$$

hence

$$\sum_{l=0}^{n-1} \left(\tilde{r}_{t_l}\right)^{2\gamma} \left(\tilde{r}_{t_{l+1}} - a\Delta t - (1 - b\Delta t)\tilde{r}_{t_l}\right)^2 - \sigma^2 \Delta t \sum_{l=0}^{n-1} \left(\tilde{r}_{t_l}\right)^{4\gamma} = 0,$$

which yields

$$\widehat{\sigma}^2 = \frac{\displaystyle\sum_{l=0}^{n-1} \left(\tilde{r}_{t_l}\right)^{2\gamma} \left(\tilde{r}_{t_{l+1}} - a\Delta t - (1 - b\Delta t)\tilde{r}_{t_l}\right)^2}{\displaystyle\Delta t \sum_{l=0}^{n-1} \left(\tilde{r}_{t_l}\right)^{4\gamma}}. \tag{S.5}$$

We also have

$$\frac{\partial}{\partial \gamma} \sum_{l=0}^{n-1} \left(\left(\tilde{r}_{t_{l+1}} - a\Delta t - (1 - b\Delta t)\tilde{r}_{t_l}\right)^2 - \sigma^2 \Delta t \left(\tilde{r}_{t_l}\right)^{2\gamma} \right)^2$$

$$= -4\sigma^2 \Delta t \sum_{l=0}^{n-1} \left(\tilde{r}_{t_l}\right)^{2\gamma} \left(\left(\tilde{r}_{t_{l+1}} - a\Delta t - (1 - b\Delta t)\tilde{r}_{t_l}\right)^2 - \sigma^2 \Delta t \left(\tilde{r}_{t_l}\right)^{2\gamma} \right) \log \tilde{r}_{t_l}$$

$$= 0,$$

which yields

$$\widehat{\sigma}^2 = \frac{\displaystyle\sum_{l=0}^{n-1} \left(\tilde{r}_{t_l}\right)^{2\gamma} \left(\tilde{r}_{t_{l+1}} - a\Delta t - (1 - b\Delta t)\tilde{r}_{t_l}\right)^2 \log \tilde{r}_{t_l}}{\displaystyle\Delta t \sum_{l=0}^{n-1} \left(\tilde{r}_{t_l}\right)^{4\gamma} \log \tilde{r}_{t_l}}, \tag{S.6}$$

and shows that γ can be estimated by matching Relations (S.5) and (S.6), *i.e.*

$$\frac{\displaystyle\sum_{l=0}^{n-1} \left(\tilde{r}_{t_l}\right)^{2\gamma} \left(\tilde{r}_{t_{l+1}} - a\Delta t - (1 - b\Delta t)\tilde{r}_{t_l}\right)^2}{\displaystyle\sum_{l=0}^{n-1} \left(\tilde{r}_{t_l}\right)^{4\gamma}}$$

$$= \frac{\displaystyle\sum_{l=0}^{n-1} \left(\tilde{r}_{t_l}\right)^{2\gamma} \left(\tilde{r}_{t_{l+1}} - a\Delta t - (1 - b\Delta t)\tilde{r}_{t_l}\right)^2 \log \tilde{r}_{t_l}}{\displaystyle\sum_{l=0}^{n-1} \left(\tilde{r}_{t_l}\right)^{4\gamma} \log \tilde{r}_{t_l}}.$$

Regarding the estimation of γ, we can combine the above relation with the second orthogonality relation

$$\sum_{l=0}^{n-1}\left(\left(\tilde{r}_{t_{l+1}}-a\Delta t-(1-b\Delta t)\tilde{r}_{t_l}\right)^2-\sigma^2\Delta t\left(\tilde{r}_{t_l}\right)^{2\gamma}\right)\tilde{r}_{t_l}$$

$$=\sigma^2\sum_{l=0}^{n-1}\left(\tilde{r}_{t_l}\right)^{2\gamma+1}\left((Z_l)^2-\Delta t\right)$$

$$\simeq 0,$$

cf. § 2.2 of Faff and Gray (2006).

Remarks.

(1) Estimators of a and b can be obtained by minimizing the residual

$$\sum_{l=0}^{n-1}\left(\tilde{r}_{t_{l+1}}-a\Delta t-(1-b\Delta t)\tilde{r}_{t_l}\right)^2$$

as in the Vasicek model, *i.e.* from the equations

$$\sum_{l=0}^{n-1}\left(\tilde{r}_{t_{l+1}}-a\Delta t-(1-b\Delta t)\tilde{r}_{t_l}\right)=0$$

and

$$\sum_{l=0}^{n-1}\left(\tilde{r}_{t_{l+1}}-a\Delta t-(1-b\Delta t)\tilde{r}_{t_l}\right)\tilde{r}_{t_l}=0.$$

(2) Instead of using the (generalised) method of moments, parameter estimation for stochastic differential equations can be achieved by maximum likelihood estimation, see *e.g.* Lindström (2007) and references therein.

Exercise 3.7.

(1) The discretization $r_{t_{k+1}}:=r_{t_k}+(a-br_{t_k})\Delta t\pm\sigma\sqrt{\Delta t}$, does not lead to a binomial tree as r_{t_2} could be obtained in *four* different ways from r_{t_0} as

$$r_{t_2}=r_{t_1}(1-b\Delta t)+a\Delta t\pm\sigma\sqrt{\Delta t}$$

$$=\begin{cases}\left(r_{t_0}(1-b\Delta t)+a\Delta t+\sigma\sqrt{\Delta t}\right)(1-b\Delta t)+a\Delta t+\sigma\sqrt{\Delta t}\\[2mm]\left(r_{t_0}(1-b\Delta t)+a\Delta t+\sigma\sqrt{\Delta t}\right)(1-b\Delta t)+a\Delta t-\sigma\sqrt{\Delta t}\\[2mm]\left(r_{t_0}(1-b\Delta t)+a\Delta t-\sigma\sqrt{\Delta t}\right)(1-b\Delta t)+a\Delta t+\sigma\sqrt{\Delta t}\\[2mm]\left(r_{t_0}(1-b\Delta t)+a\Delta t-\sigma\sqrt{\Delta t}\right)(1-b\Delta t)+a\Delta t-\sigma\sqrt{\Delta t}.\end{cases}$$

(2) By the Girsanov Theorem 2.1, the process $(r_t/\sigma)_{t\in[0,T]}$ with

$$\frac{dr_t}{\sigma} = \frac{a - br_t}{\sigma}dt + dB_t$$

is a standard Brownian motion under the probability measure $\mathbb{Q}$ with Radon-Nikodym density

$$\frac{d\mathbb{Q}}{d\mathbb{P}} = \exp\left(-\frac{1}{\sigma}\int_0^T (a - br_t)dB_t - \frac{1}{2\sigma^2}\int_0^T (a - br_t)^2 dt\right)$$

$$\simeq \exp\left(-\frac{1}{\sigma^2}\int_0^T (a - br_t)(dr_t - (a - br_t)dt) - \frac{1}{2\sigma^2}\int_0^T (a - br_t)^2 dt\right)$$

$$= \exp\left(-\frac{1}{\sigma^2}\int_0^T (a - br_t)dr_t + \frac{1}{2\sigma^2}\int_0^T (a - br_t)^2 dt\right).$$

In other words, if we generate $(r_t/\sigma)_{t\in[0,T]}$ and the increments $\sigma^{-1}dr_t \simeq \pm\sqrt{\Delta t}$ as a standard Brownian motion under $\mathbb{Q}$, then, under the probability measure $\mathbb{P}$ such that

$$\frac{d\mathbb{P}}{d\mathbb{Q}} = \exp\left(\frac{1}{\sigma^2}\int_0^T (a - br_t)dr_t - \frac{1}{2\sigma^2}\int_0^T (a - br_t)^2 dt\right)$$

$$\simeq 2^{T/\Delta T} \prod_{0<t<T}\left(\frac{1}{2} \pm \frac{a - br_t}{2\sigma}\sqrt{\Delta t}\right),$$

the process

$$dB_t = \frac{dr_t}{\sigma} - \frac{a - br_t}{\sigma}dt$$

will be a standard Brownian motion under $\mathbb{P}$, and the samples

$$dr_t = (a - br_t)dt + \sigma dB_t$$

of $(r_t)_{t\in[0,T]}$ will be distributed as those of a Vasicek process under $\mathbb{P}$.

(3) We check that

$$\mathbb{E}[\Delta r_{t_1} \mid r_{t_0}] = (a - br_{t_0})\Delta t$$
$$= p(r_{t_0})\sigma\sqrt{\Delta t} - (1 - p(r_{t_0}))\sigma\sqrt{\Delta t}$$
$$= \sigma p(r_{t_0})\sqrt{\Delta t} - \sigma q(r_{t_0})\sqrt{\Delta t},$$

with

$$p(r_{t_0}) = \frac{1}{2} + \frac{a - br_{t_0}}{2\sigma}\sqrt{\Delta t} \quad \text{and} \quad q(r_{t_0}) = \frac{1}{2} - \frac{a - br_{t_0}}{2\sigma}\sqrt{\Delta t}.$$

Similarly, we have

$$\mathbb{E}[\Delta r_{t_2} \mid r_{t_1}] = (a - br_{t_1})\Delta t$$

$$= p(r_{t_1})\sigma\sqrt{\Delta t} - (1 - p(r_{t_1}))\sigma\sqrt{\Delta t}$$
$$= \sigma p(r_{t_1})\sqrt{\Delta t} - \sigma q(r_{t_1})\sqrt{\Delta t},$$

with

$$p(r_{t_1}) = \frac{1}{2} + \frac{a - br_{t_1}}{2\sigma}\sqrt{\Delta t}, \qquad q(r_{t_1}) = \frac{1}{2} - \frac{a - br_{t_1}}{2\sigma}\sqrt{\Delta t}.$$

Chapter 4

Exercise 4.1.

(1) Consider the function $F(t, x)$ defined via

$$F(t, x) = \mathbb{E}^*\left[e^{-\int_t^T r_s ds} \,\middle|\, r_t = x\right], \qquad 0 \leqslant t \leqslant T.$$

We have

$$F(t, r_t) = F(t, r_0 + \theta t + \sigma B_t),$$

and by standard arbitrage arguments as in the proof of Proposition 4.2, the PDE satisfied by $F(t, x)$ is

$$-xF(t, x) + \frac{\partial F}{\partial t}(t, x) + \theta\frac{\partial F}{\partial x}(t, x) + \frac{\sigma^2}{2}\frac{\partial^2 F}{\partial x^2}(t, x) = 0,$$

with terminal condition $F(T, x) = 1$.

(2) We have

$$F(t, r_t) = \mathbb{E}^*\left[\exp\left(-\int_t^T (r_0 + \theta s + \sigma B_s)ds\right)\,\middle|\,\mathcal{F}_t\right]$$

$$= \mathbb{E}^*\left[\exp\left(-\int_t^T (r_t + \theta(s - t) + \sigma(B_s - B_t))ds\right)\,\middle|\,\mathcal{F}_t\right]$$

$$= \exp\left(-(T - t)r_t - \frac{\theta}{2}(T - t)^2\right)\mathbb{E}^*\left[e^{-\sigma\int_t^T (B_s - B_t)ds}\,\middle|\,\mathcal{F}_t\right]$$

$$= \exp\left(-(T - t)r_t - \frac{\theta}{2}(T - t)^2\right)\mathbb{E}^*\left[e^{-\sigma\int_t^T (B_s - B_t)ds}\right]$$

$$= \exp\left(-(T - t)r_t - \frac{\theta}{2}(T - t)^2\right)\mathbb{E}^*\left[e^{-\sigma\int_0^{T-t} B_s ds}\right],$$

$0 \leqslant t \leqslant T$. Next, we have

$$\mathbb{E}^* \left[e^{-\sigma \int_0^{T-t} B_s ds} \right] = \mathbb{E}^* \left[\exp \left(-\sigma \int_0^{T-t} \int_0^{T-t} \mathbb{1}_{\{u \leqslant s\}} dB_u ds \right) \right]$$

$$= \mathbb{E}^* \left[\exp \left(-\sigma \int_0^{T-t} \int_0^{T-t} \mathbb{1}_{\{u \leqslant s\}} ds dB_u \right) \right]$$

$$= \mathbb{E}^* \left[\exp \left(-\sigma \int_0^{T-t} \int_0^{T-t} \mathbb{1}_{\{u \leqslant s\}} ds dB_u \right) \right]$$

$$= \mathbb{E}^* \left[\exp \left(-\sigma \int_0^{T-t} \int_u^{T-t} ds dB_u \right) \right]$$

$$= \mathbb{E}^* \left[e^{-\sigma \int_0^{T-t} (T-t-u) dB_u} \right]$$

$$= \exp \left(\frac{\sigma^2}{2} \int_0^{T-t} (T-t-u)^2 du \right)$$

$$= \exp \left(\frac{\sigma^2}{2} \int_0^{T-t} u^2 du \right)$$

$$= e^{\sigma^2 (T-t)^3 / 6},$$

hence

$$F(t, r_t) = \exp \left(-r_t(T-t) - \frac{\theta}{2}(T-t)^2 + \frac{\sigma^2}{6}(T-t)^3 \right),$$

and

$$F(t, x) = \exp \left(-x(T-t) - \frac{\theta}{2}(T-t)^2 + \frac{\sigma^2}{6}(T-t)^3 \right).$$

(3) We have

$$\frac{\partial F}{\partial t}(t, x) = \left(x - \frac{\sigma^2}{2}(T-t)^2 + \theta(T-t) \right) F(t, x),$$

$$\frac{\partial F}{\partial x}(t, x) = -(T-t) F(t, x),$$

and

$$\frac{\partial^2 F}{\partial x^2}(t, x) = (T-t)^2 F(t, x),$$

which shows by addition that $F(t, x)$ satisfy the PDE with the terminal condition $F(T, x) = 1$.

Exercise 4.2.

(1) Applying the Itô formula

$$df(r_t) = f'(r_t)dr_t + \frac{1}{2}f''(r_t)(dr_t)^2$$

to the function $f(x) = x^{2-\gamma}$ with

$$f'(x) = (2-\gamma)x^{1-\gamma} \quad \text{and} \quad f''(x) = (2-\gamma)(1-\gamma)x^{-\gamma},$$

we have

$$
\begin{aligned}
dR_t &= dr_t^{2-\gamma} \\
&= df(r_t) \\
&= f'(r_t)dr_t + \frac{1}{2}f''(r_t)(dr_t)^2 \\
&= f'(r_t)\left((\beta r_t^{\gamma-1} + \alpha r_t)dt + \sigma r_t^{\gamma/2}dB_t\right)dr_t + \\
&\quad \frac{1}{2}f''(r_t)\left((\beta r_t^{\gamma-1} + \alpha r_t)dt + \sigma r_t^{\gamma/2}dB_t\right)^2 \\
&= f'(r_t)\left((\beta r_t^{\gamma-1} + \alpha r_t)dt + \sigma r_t^{\gamma/2}dB_t\right)dr_t + \frac{\sigma^2}{2}f''(r_t)r_t^{\gamma}dt \\
&= (2-\gamma)r_t^{1-\gamma}\left((\beta r_t^{\gamma-1} + \alpha r_t)dt + \sigma r_t^{\gamma/2}dB_t\right) \\
&\quad + \frac{\sigma^2}{2}(2-\gamma)(1-\gamma)r_t^{\gamma}r_t^{-\gamma}dt \\
&= (2-\gamma)(\beta + \alpha r_t^{2-\gamma})dt + \frac{\sigma^2}{2}(2-\gamma)(1-\gamma)dt + \sigma(2-\gamma)r_t^{1-\gamma/2}dB_t \\
&= (2-\gamma)\left(\beta + \frac{\sigma^2}{2}(1-\gamma) + \alpha R_t\right)dt + (2-\gamma)\sigma\sqrt{R_t}dB_t.
\end{aligned}
$$

We conclude that the process $R_t = r_t^{2-\gamma}$ follows the CIR equation

$$dR_t = b(a - R_t)dt + \eta\sqrt{R_t}dB_t$$

with initial condition $R_0 = r_0^{2-\gamma}$ and coefficients

$$b = (2-\gamma)\alpha, \quad a = \frac{1}{\alpha}\left(\beta + (1-\gamma)\frac{\sigma^2}{2}\right), \quad \text{and} \quad \eta = (2-\gamma)\sigma.$$

(2) By Itô's formula and the relation $P(t,T) = F(t,r_t)$, $t \in [0,T]$, we have

$$
\begin{aligned}
d\left(e^{-\int_0^t r_s ds}P(t,T)\right) &= -r_t e^{-\int_0^t r_s ds}P(t,T)dt + e^{-\int_0^t r_s ds}dP(t,T) \\
&= -r_t e^{-\int_0^t r_s ds}F(t,r_t)dt + e^{-\int_0^t r_s ds}dF(t,r_t) \\
&= -r_t e^{-\int_0^t r_s ds}F(t,r_t)dt + e^{-\int_0^t r_s ds}\frac{\partial F}{\partial t}(t,r_t)dt
\end{aligned}
$$

$$+ \, \mathrm{e}^{-\int_0^t r_s ds} \frac{\partial F}{\partial x}(t, r_t) dr_t + \frac{1}{2} \mathrm{e}^{-\int_0^t r_s ds} \frac{\partial^2 F}{\partial x^2}(t, r_t)(dr_t)^2$$

$$= -r_t \mathrm{e}^{-\int_0^t r_s ds} F(t, r_t) dt$$

$$+ \, \mathrm{e}^{-\int_0^t r_s ds} \frac{\partial F}{\partial x}(t, r_t) \left((\beta r_t^{-(1-\gamma)} + \alpha r_t) dt + \sigma r_t^{\gamma/2} dB_t \right)$$

$$+ \, \mathrm{e}^{-\int_0^t r_s ds} \left(\frac{\partial F}{\partial t}(t, r_t) + \frac{\sigma^2}{2} r_t^\gamma \frac{\partial^2 F}{\partial x^2}(t, r_t) \right) dt$$

$$= \mathrm{e}^{-\int_0^t r_s ds} \sigma r_t^{\gamma/2} \frac{\partial F}{\partial x}(t, r_t) dB_t$$

$$+ \, \mathrm{e}^{-\int_0^t r_s ds} \left(-r_t F(t, r_t) + (\beta r_t^{-(1-\gamma)} + \alpha r_t) \frac{\partial F}{\partial x}(t, r_t) \right.$$

$$\left. + \frac{\sigma^2}{2} r_t^\gamma \frac{\partial^2 F}{\partial x^2}(t, r_t) + \frac{\partial F}{\partial t}(t, r_t) \right) dt. \tag{S.7}$$

Given that $t \mapsto \mathrm{e}^{-\int_0^t r_s ds} P(t, T)$ is a martingale, the above expression (S.7) should only contain terms in dB_t and all terms in dt should vanish inside (S.7). This leads to the identities

$$\begin{cases} r_t F(t, r_t) = (\beta r_t^{-(1-\gamma)} + \alpha r_t) \frac{\partial F}{\partial x}(t, r_t) + \frac{\sigma^2}{2} r_t^\gamma \frac{\partial^2 F}{\partial x^2}(t, r_t) + \frac{\partial F}{\partial t}(t, r_t) \\[2ex] d\left(\mathrm{e}^{-\int_0^t r_s ds} P(t, T) \right) = \sigma \mathrm{e}^{-\int_0^t r_s ds} r_t^{\gamma/2} \frac{\partial F}{\partial x}(t, r_t) dB_t, \end{cases}$$

and to the PDE

$$x F(t, x) = \frac{\partial F}{\partial t}(t, x) + (\beta x^{-(1-\gamma)} + \alpha x) \frac{\partial F}{\partial x}(t, x) + \frac{\sigma^2}{2} x^\gamma \frac{\partial^2 F}{\partial x^2}(t, x).$$

Exercise 4.3. (Exercise 3.5 continued). From (1.17) and Proposition 4.2, the bond pricing PDE is given by

$$\begin{cases} \dfrac{\partial F}{\partial t}(t, x) = x F(t, x) - (\alpha - \beta x) \dfrac{\partial F}{\partial x}(t, x) - \dfrac{\sigma^2}{2} x^2 \dfrac{\partial^2 F}{\partial x^2}(t, x) \\[2ex] F(T, x) = 1. \end{cases}$$

When $\alpha = 0$, we search for a solution of the form

$$F(t, x) = \mathrm{e}^{A(T-t) - x B(T-t)},$$

with $A(0) = B(0) = 0$, which implies

$$\begin{cases} A'(s) = 0 \\[2ex] B'(s) + \beta B(s) + \dfrac{\sigma^2}{2} B^2(s) = 1, \end{cases}$$

hence in particular $A(s) = 0$, $s \in \mathbb{R}$, and $B(s)$ solves a Riccati equation, whose solution is easily checked to be

$$B(s) = \frac{2(e^{\gamma s} - 1)}{2\gamma + (\beta + \gamma)(e^{\gamma s} - 1)},$$

with $\gamma = \sqrt{\beta^2 + 2\sigma^2}$.

Exercise 4.4.

(1) The process $e^{-\int_0^t r_s ds} F(t, r_t)$ is a martingale, and we have

$$d\left(e^{-\int_0^t r_s ds} F(t, r_t)\right)$$

$$= e^{-\int_0^t r_s ds} \left(-r_t F(t, r_t) dt + \frac{\partial F}{\partial t}(t, r_t) dt + \frac{\partial F}{\partial x}(t, r_t) dr_t\right)$$

$$+ \frac{1}{2} e^{-\int_0^t r_s ds} \frac{\partial^2 F}{\partial x^2}(t, r_t) dr_t \cdot dr_t$$

$$= -r_t e^{-\int_0^t r_s ds} F(t, r_t) dt + e^{-\int_0^t r_s ds} \frac{\partial F}{\partial t}(t, r_t) dt$$

$$+ e^{-\int_0^t r_s ds} \frac{\partial F}{\partial x}(t, r_t)\left(-ar_t dt + \sigma\sqrt{r_t} dB_t\right)$$

$$+ r_t e^{-\int_0^t r_s ds} \frac{\sigma^2}{2} \frac{\partial^2 F}{\partial x^2}(t, r_t) dt,$$

hence

$$- xF(t, x) + \frac{\partial F}{\partial t}(t, x) - ax\frac{\partial F}{\partial x}(t, x) + x\frac{\sigma^2}{2}\frac{\partial^2 F}{\partial x^2}(t, x) = 0. \qquad \text{(S.8)}$$

(2) Plugging $F(t, x) = e^{A(T-t) + xC(T-t)}$ into the PDE (S.8) shows that

$$-x - A'(T-t) - xC'(T-t) - axC(T-t) + \frac{\sigma^2 x}{2}C^2(T-t) = 0.$$

Taking successively $x = 0$ and $x = 1$ in the above relation then yields the differential equations

$$\begin{cases} A'(s) = 0, \\ -1 - C'(s) - aC(s) + \dfrac{\sigma^2}{2}C^2(s) = 0, \qquad 0 \leqslant s \leqslant T. \end{cases}$$

Remark: The initial condition $A(0) = 0$ shows that $A(s) = 1$, and it can be shown from the condition $C(0) = 0$ that

$$C(s) = \frac{2(1 - e^{\gamma s})}{2\gamma + (a + \gamma)(e^{\gamma s} - 1)}, \qquad 0 \leqslant s \leqslant T,$$

with $\gamma = \sqrt{a^2 + 2\sigma^2}$, see *e.g.* Equation (3.25) page 66 of Brigo and Mercurio (2006).

Exercise 4.5.

(1) We have

$$
\begin{cases}
y_{0,1} = -\dfrac{1}{T_1} \log P(0, T_1) = 9.53\%, \\[2mm]
y_{0,2} = -\dfrac{1}{T_2} \log P(0, T_2) = 9.1\%, \\[2mm]
y_{1,2} = -\dfrac{1}{T_2 - T_1} \log \dfrac{P(0, T_2)}{P(T_1, T_2)} = 8.6\%,
\end{cases}
$$

with $T_1 = 1$ and $T_2 = 2$.

(2) We have

$$
P_c(1,2) = (\$1 + \$0.1) \times P_0(1,2) = (\$1 + \$0.1) \times e^{-(T_2 - T_2)y_{1,2}} = \$1.00914,
$$

and

$$
\begin{aligned}
P_c(0,2) &= (\$1 + \$0.1) \times P_0(0,2) + \$0.1 \times P_0(0,1) \\
&= (\$1 + \$0.1) \times e^{-(T_2 - T_2)y_{0,2}} + \$0.1 \times e^{-(T_2 - T_2)y_{0,1}} \\
&= \$1.00831.
\end{aligned}
$$

Exercise 4.6.

(1) We have

$$
P(1,2) = \mathbb{E}^* \left[\frac{100}{1 + r_1} \right] = \frac{100}{2(1 + r_1^u)} + \frac{100}{2(1 + r_1^d)}.
$$

(2) We have

$$
P(0,2) = \frac{100}{2(1 + r_0)(1 + r_1^u)} + \frac{100}{2(1 + r_0)(1 + r_1^d)}.
$$

(3) We have $P(0,1) = 91.74 = 100/(1 + r_0)$, hence

$$
r_0 = \frac{100 - P(0,1)}{P(0,1)} = \frac{100}{91.74} - 1 = 0.090037061 \simeq 9\%.
$$

(4) We have

$$
83.40 = P(0,2) = \frac{P(0,1)}{2(1 + r_1^u)} + \frac{P(0,1)}{2(1 + r_1^d)}
$$

and $r_1^u / r_1^d = e^{2\sigma\sqrt{\Delta t}}$, hence

$$
83.40 = P(0,2) = \frac{P(0,1)}{2(1 + r_1^d e^{2\sigma\sqrt{\Delta t}})} + \frac{P(0,1)}{2(1 + r_1^d)}
$$

or

$$e^{2\sigma\sqrt{\Delta t}}(r_1^d)^2 + 2r_1^d\, e^{\sigma\sqrt{\Delta t}} \cosh\left(\sigma\sqrt{\Delta t}\right)\left(1 - \frac{P(0,1)}{2P(0,2)}\right) + 1 - \frac{P(0,1)}{P(0,2)} = 0,$$

and

$$r_1^d = e^{-\sigma\sqrt{\Delta t}}\left(\cosh\left(\sigma\sqrt{\Delta t}\right)\left(\frac{P(0,1)}{2P(0,2)} - 1\right)\right.$$

$$\left. \pm\sqrt{\left(\frac{P(0,1)}{2P(0,2)} - 1\right)^2 \cosh^2\left(\sigma\sqrt{\Delta t}\right) + \frac{P(0,1)}{P(0,2)} - 1}\right)$$

$$= 0.078684844 \simeq 7.87\%,$$

and

$$r_1^u = r_1^d\, e^{2\sigma\sqrt{\Delta t}}$$

$$= e^{\sigma\sqrt{\Delta t}}\left(\cosh\left(\sigma\sqrt{\Delta t}\right)\left(\frac{P(0,1)}{2P(0,2)} - 1\right)\right.$$

$$\left. \pm\sqrt{\left(\frac{P(0,1)}{2P(0,2)} - 1\right)^2 \cosh^2\left(\sigma\sqrt{\Delta t}\right) + \frac{P(0,1)}{P(0,2)} - 1}\right)$$

$$= 0.122174525 \simeq 12.22\%.$$

We also find

$$\mu = \frac{1}{\Delta t}\left(\sigma\sqrt{\Delta t} + \log\frac{r_1^d}{r_0}\right)$$

$$= \frac{1}{\Delta t}\left(-\sigma\sqrt{\Delta t} + \log\frac{r_1^u}{r_0}\right) = 0.085229181 \simeq 8.52\%.$$

Exercise 4.7. We have

$$\frac{\partial}{\partial r}P_c(0,n) = \frac{\partial}{\partial r}\left(\frac{1}{(1+r)^n} + \frac{c}{r}\left(1 - \frac{1}{(1+r)^n}\right)\right)$$

$$= -\frac{n}{(1+r)^{n+1}} - \frac{c}{r^2}\left(1 - \frac{1}{(1+r)^n}\right) + \frac{nc}{r(1+r)^{n+1}},$$

hence

$$D_c(0,n) = -\frac{1+r}{P_c(0,n)}\frac{\partial}{\partial r}P_c(0,n)$$

$$= -\frac{-\dfrac{n}{(1+r)^n} - \dfrac{(1+r)c}{r^2}\left(1 - \dfrac{1}{(1+r)^n}\right) + \dfrac{nc}{r(1+r)^n}}{\dfrac{1}{(1+r)^n} + \dfrac{c}{r}\left(1 - \dfrac{1}{(1+r)^n}\right)}$$

$$= -\frac{-nr - \dfrac{1+r}{r}\left(c((1+r)^n - 1)\right) + nc}{r + c\left((1+r)^n - 1\right)}$$

$$= -\frac{1+r - nr - \dfrac{1+r}{r}\left(r + c((1+r)^n - 1)\right) + nc}{r + c\left((1+r)^n - 1\right)}$$

$$= \frac{1+r}{r} - \frac{1+r+n(c-r)}{r + c\left((1+r)^n - 1\right)}$$

$$= \frac{(1 - c/r)n}{1 + c\left((1+r)^n - 1\right)/r} + \frac{1+r}{r}\left(\frac{c\left((1+r)^n - 1\right)}{r + c\left((1+r)^n - 1\right)}\right),$$

with $D_0(0, n) = n$. We note that

$$\lim_{n \to \infty} D_c(0, n) = \lim_{n \to \infty}\left(\frac{1+r}{r} - \frac{1+r+n(c-r)}{r + c\left((1+r)^n - 1\right)}\right) = 1 + \frac{1}{r}.$$

When n becomes large, the duration (or relative sensitivity) of the bond price converges to $1 + 1/r$ whenever the (nonnegative) coupon amount c is nonzero, otherwise the bond duration of $P_c(0, n)$ is n. In particular, the presence of a nonzero coupon makes the duration (or relative sensitivity) of the bond price bounded as n increases, whereas the duration n of $P_0(0, n)$ goes to infinity as n increases.

As a consequence, the presence of the coupon tends to put an upper limit the risk and sensitivity of bond prices with respect to the market interest rate r as n becomes large, which can be used for bond *immunization*. Note that the duration $D_c(0, n)$ can also be written as the relative average

$$D_c(0, n) = \frac{1}{P_c(0, n)}\left(\frac{n}{(1+r)^n} + c\sum_{k=1}^{n}\frac{k}{(1+r)^k}\right)$$

of zero coupon bond durations, weighted by their respective zero-coupon prices.

Problem 4.8.

(1) We have

$$d\left(e^{-\int_0^t r_s ds} P(t, T)\right) = -e^{-\int_0^t r_s ds} P(t, T) r_t dt + e^{-\int_0^t r_s ds} dP(t, T)$$

$$= -e^{-\int_0^t r_s ds} P(t, T) r_t dt + e^{-\int_0^t r_s ds} dF(t, X_t)$$

$$= -e^{-\int_0^t r_s ds} P(t, T) r_t dt + e^{-\int_0^t r_s ds}\frac{\partial F}{\partial t}(t, X_t) dt$$

$$+ e^{-\int_0^t r_s ds}\frac{\partial F}{\partial x}(t, X_t) dX_t + \frac{\sigma^2}{2}e^{-\int_0^t r_s ds}\frac{\partial^2 F}{\partial x^2}(t, X_t) dt$$

$$= \sigma e^{-\int_0^t r_s ds} \frac{\partial F}{\partial x}(t, X_t) dB_t$$

$$+ e^{-\int_0^t r_s ds} \left(-r_t P(t, T) + \frac{\partial F}{\partial t}(t, X_t) - bX_t \frac{\partial F}{\partial x}(t, X_t) \right) dt$$

$$- \frac{\sigma^2}{2} e^{-\int_0^t r_s ds} \frac{\partial^2 F}{\partial x^2}(t, X_t) dt.$$

Since

$$t \mapsto e^{-\int_0^t r_s ds} P(t, T) = e^{-\int_0^t r_s ds} \mathbb{E}^* \left[e^{-\int_t^T r_s ds} \,\Big|\, \mathcal{F}_t \right]$$

$$= \mathbb{E}^* \left[e^{-\int_0^T r_s ds} \,\Big|\, \mathcal{F}_t \right],$$

is a martingale we get that

$$-r_t P(t, T) + \frac{\partial F}{\partial t}(t, X_t) - bX_t \frac{\partial F}{\partial x}(t, X_t) + \frac{\sigma^2}{2} \frac{\partial^2 F}{\partial x^2}(t, X_t) = 0,$$

and the PDE

$$-(r+x) F(t, x) + \frac{\partial F}{\partial t}(t, x) - bx \frac{\partial F}{\partial x}(t, x) + \frac{\sigma^2}{2} \frac{\partial^2 F}{\partial x^2}(t, x) = 0.$$

(2) We have

$$X_t = \sigma \int_0^t e^{-(t-s)b} dB_s, \qquad t \geqslant 0.$$

(3) Integrating both sides of the stochastic differential equation defining $(X_t)_{t \in \mathbb{R}_+}$, we get

$$X_t = -b \int_0^t X_s ds + \sigma B_t,$$

hence

$$\int_0^t X_s ds = \frac{1}{b} (\sigma B_t - X_t)$$

$$= \frac{\sigma}{b} \left(B_t - \int_0^t e^{-(t-s)b} dB_s \right)$$

$$= \frac{\sigma}{b} \int_0^t \left(1 - e^{-(t-s)b} \right) dB_s, \qquad t \geqslant 0.$$

(4) We have

$$\int_t^T X_s ds = \int_0^T X_s ds - \int_0^t X_s ds$$

$$= \frac{\sigma}{b} \int_0^T \left(1 - e^{-(T-s)b} \right) dB_s - \frac{\sigma}{b} \int_0^t \left(1 - e^{-(t-s)b} \right) dB_s$$

$$= -\frac{\sigma}{b} \left(\int_0^t \left(e^{-(T-s)b} - e^{-(t-s)b} \right) dB_s + \int_t^T \left(e^{-(T-s)b} - 1 \right) dB_s \right).$$

(5) Applying Corollary 1.3, we find

$$
\mathbb{E}^* \left[\int_t^T X_s ds \,\middle|\, \mathcal{F}_t \right] =
$$

$$
- \frac{\sigma}{b} \mathbb{E}^* \left[\int_0^t \left(e^{-(T-s)b} - e^{-(t-s)b} \right) dB_s + \int_t^T \left(e^{-(T-s)b} - 1 \right) dB_s \,\middle|\, \mathcal{F}_t \right]
$$

$$
= - \frac{\sigma}{b} \mathbb{E}^* \left[\int_0^t \left(e^{-(T-s)b} - e^{-(t-s)b} \right) dB_s \,\middle|\, \mathcal{F}_t \right]
$$

$$
- \frac{\sigma}{b} \mathbb{E}^* \left[\int_t^T \left(e^{-(T-s)b} - 1 \right) dB_s \,\middle|\, \mathcal{F}_t \right]
$$

$$
= - \frac{\sigma}{b} \mathbb{E}^* \left[\int_0^t \left(e^{-(T-s)b} - e^{-(t-s)b} \right) dB_s \,\middle|\, \mathcal{F}_t \right]
$$

$$
= - \frac{\sigma}{b} \int_0^t \left(e^{-(T-s)b} - e^{-(t-s)b} \right) dB_s, \qquad t \geqslant 0.
$$

(6) We have

$$
\mathbb{E}^* \left[\int_t^T X_s ds \,\middle|\, \mathcal{F}_t \right] = - \frac{\sigma}{b} \int_0^t \left(e^{-(T-s)b} - e^{-(t-s)b} \right) dB_s
$$

$$
= - \frac{\sigma}{b} \left(e^{-(T-t)b} - 1 \right) \int_0^t e^{-(t-s)b} dB_s
$$

$$
= \frac{X_t}{b} \left(1 - e^{-(T-t)b} \right), \qquad 0 \leqslant t \leqslant T.
$$

(7) From the properties of variance and conditional variance stated in the appendix, we have

$$
\mathrm{Var} \left[\int_t^T X_s ds \,\middle|\, \mathcal{F}_t \right]
$$

$$
= \frac{\sigma^2}{b^2}
$$

$$
\times \mathrm{Var} \left[\int_0^t \left(e^{-(T-s)b} - e^{-(t-s)b} \right) dB_s + \int_t^T \left(e^{-(T-s)b} - 1 \right) dB_s \,\middle|\, \mathcal{F}_t \right]
$$

$$
= \frac{\sigma^2}{b^2} \mathrm{Var} \left[\int_t^T \left(e^{-(T-s)b} - 1 \right) dB_s \,\middle|\, \mathcal{F}_t \right]
$$

$$
= \frac{\sigma^2}{b^2} \mathrm{Var} \left[\int_t^T \left(e^{-(T-s)b} - 1 \right) dB_s \right]
$$

$$
= \frac{\sigma^2}{b^2} \int_t^T \left| 1 - e^{-(T-s)b} \right|^2 ds, \qquad 0 \leqslant t \leqslant T.
$$

(8) Given $\mathcal{F}_t$, the random variable $\int_t^T X_s ds$ has a Gaussian distribution with conditional mean

$$\mathbb{E}^*\left[\int_t^T X_s ds \,\bigg|\, \mathcal{F}_t\right] = \frac{X_t}{b}\left(1 - e^{-(T-t)b}\right)$$

and conditional variance

$$\text{Var}\left[\int_t^T X_s ds \,\bigg|\, \mathcal{F}_t\right] = \frac{\sigma^2}{b^2}\int_t^T \left|1 - e^{-(T-s)b}\right|^2 ds.$$

(9) We have

$$P(t,T) = \mathbb{E}^*\left[e^{-\int_t^T r_s ds} \,\bigg|\, \mathcal{F}_t\right]$$

$$= \exp\left(-(T-t)r - \mathbb{E}^*\left[\int_t^T X_s ds \,\bigg|\, \mathcal{F}_t\right] + \frac{1}{2}\text{Var}\left[\int_t^T X_s ds \,\bigg|\, \mathcal{F}_t\right]\right)$$

$$= \exp\left(-(T-t)r - \frac{X_t}{b}\left(1 - e^{-(T-t)b}\right) + \frac{\sigma^2}{2b^2}\int_t^T \left|1 - e^{-(T-s)b}\right|^2 ds\right).$$

(10) We have

$$F(t,x) = \exp\left(-(T-t)r - \frac{x}{b}\left(1 - e^{-(T-t)b}\right) + \frac{\sigma^2}{2b^2}\int_t^T \left|1 - e^{-(T-s)b}\right|^2 ds\right),$$

hence

$$\frac{\partial F}{\partial t}(t,x) = \left(r + xe^{-(T-t)b} - \frac{\sigma^2}{2b^2}\left(1 - e^{-(T-t)b}\right)^2\right)F(t,x),$$

$$\frac{\partial F}{\partial x}(t,x) = -\frac{1}{b}\left(1 - e^{-(T-t)b}\right)F(t,x),$$

and

$$\frac{\partial^2 F}{\partial x^2}(t,x) = \frac{1}{b^2}\left(1 - e^{-(T-t)b}\right)^2 F(t,x),$$

which implies

$$-(r+x)F(t,x) + \frac{\partial F}{\partial t}(t,x) - bx\frac{\partial F}{\partial x}(t,x) + \frac{\sigma^2}{2}\frac{\partial^2 F}{\partial x^2}(t,x) = 0.$$

Chapter 5

Exercise 5.1. (Exercise 4.1 continued).

(1) We have

$$\log P(t,T) = -(T-t)r_0 - (T^2-t^2)\frac{\theta}{2} + \frac{\sigma^2}{6}(T-t)^3 - (T-t)(r_t - \theta t - r_0)$$

and

$$\log P(t,S) = -(S-t)r_0 - (S^2-t^2)\frac{\theta}{2} + (S-t)^3\frac{\sigma^2}{6} - (S-t)(r_t - \theta t - r_0),$$

hence

$$\log P(t,T) - \log P(t,S)$$
$$= \frac{\theta}{2}(S^2 - T^2) + \frac{\sigma^2}{6}(T-t)^3 + (S-T)(r_t - \theta t) + \frac{\sigma^2}{6}(S-t)^3,$$

and

$$f(t,T,S) = \frac{1}{S-T}(\log P(t,T) - \log P(t,S))$$
$$= \frac{1}{S-T}$$
$$\times \left((S^2 - T^2)\frac{\theta}{2} + \frac{\sigma^2}{6}(T-t)^3 + (S-T)(r_t - \theta t) + \frac{\sigma^2}{6}(S-t)^3 \right).$$

(2) We have

$$f(t,T) = -\frac{\partial}{\partial T}\log P(t,T)$$

$$= r_0 + T\theta - \frac{\sigma^2}{2}(T-t)^2 + (r_t - \theta t - r_0)$$

$$= (T-t)\theta - \frac{\sigma^2}{2}(T-t)^2 + r_t$$

$$= f(0,T) + \sigma^2 t\left(T - \frac{t}{2}\right) + \sigma B_t. \tag{S.9}$$

Exercise 5.2. (Problem 4.8 continued).

(1) We have

$$f(t,T,S) = -\frac{\log P(t,S) - \log P(t,T)}{S-T}$$
$$= \frac{r(S-t) + X_t(1 - e^{-(S-t)b})/b - \sigma^2 \int_t^S |1 - e^{-(S-s)b}|^2 ds/(2b^2)}{S-T}$$

$$-\frac{(T-t)r + X_t\big(1 - e^{-(T-t)b}\big)/b - \sigma^2 \int_t^T \big|1 - e^{-(T-s)b}\big|^2 ds/(2b^2)}{S - T}$$

$$= r - \frac{X_t}{b} \times \frac{e^{-(S-t)b} - e^{-(T-t)b}}{S - T}$$

$$+ \frac{\sigma^2}{2b^2} \int_t^T \frac{\big|1 - e^{-(T-s)b}\big|^2}{S - T} ds - \frac{\sigma^2}{2b^2} \int_t^S \frac{\big|1 - e^{-(S-s)b}\big|^2}{S - T} ds.$$

(2) We have

$$f(t,T) = \lim_{S \searrow T} f(t,T,S)$$

$$= r + X_t e^{-(T-t)b} - \frac{\sigma^2}{b^2} \int_t^T e^{-(T-s)b}\big(e^{-(T-s)b} - 1\big) ds$$

$$= r + X_t e^{-(T-t)b} - \frac{\sigma^2}{2b^2}\big(1 - e^{-(T-t)b}\big)^2. \tag{S.10}$$

Exercise 5.3. We have

$$P(0,T_2) = e^{-\int_0^{T_2} f(t,s)ds} = e^{-r_1 T_1 - r_2(T_2 - T_1)}, \qquad 0 \leqslant t \leqslant T_2,$$

and

$$P(T_1,T_2) = e^{-\int_{T_1}^{T_2} f(t,s)ds} = e^{-r_2(T_2 - T_1)}, \qquad 0 \leqslant t \leqslant T_2,$$

from which we deduce

$$r_2 = -\frac{1}{T_2 - T_1} \log P(T_1,T_2),$$

and

$$r_1 = -r_2 \frac{T_2 - T_1}{T_1} - \frac{1}{T_1} \log P(0,T_2)$$

$$= \frac{1}{T_1} \log P(T_1,T_2) - \frac{1}{T_1} \log P(0,T_2)$$

$$= -\frac{1}{T_1} \log \frac{P(0,T_2)}{P(T_1,T_2)}.$$

Exercise 5.4. (Exercise 4.1 continued).

(1) By Definition 5.1, we have

$$f(t,T,S) = \frac{1}{S-T}(\log P(t,T) - \log P(t,S))$$

$$= \frac{1}{S-T}\left(\left(-(T-t)r_t + \frac{\sigma^2}{6}(T-t)^3\right) - \left(-(S-t)r_t + \frac{\sigma^2}{6}(S-t)^3\right)\right)$$

$$= r_t + \frac{1}{S-T}\frac{\sigma^2}{6}\big((T-t)^3 - (S-t)^3\big).$$

(2) We have

$$f(t,T) = -\frac{\partial}{\partial T}\log P(t,T) = r_t - \frac{\sigma^2}{2}(T-t)^2.$$

(3) We have

$$d_t f(t,T) = (T-t)\sigma^2 dt + a dt + \sigma dB_t.$$

Exercise 5.5. Bridge model. (Exercise 1.8 continued).

(1) By Exercise 1.8-(4) we check that $P(T,T) = e^{X_T^T} = 1$.
(2) We have

$$\begin{aligned}
f(t,T,S) &= -\frac{1}{S-T}\left(X_t^S - X_t^T - \mu(S-T)\right)\\
&= \mu - \frac{\sigma}{S-T}\left(\int_0^t \frac{S-t}{S-s}dB_s - \int_0^t \frac{T-t}{T-s}dB_s\right)\\
&= \mu - \frac{\sigma}{S-T}\int_0^t \left(\frac{S-t}{S-s} - \frac{T-t}{T-s}\right)dB_s\\
&= \mu - \frac{\sigma}{S-T}\int_0^t \frac{(T-s)(S-t)-(T-t)(S-s)}{(S-s)(T-s)}dB_s\\
&= \mu + \sigma\int_0^t \frac{s-t}{(S-s)(T-s)}dB_s.
\end{aligned}$$

(3) We have

$$f(t,T) = \mu - \sigma\int_0^t \frac{t-s}{(T-s)^2}dB_s.$$

(4) We note that according to the Itô isometry (1.4), the limit

$$\lim_{T\searrow t} f(t,T) = \mu - \sigma\int_0^t \frac{dB_s}{t-s}$$

does not exist in $L^2(\Omega)$, as we have

$$\int_0^t \frac{1}{(t-s)^2}ds = +\infty.$$

(5) By Itô's calculus, we have

$$\begin{aligned}
\frac{dP(t,T)}{P(t,T)} &= \sigma dB_t + \frac{\sigma^2}{2}dt + \mu dt - \frac{X_t^T}{T-t}dt\\
&= \sigma dB_t + \frac{\sigma^2}{2}dt - \frac{\log P(t,T)}{T-t}dt, \qquad 0 \leqslant t < T.
\end{aligned}$$

(6) Letting

$$r_t^{\mathrm{T}} := \mu + \frac{\sigma^2}{2} - \frac{X_t^T}{T-t}$$

$$= \mu + \frac{\sigma^2}{2} - \sigma \int_0^t \frac{dB_s}{T-s},$$

by Question (5) we find that

$$\frac{dP(t,T)}{P(t,T)} = \sigma dB_t + r_t^{\mathrm{T}} dt, \qquad 0 \leqslant t \leqslant T.$$

(7) The equation of Question (6) can be solved as

$$P(t,T) = P(0,T) \exp\left(\sigma B_t - \frac{\sigma^2}{2}t + \int_0^t r_s^{\mathrm{T}} ds\right), \qquad 0 \leqslant t \leqslant T,$$

hence the process

$$P(t,T) \exp\left(-\int_0^t r_s^{\mathrm{T}} ds\right) = P(0,T) \exp\left(\sigma B_t - \frac{\sigma^2}{2}t\right), \qquad 0 \leqslant t \leqslant T,$$

is a martingale under $\mathbb{P}^*$, with the relation

$$P(t,T) e^{-\int_0^t r_s^{\mathrm{T}} ds} = \mathbb{E}^*\left[P(T,T) e^{-\int_0^T r_s^{\mathrm{T}}} \,\middle|\, \mathcal{F}_t\right]$$

$$= \mathbb{E}^*\left[e^{-\int_0^T r_s^{\mathrm{T}} ds} \,\middle|\, \mathcal{F}_t\right], \qquad 0 \leqslant t \leqslant T,$$

showing that

$$P(t,T) = e^{\int_0^t r_s^{\mathrm{T}} ds} \, \mathbb{E}^*\left[e^{-\int_0^T r_s^{\mathrm{T}} ds} \,\middle|\, \mathcal{F}_t\right]$$

$$= \mathbb{E}^*\left[e^{\int_0^t r_s^{\mathrm{T}} ds} e^{-\int_0^T r_s^{\mathrm{T}} ds} \,\middle|\, \mathcal{F}_t\right]$$

$$= \mathbb{E}^*\left[e^{-\int_t^T r_s^{\mathrm{T}} ds} \,\middle|\, \mathcal{F}_t\right], \qquad 0 \leqslant t \leqslant T.$$

Exercise 5.6.

(1) We have

$$P(t,T) = \mathbb{E}^*\left[e^{-\int_t^T r_s ds}\right]$$

$$= \mathbb{E}^*\left[\exp\left(-\int_t^T h(s)ds - \int_t^T X_s ds\right)\right]$$

$$= e^{-\int_t^T h(s)ds} \, \mathbb{E}^*\left[e^{-\int_t^T X_s ds}\right]$$

$$= \exp\left(-\int_t^T h(s)ds + A(T-t) + X_t C(T-t)\right),$$

hence, since $X_0 = 0$ we find $P(0,T) = \exp\left(-\int_0^T h(s)ds + A(T)\right)$.

(2) By the identification

$$P(t,T) = \exp\left(-\int_t^T h(s)ds + A(T-t) + X_t C(T-t)\right) = e^{-\int_t^T f(t,s)ds}$$

we find

$$\int_t^T h(s)ds = \int_t^T f(t,s)ds + A(T-t) + X_t C(T-t),$$

and, by differentiation with respect to T, this yields

$$h(T) = f(t,T) + A'(T-t) + X_t C'(T-t), \qquad 0 \leqslant t \leqslant T,$$

where

$$A(T-t) = \frac{4ab - 3\sigma^2}{4b^3} + \frac{\sigma^2 - 2ab}{2b^2}(T-t) + \frac{\sigma^2 - ab}{b^3}e^{-(T-t)b} - \frac{\sigma^2}{4b^3}e^{-2(T-t)b}.$$

Given an initial market data curve $f^M(0,T)$, the matching $f^M(0,T) = f(0,T)$ can be achieved at time $t = 0$ by letting

$$h(T) := f^M(0,T) + A'(T)$$

$$= f^M(0,T) + \frac{\sigma^2 - 2ab}{2b^2} - \frac{\sigma^2 - ab}{b^2}e^{-bT} + \frac{\sigma^2}{2b^2}e^{-2bT}, \quad T > 0.$$

Note however that in general, at time $t \in (0,T]$ we will have

$$h(T) = f(t,T) + A'(T-t) + X_t C'(T-t) = f^M(0,T) + A'(T),$$

and the relation

$$f(t,T) = f^M(0,T) + A'(T) - A'(T-t) - X_t C'(T-t), \quad 0 \leqslant t \leqslant T,$$

will allow us to match market data at time $t = 0$ only, *i.e.* for the initial curve. In any case, model calibration is to be done at time $t = 0$.

Exercise 5.7.

(1) From the definition

$$L(t,t,T) = \frac{1}{T-t}\left(\frac{1}{P(t,T)} - 1\right),$$

of the yield $L(t,T,T)$, we have

$$P(t,T) = \frac{1}{1 + (T-t)L(t,t,T)},$$

and similarly

$$P(t,S) = \frac{1}{1 + (S-t)L(t,t,S)}.$$

Hence, we get

$$L(t, T, S) = \frac{1}{S-T}\left(\frac{P(t,T)}{P(t,S)} - 1\right)$$

$$= \frac{1}{S-T}\left(\frac{1 + (S-t)L(t,t,S)}{1 + (T-t)L(t,t,T)} - 1\right)$$

$$= \frac{1}{S-T}\left(\frac{(S-t)L(t,t,S) - (T-t)L(t,t,T)}{1 + (T-t)L(t,t,T)}\right).$$

(2) When $T =$ one year and $L(0,0,T) = 2\%$, $L(0,0,2T) = 2.5\%$, we find

$$L(t, T, S) = \frac{1}{T}\left(\frac{2TL(0,0,2T) - TL(0,0,T)}{1 + TL(0,0,T)}\right)$$

$$= \frac{2 \times 0.025 - 0.02}{1 + 0.02} = 2.94\%,$$

so that we would not prefer a spot rate at $L(T, T, 2T) = 2\%$ over a forward contract with rate $L(0, T, 2T) = 2.94\%$.

Exercise 5.8.

(1) When $n = 1$, Relation (5.19) shows that $\tilde{f}(t, t, T_1) = f(t, t, T_1)$ with $F(t, x) = c_1 e^{-(T_1-1)x}$ and $P(t, T_1) = c_1 e^{f(t,t,T_1)}$, hence

$$D(t, T_1) := -\frac{1}{P(t,T_1)}\frac{\partial F}{\partial x}(t, f(t,t,T_1)) = T_1 - t, \qquad 0 \leqslant t \leqslant T_1.$$

(2) In general, we have

$$D(t, T_n) = -\frac{1}{P(t,T_n)}\frac{\partial F}{\partial x}\left(t, \tilde{f}(t,t,T_n)\right)$$

$$= \frac{1}{P(t,T_n)}\sum_{k=1}^{n}(T_k - t)c_k e^{-(T_k-t)\tilde{f}(t,t,T_n)}$$

$$= \sum_{k=1}^{n}(T_k - t)w_k,$$

where

$$w_k := \frac{c_k}{P(t,T_n)} e^{-(T_k-t)\tilde{f}(t,t,T_n)}$$

$$= \frac{c_k e^{-(T_k-t)\tilde{f}(t,t,T_n)}}{\displaystyle\sum_{l=1}^{n}c_l e^{-(T_l-t)f(t,t,T_l)}}$$

$$= \frac{c_k \, e^{-(T_k-t)\tilde{f}(t,t,T_n)}}{\displaystyle\sum_{l=1}^{n} c_l \, e^{-(T_l-t)\tilde{f}(t,t,T_n)}}, \qquad k = 1, 2, \ldots, n,$$

and the weights $w_1, w_2, \ldots, w_n$ satisfy $\displaystyle\sum_{k=1}^{n} w_k = 1$.

(3) We have

$$C(t, T_n) = \frac{1}{P(t, T_n)} \frac{\partial^2 F}{\partial x^2}\left(t, \tilde{f}(t, t, T_n)\right)$$

$$= \sum_{k=1}^{n} (T_k - t)^2 w_k$$

$$= \sum_{k=1}^{n} (T_k - t - D(t, T_n))^2 w_k + 2D(t, T_n) \sum_{k=1}^{n} (T_k - t) w_k - (D(t, T_n))^2$$

$$= \sum_{k=1}^{n} (T_k - t - D(t, T_n))^2 w_k + (D(t, T_n))^2$$

$$= (S(t, T_n))^2 + (D(t, T_n))^2,$$

with

$$(S(t, T_n))^2 := \sum_{k=1}^{n} (T_k - t - D(t, T_n))^2 w_k.$$

(4) We have

$$D(t, T_n) = \frac{1}{P(t, T_n)} \sum_{k=1}^{n} c_k B(T_k - t) \, e^{A(T_k-t)+B(T_k-t)f_\alpha(t,t,T_n)}$$

$$= e^{-A(T_n-t)-B(T_n-t)f_\alpha(t,t,T_n)} \sum_{k=1}^{n} c_k B(T_k - t) \, e^{A(T_k-t)+B(T_k-t)f_\alpha(t,t,T_n)}$$

$$= \sum_{k=1}^{n} c_k B(T_k - t) \, e^{A(T_k-t)-A(T_n-t)+(B(T_k-t)-B(T_n-t))f_\alpha(t,t,T_n)}.$$

(5) We have

$$D(t, T_n)$$

$$= \frac{1}{b} \sum_{k=1}^{n} c_k \left(1 - e^{-(T_k-t)b}\right)$$

$$\times e^{A(T_k-t)-A(T_n-t)+\left(e^{-(T_n-t)b}-e^{-(T_k-t)b}\right)f_\alpha(t,t,T_n)/b}$$

$$= \frac{1}{b} \sum_{k=1}^{n} c_k \left(1 - e^{-(T_k - t)b} \right)$$

$$\times e^{A(T_k - t) - A(T_n - t)} \left(P(t, t + \alpha(T_n - t)) \right)^{\frac{e^{-(T_n - t)b} - e^{-(T_k - t)b}}{(T_n - t)\alpha b}}.$$

Chapter 6

Exercise 6.1. Using the decomposition

$$P(t, T) = F_1(t, X_t) F_2(t, Y_t) \exp \left(- \int_t^T \varphi(s) ds + U(t, T) \right)$$

we have, by (6.13),

$$\frac{dP(t, T)}{P(t, T)} = r_t dt + \sigma C_1(t, T) dB_t^{(1)} + \eta C_2(t, T) dB_t^{(2)},$$

where

$$C_1(t, T) = \frac{e^{-(T-t)a} - 1}{a} \quad \text{and} \quad C_2(t, T) = \frac{e^{-(T-t)b} - 1}{b}, \qquad 0 \leqslant t \leqslant T.$$

Exercise 6.2.

(1) We have

$$X_t = X_0 e^{-bt} + \sigma \int_0^t e^{-(t-s)b} dB_s^{(1)}$$

and

$$Y_t = Y_0 e^{-bt} + \sigma \int_0^t e^{-(t-s)b} dB_s^{(2)}, \qquad t \geqslant 0,$$

see (3.2).

(2) We have

$$\mathrm{Var}[X_t] = \mathrm{Var}[Y_t] = \frac{\sigma^2}{2b} \left(1 - e^{-2bt} \right), \qquad t \geqslant 0,$$

see page 40, and therefore

$$\mathrm{Cov}(X_t, Y_t)$$

$$= \mathrm{Cov} \left(X_0 e^{-bt} + \sigma \int_0^t e^{-(t-s)b} dB_s^{(1)}, Y_0 e^{-bt} + \sigma \int_0^t e^{-(t-s)b} dB_s^{(2)} \right)$$

$$= \sigma^2 \operatorname{Cov}\left(\int_0^t e^{-(t-s)b} dB_s^{(1)}, \int_0^t e^{-(t-s)b} dB_s^{(2)}\right)$$

$$= \sigma^2 \mathbb{E}^*\left[\int_0^t e^{-(t-s)b} dB_s^{(1)} \int_0^t e^{-(t-s)b} dB_s^{(2)}\right]$$

$$= \rho\sigma^2 \int_0^t e^{-2(t-s)b} ds$$

$$= \rho\frac{\sigma^2}{2b}\left(1 - e^{-2bt}\right), \qquad t \geqslant 0.$$

(3) We have

$$\operatorname{Cov}(\log P(t, T_1), \log P(t, T_2))$$

$$= \operatorname{Cov}\left(\log\left(F_1(t, X_t, T_1)F_2(t, Y_t, T_2)e^{\rho U(t,T_1)}\right),\right.$$

$$\left.\log\left(F_1(t, X_t, T_1)F_2(t, Y_t, T_2)e^{\rho U(t,T_2)}\right)\right)$$

$$= \operatorname{Cov}\left(C_1^{T_1} + X_t A_1^{T_1} + C_2^{T_1} + Y_t A_2^{T_1}, C_1^{T_2} + X_t A_1^{T_2} + C_2^{T_2} + Y_t A_2^{T_2}\right)$$

$$= \operatorname{Cov}\left(X_t A_1^{T_1} + Y_t A_2^{T_1}, X_t A_1^{T_2} + Y_t A_2^{T_2}\right)$$

$$= A_1^{T_1} A_1^{T_2} \operatorname{Var}[X_t] + A_2^{T_1} A_2^{T_2} \operatorname{Var}[Y_t] + \left(A_1^{T_1} A_2^{T_2} + A_1^{T_2} A_2^{T_1}\right) \operatorname{Cov}(X_t, Y_t)$$

$$= \left(A_1^{T_1} A_1^{T_2} + A_2^{T_1} A_2^{T_2} + \rho\left(A_1^{T_1} A_2^{T_2} + A_1^{T_2} A_2^{T_1}\right)\right) \operatorname{Var}[X_t].$$

When $\operatorname{Cov}(X_t, Y_t) = \rho \operatorname{Var}[X_t] = \rho \operatorname{Var}[Y_t]$, we find the correlation

$$\operatorname{Cov}(\log P(t, T_1), \log P(t, T_2)) = \frac{\operatorname{Cov}(\log P(t, T_1), \log P(t, T_2))}{\sqrt{\operatorname{Var}[\log P(t, T_1)]\operatorname{Var}[\log P(t, T_2)]}}$$

$$= \frac{A_1^{T_1} A_1^{T_2} + A_2^{T_1} A_2^{T_2} + \rho\left(A_1^{T_1} A_2^{T_2} + A_1^{T_2} A_2^{T_1}\right)}{\sqrt{\left(A_1^{T_1}\right)^2 + \left(A_2^{T_1}\right)^2 + \rho\left(A_1^{T_1} A_2^{T_1} + A_1^{T_1} A_2^{T_1}\right)}}$$

$$\times \frac{1}{\sqrt{\left(A_1^{T_2}\right)^2 + \left(A_2^{T_2}\right)^2 + \rho\left(A_1^{T_2} A_2^{T_2} + A_1^{T_2} A_2^{T_2}\right)}}.$$

When $\rho = 1$, we find

$$\operatorname{Cov}(\log P(t, T_1), \log P(t, T_2))$$

$$= \frac{1}{\sqrt{\left(A_1^{T_1}\right)^2 + \left(A_2^{T_1}\right)^2 + A_1^{T_1} A_2^{T_1} + A_1^{T_1} A_2^{T_1}}}$$

$$\times \frac{A_1^{T_1} A_1^{T_2} + A_2^{T_1} A_2^{T_2} + A_1^{T_1} A_2^{T_2} + A_1^{T_2} A_2^{T_1}}{\sqrt{\left(A_1^{T_2}\right)^2 + \left(A_2^{T_2}\right)^2 + A_1^{T_2} A_2^{T_2} + A_1^{T_2} A_2^{T_2}}}$$

$$= \frac{\left(A_1^{T_1} + A_2^{T_1}\right)\left(A_1^{T_2} + A_2^{T_2}\right)}{\left|A_1^{T_1} + A_2^{T_1}\right|\left|A_1^{T_2} + A_2^{T_2}\right|}$$

$$= \pm 1.$$

For example, if $A_1^{T_1} = 4$, $A_1^{T_2} = 1$, $A_2^{T_1} = 1$ and $A_2^{T_2} = 4$, we find

$$\text{Cov}(\log P(t, T_1), \log P(t, T_2)) = \frac{8 + 17\rho}{17 + 8\rho}.$$

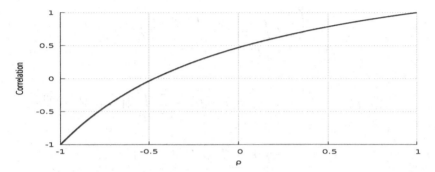

Fig. S.1: Log bond prices correlation graph in the two-factor model.

Exercise 6.3.

(1) We have

$$dr_t = -ar_0 e^{-at}dt + \varphi'(t)dt + dX_t$$
$$= -ar_0 e^{-at}dt + \theta(t)dt - a\int_0^t \theta(u) e^{-(t-u)a}dudt - aX_t dt + \sigma dB_t$$
$$= -ar_0 e^{-at}dt + \theta(t)dt - a\varphi(t)dt - aX_t dt + \sigma dB_t$$
$$= (\theta(t) - ar_t)dt + \sigma dB_t.$$

(2) By standard arguments, we obtain

$$-xF(t,x) + (\theta(t) - ax)\frac{\partial F}{\partial x}(t,x) + \frac{\sigma^2}{2}\frac{\partial^2 F}{\partial x^2}(t,x) + \frac{\partial F}{\partial t}(t,x) = 0, \quad \text{(S.11)}$$

under the terminal condition $F(T, x) = 1$, $x \in \mathbb{R}$.

(3) We have

$$P(t, T) = \mathbb{E}^*\left[e^{-\int_t^T r_s ds} \,\Big|\, \mathcal{F}_t \right]$$
$$= \mathbb{E}^*\left[e^{-\int_t^T (r_0 e^{-as} + \varphi(s) + X_s)ds} \,\Big|\, \mathcal{F}_t \right]$$
$$= e^{-\int_t^T (r_0 e^{-as} + \varphi(s))ds} \mathbb{E}^*\left[e^{-\int_t^T X_s ds} \,\Big|\, \mathcal{F}_t \right]$$
$$= e^{-\int_t^T (r_0 e^{-as} + \varphi(s))ds}$$

$$\times \exp\left(-\mathbb{E}^*\left[\int_t^T X_s ds \,\Big|\, \mathcal{F}_t\right] + \frac{1}{2}\operatorname{Var}\left[\int_t^T X_s ds \,\Big|\, \mathcal{F}_t\right]\right)$$

$$= \exp\left(-\int_t^T (r_0 e^{-as} + \varphi(s))ds - \frac{X_t}{a}\left(1 - e^{-(T-t)a}\right)\right)$$

$$\times \exp\left(\frac{\sigma^2}{2a^2}\int_t^T \left|1 - e^{-(T-s)a}\right|^2 ds\right)$$

$$= e^{A(t,T) + X_t C(t,T)},$$

where

$$A(t,T) = -\int_t^T (r_0 e^{-as} + \varphi(s))ds + \frac{\sigma^2}{2a^2}\int_t^T \left|1 - e^{-(T-s)a}\right|^2 ds,$$

and

$$C(t,T) = \frac{1}{a}\left(e^{-(T-t)a} - 1\right).$$

(4) We have

$$f(t,T) = -\frac{\partial}{\partial T}\log P(t,T)$$

$$= r_0 e^{-aT} + \varphi(T) + X_t e^{-(T-t)a}$$

$$\quad - \frac{\sigma^2}{2a^2}\frac{\partial}{\partial T}\int_t^T \left|1 - e^{-(T-s)a}\right|^2 ds$$

$$= r_0 e^{-aT} + \varphi(T) + X_t e^{-(T-t)a} + \frac{\sigma^2}{2a^2}\left(1 - e^{-(T-s)a}\right)^2\Big|_{s=T}$$

$$\quad - \frac{\sigma^2}{2a^2}\int_t^T \frac{\partial}{\partial T}\left(1 - e^{-(T-s)a}\right)^2 ds$$

$$= r_0 e^{-aT} + \varphi(T) + X_t e^{-(T-t)a}$$

$$\quad - \frac{\sigma^2}{a^2}\int_t^T e^{-(T-s)a}\left(1 - e^{-(T-s)a}\right)ds$$

$$= r_0 e^{-aT} + \varphi(T) + X_t e^{-(T-t)a}$$

$$\quad - \frac{\sigma^2}{a^2}\int_t^T e^{-(T-s)a}ds + \frac{\sigma^2}{a^2}\int_t^T e^{-2(T-s)a}ds$$

$$= r_0 e^{-aT} + \varphi(T) + X_t e^{-(T-t)a}$$

$$\quad - \frac{\sigma^2}{2a^2}\left(1 - e^{-(T-t)a}\right) + \frac{\sigma^2}{a^2}\left(1 - e^{-2(T-t)a}\right)$$

$$= r_0 e^{-aT} + \varphi(T) + X_t e^{-(T-t)a} - \frac{\sigma^2}{2a^2} + \frac{\sigma^2}{a^2}e^{-(T-t)a}$$

$$\quad - \frac{\sigma^2}{2a^2}e^{-2(T-t)a}$$

$$= r_0 \, e^{-aT} + \varphi(T) + X_t \, e^{-(T-t)a} - \frac{\sigma^2}{2a^2} \left(1 - e^{-(T-t)a} \right)^2,$$

$0 \leqslant t \leqslant T.$

(5) Since $t = 0$, it suffices to let

$$\varphi(T) = -r_0 \, e^{-aT} + f^M(0, T) + \frac{\sigma^2}{2a^2} \left(1 - e^{-aT} \right)^2, \quad T > 0,$$

to obtain $f(0, T) = f^M(0, T)$, $T > 0$.

(6) By differentiating the relation

$$\varphi(T) = \int_0^T \theta(t) \, e^{-(T-t)a} dt = -r_0 \, e^{-aT} + f^M(0, T) + \frac{\sigma^2}{2a^2} \left(1 - e^{-aT} \right)^2,$$

$T > 0$, we get

$$\theta(T) - a\varphi(T) = ar_0 \, e^{-aT} + \frac{\partial f^M}{\partial t}(0, T) + \frac{\sigma^2}{a} e^{-aT} \left(1 - e^{-aT} \right), \quad T > 0,$$

hence

$$\theta(t) = a\varphi(t) + ar_0 \, e^{-at} + \frac{\partial f^M}{\partial t}(0, t) + \frac{\sigma^2}{a} e^{-at} \left(1 - e^{-at} \right)$$

$$= af(0, t) + \frac{\partial f^M}{\partial t}(0, t) + \frac{\sigma^2}{2a} \left(1 - e^{-2at} \right)$$

$$= af^M(0, t) + \frac{\partial f^M}{\partial t}(0, t) + \frac{\sigma^2}{2a} \left(1 - e^{-2at} \right), \quad t \geqslant 0.$$

(7) From the Itô formula, the PDE (S.11) and the martingale property of $t \mapsto e^{-\int_0^t r_s ds} P(t, T)$, we have

$$d\left(e^{-\int_0^t r_s ds} P(t, T) \right) = d\left(e^{-\int_0^t r_s ds} F(t, r_t) \right)$$

$$= \sigma e^{-\int_0^t r_s ds} \frac{\partial F}{\partial x}(t, r_t) dB_t$$

$$= \sigma e^{-\int_0^t r_s ds} P(t, T) \frac{\partial}{\partial x} \log F(t, r_t) dB_t,$$

hence

$$\zeta_t = \sigma \frac{\partial}{\partial x} \log F(t, r_t) = \sigma C(t, T),$$

and we have

$$dP(t, T) = e^{\int_0^t r_s ds} d\left(e^{-\int_0^t r_s ds} P(t, T) \right) + r_t P(t, T) dt$$

$$= r_t P(t, T) dt + \zeta_t P(t, T) dB_t.$$

Chapter 7

Exercise 7.1. (Exercise 5.1 continued).

(1) By (S.9), we have
$$d_t f(t, T) = (T - t)\sigma^2 dt - \theta dt + dr_t = (T - t)\sigma^2 dt + \sigma dB_t.$$

(2) The HJM Condition (7.6) is satisfied since the drift of $d_t f(t, T)$ equals $\sigma \int_t^T \sigma ds = (T - t)\sigma^2$.

Exercise 7.2. (Exercise 5.2 continued).

(1) By (S.10), we have
$$
\begin{aligned}
d_t f(t, T) &= e^{-(T-t)b} dX_t + bX_t e^{-(T-t)b} dt \\
&\quad + \frac{\sigma^2}{b} e^{-(T-t)b} \left(1 - e^{-(T-t)b}\right) dt \\
&= \frac{\sigma^2}{b} e^{-(T-t)b} \left(1 - e^{-(T-t)b}\right) dt + \sigma e^{-(T-t)b} dB_t.
\end{aligned}
$$

(2) We have
$$\frac{\sigma^2}{b} e^{-(T-t)b} \left(1 - e^{-(T-t)b}\right) = \sigma e^{-(T-t)b} \int_t^T \sigma e^{-(s-t)b} ds,$$
which is the HJM absence of arbitrage Condition (7.6).

Exercise 7.3. (Exercise 5.4 continued). We have
$$d_t f(t, T) = (T - t)\sigma^2 dt + a dt + \sigma dB_t,$$
hence the HJM condition (7.6) is satisfied since the drift of $d_t f(t, T)$ equals $\sigma \int_t^T \sigma ds$.

Exercise 7.4. (Exercise 6.3 continued).

(1) We have
$$
\begin{aligned}
d_t f(t, T) &= aX_t e^{-(T-t)a} dt + e^{-(T-t)a} dX_t \\
&\quad + \frac{\sigma^2}{a} e^{-(T-t)a} \left(1 - e^{-(T-t)a}\right) dt \\
&= aX_t e^{-(T-t)a} dt + e^{-(T-t)a} \left(-aX_t + \sigma dB_t\right) \\
&\quad + \frac{\sigma^2}{a} e^{-(T-t)a} \left(1 - e^{-(T-t)a}\right) dt \\
&= e^{-(T-t)a} \sigma dB_t + \frac{\sigma^2}{a} e^{-(T-t)a} \left(1 - e^{-(T-t)a}\right) dt.
\end{aligned}
$$

(2) We check that the equality (7.6) holds, as

$$\sigma^2 e^{-(T-t)a} \int_t^T e^{-(T-s)a} = \frac{\sigma^2}{a} e^{-(T-t)a}\left(1 - e^{-(T-t)a}\right), \quad 0 \leqslant t \leqslant T.$$

Exercise 7.5.

(1) We have $\sigma(t, s) = \sigma s$, and we check that

$$\alpha(t, T) = \frac{\sigma^2}{2} T(T^2 - t^2) = \sigma T \int_t^T \sigma s \, ds = \sigma(t, T) \int_t^T \sigma(t, s) \, ds.$$

(2) We have

$$
\begin{aligned}
f(t, T) &= f(0, T) + \int_0^t d_s f(s, T) \\
&= f(0, T) + \frac{\sigma^2}{2} T \int_0^t (T^2 - s^2) \, ds + \sigma T \int_0^t dB_s \\
&= f(0, T) + \frac{\sigma^2}{2} T^3 \int_0^t ds - \frac{\sigma^2}{2} T \int_0^t s^2 \, ds + \sigma T \int_0^t dB_s \\
&= f(0, T) + \frac{\sigma^2}{2} T^3 t - \frac{\sigma^2}{6} T t^3 + \sigma T B_t \\
&= f(0, T) + \sigma^2 T t \left(\frac{T^2}{2} - \frac{t^2}{6}\right) + \sigma T B_t.
\end{aligned}
$$

(3) We have

$$r_t = f(t, t) = f(0, t) + \sigma^2 t^2 \left(\frac{t^2}{2} - \frac{t^2}{6}\right) + \sigma t B_t = f(0, t) + \frac{\sigma^2}{3} t^4 + \sigma t B_t.$$

(4) We have

$$
\begin{aligned}
dr_t &= \frac{4}{3} \sigma^2 t^3 dt + \sigma B_t dt + \sigma t dB_t \\
&= \frac{4}{3} \sigma^2 t^3 dt + \frac{1}{t}\left(r_t - f(0, t) - \frac{\sigma^2}{3} t^4\right) dt + \sigma t dB_t \\
&= \frac{1}{t}(r_t - f(0, t) + \sigma^2 t^4) dt + \sigma t dB_t \\
&= \sigma^2 t^3 dt + \frac{1}{t}(r_t - f(0, t)) dt + \sigma t dB_t,
\end{aligned}
$$

which is a Hull-White-type short-term interest rate model with the time-dependent deterministic coefficients $\eta(t) = \sigma^2 t^3$, $\psi(t) = 1/t$ and $\xi(t) = \sigma t$. Note that $t \mapsto f(0, t)$ is the *initial* rate curve data.

Exercise 7.6.

(1) We have

$$f(t,x) = f(0,x) + \alpha \int_0^t x^2 ds + \sigma \int_0^t d_s B(s,x) = r + \alpha t x^2 + \sigma B(t,x).$$

(2) We have

$$r_t = f(t,0) = r + B(t,0) = r.$$

(3) We have

$$P(t,T) = e^{-\int_t^T f(t,s)ds}$$

$$= \exp\left(-(T-t)r - \alpha t \int_0^{T-t} s^2 ds - \sigma \int_0^{T-t} B(t,x)dx \right)$$

$$= \exp\left(-(T-t)r - \frac{\alpha}{3} t(T-t)^3 - \sigma \int_0^{T-t} B(t,x)dx \right), \qquad 0 \leqslant t \leqslant T.$$

(4) Using (7.23), we have

$$\mathbb{E}^*\left[\left(\int_0^{T-t} B(t,x)dx \right)^2 \right] = \int_0^{T-t} \int_0^{T-t} \mathbb{E}^*[B(t,x)B(t,y)]dxdy$$

$$= t \int_0^{T-t} \int_0^{T-t} \min(x,y)dxdy$$

$$= 2t \int_0^{T-t} \int_0^y xdxdy = \frac{1}{3} t(T-t)^3.$$

(5) By Question (4), we have

$$\mathbb{E}^*[P(t,T)]$$

$$= \mathbb{E}^*\left[\exp\left(-(T-t)r - \frac{\alpha}{3} t(T-t)^3 - \sigma \int_0^{T-t} B(t,x)dx \right) \right]$$

$$= \exp\left(-(T-t)r - \frac{\alpha}{3} t(T-t)^3 \right) \mathbb{E}^*\left[e^{-\sigma \int_0^{T-t} B(t,x)dx} \right]$$

$$= \exp\left(-(T-t)r - \frac{\alpha}{3} t(T-t)^3 \right) \exp\left(\frac{\sigma^2}{2} \operatorname{Var}\left[\int_0^{T-t} B(t,x)dx \right] \right)$$

$$= \exp\left(-(T-t)r - \frac{\alpha}{3} t(T-t)^3 + \frac{\sigma^2}{6} t(T-t)^3 \right), \qquad 0 \leqslant t \leqslant T.$$

(6) By Question (5), we check that the required relation is satisfied if

$$-\frac{\alpha}{3} t(T-t)^3 + \frac{\sigma^2}{6} t(T-t)^3 = 0,$$

i.e. $\alpha = \sigma^2/2$.

Remark: In order to derive an analog of the HJM absence of arbitrage condition in this stochastic string model, one would have to check whether the discounted bond price $\mathrm{e}^{-rt}P(t,T)$ can be a martingale, which would require doing stochastic calculus with respect to the Brownian sheet $B(t,x)$.

Chapter 8

Exercise 8.1.

(1) By partial differentiation with respect to T under the expectation $\widehat{\mathbb{E}}$, and applying Proposition 8.1, we have

$$
\begin{aligned}
\frac{\partial P}{\partial T}(t,T) &= \frac{\partial}{\partial T}\,\mathbb{E}^*\left[\,\mathrm{e}^{-\int_t^T r_s ds}\;\Big|\;\mathcal{F}_t\right] \\
&= \mathbb{E}^*\left[\,-r_T\,\mathrm{e}^{-\int_t^T r_s ds}\;\Big|\;\mathcal{F}_t\right] \\
&= -P(t,T)\widehat{\mathbb{E}}[r_T\mid\mathcal{F}_t].
\end{aligned}
$$

(2) As a consequence of Question (1), we find

$$
f(t,T) = -\frac{1}{P(t,T)}\frac{\partial P}{\partial T}(t,T) = \widehat{\mathbb{E}}[r_T\mid\mathcal{F}_t], \quad 0\leqslant t\leqslant T,
$$

see Relation (22) page 10 of Mamon (2004).

Remark. In the Vasicek model, by (3.2) we have

$$
\begin{aligned}
r_T &= r_0\,\mathrm{e}^{-bT} + \frac{a}{b}\big(1 - \mathrm{e}^{-bT}\big) + \sigma\int_0^T \mathrm{e}^{-(T-s)b}dB_s \\
&= r_0\,\mathrm{e}^{-bT} + \frac{a}{b}\big(1 - \mathrm{e}^{-bT}\big) + \sigma\int_0^T \mathrm{e}^{-(T-s)b}d\widehat{B}_s \\
&\quad -\frac{\sigma^2}{b}\int_0^T \mathrm{e}^{-(T-s)b}\big(1 - \mathrm{e}^{-(T-s)b}\big)ds \\
&= r_0\,\mathrm{e}^{-bT} + \frac{a}{b}\big(1 - \mathrm{e}^{-bT}\big) + \sigma\int_0^T \mathrm{e}^{-(T-s)b}d\widehat{B}_s \\
&\quad -\frac{\sigma^2}{b}\int_0^T \mathrm{e}^{-(T-s)b}ds + \frac{\sigma^2}{b}\int_0^T \mathrm{e}^{-2(T-s)b}ds,
\end{aligned}
$$

hence

$$\widehat{\mathbb{E}}[r_T \mid \mathcal{F}_t] = r_0 \, e^{-bT} + \frac{a}{b}\left(1 - e^{-bT}\right) + \sigma \int_0^t e^{-(T-s)b} d\widehat{B}_s$$

$$- \frac{\sigma^2}{b} \int_0^T e^{-(T-s)b} ds + \frac{\sigma^2}{b} \int_0^T e^{-2(T-s)b} ds$$

$$= r_0 \, e^{-bT} + \frac{a}{b}\left(1 - e^{-bT}\right) + \sigma \int_0^t e^{-(T-s)b} dB_s$$

$$+ \frac{\sigma^2}{b} \int_0^t e^{-(T-s)b}\left(1 - e^{-(T-s)b}\right) ds$$

$$- \frac{\sigma^2}{b} \int_0^T e^{-(T-s)b} ds + \frac{\sigma^2}{b} \int_0^T e^{-2(T-s)b} ds$$

$$= r_0 \, e^{-bT} + \frac{a}{b}\left(1 - e^{-bT}\right) + e^{-(T-t)b}\left(r_t - r_0 \, e^{-bt} - \frac{a}{b}\left(1 - e^{-bt}\right)\right)$$

$$+ \frac{\sigma^2}{b} \int_0^t e^{-(T-s)b} ds - \frac{\sigma^2}{b} \int_0^t e^{-2(T-s)b} ds$$

$$- \frac{\sigma^2}{b} \int_0^T e^{-(T-s)b} ds + \frac{\sigma^2}{b} \int_0^T e^{-2(T-s)b} ds$$

$$= \frac{a}{b} + e^{-(T-t)b}\left(r_t - \frac{a}{b}\right) - \frac{\sigma^2}{b} \int_t^T e^{-(T-s)b} ds + \frac{\sigma^2}{b} \int_t^T e^{-2(T-s)b} ds$$

$$= \frac{a}{b} + e^{-(T-t)b}\left(r_t - \frac{a}{b}\right) - \frac{\sigma^2}{b} \int_0^{T-t} e^{-bs} ds + \frac{\sigma^2}{b} \int_0^{T-t} e^{-2bs} ds$$

$$= \frac{a}{b} + e^{-(T-t)b}\left(r_t - \frac{a}{b}\right) - \frac{\sigma^2}{b^2}\left(1 - e^{-(T-t)b}\right) + \frac{\sigma^2}{b^2}\left(1 - e^{-2(T-t)b}\right)$$

$$= \frac{a}{b} - \frac{\sigma^2}{2b^2} + e^{-(T-t)b}\left(r_t - \frac{a}{b} + \frac{\sigma^2}{b^2}\right) - \frac{\sigma^2}{b^2} e^{-2(T-t)b},$$

which recovers (6.2).

Exercise 8.2. Bond call options (1). (Exercise 7.1 continued).

(1) We have

$$dP(t, T) = dF(t, r_t) = r_t P(t, T) dt + \sigma \frac{\partial F}{\partial x}(t, r_t) dB_t,$$

where the remaining terms in factor of dt add up to zero from the martingale property of $t \mapsto e^{-\int_0^t r_s ds} P(t, T)$. Hence

$$dP(t, T) = r_t P(t, T) dt - (T - t)\sigma F(t, x) dB_t,$$

and

$$\frac{dP(t, T)}{P(t, T)} = r_t dt - (T - t)\sigma dB_t.$$

(2) From Question (1), we have

$$d\big(e^{-\int_0^t r_s ds} P(t,T)\big) = -r_t\, e^{-\int_0^t r_s ds} P(t,T)dt + e^{-\int_0^t r_s ds} dP(t,T)$$

$$= \sigma e^{-\int_0^t r_s ds} \frac{\partial F}{\partial x}(t,r_t)dB_t$$

$$= -(T-t)\sigma\, e^{-\int_0^t r_s ds} F(t,r_t)dB_t.$$

(3) We have

$$\Psi(t) = \mathbb{E}\left[\frac{d\widehat{\mathbb{P}}}{d\mathbb{P}^*}\,\Big|\,\mathcal{F}_t\right] = \frac{P(t,T)}{P(0,T)}\, e^{-\int_0^t r_s ds}, \qquad 0 \leqslant t \leqslant T.$$

(4) We have $d\Psi(t) = \Psi(t)\zeta_t dB_t$, with $\zeta_t = -(T-t)\sigma$, $t \in [0,T]$.

(5) We have

$$\Psi(t) = \exp\left(\int_0^t \zeta_s dB_s - \frac{1}{2}\int_0^t \zeta_s^2 ds\right),$$

hence

$$\mathbb{E}\left[\frac{d\widehat{\mathbb{P}}}{d\mathbb{P}^*}\,\Big|\,\mathcal{F}_T\right] = \Psi(T) = \exp\left(\int_0^T \zeta_s dB_s - \frac{1}{2}\int_0^T \zeta_s^2 ds\right).$$

(6) By the Girsanov Theorem 2.1, the process $\big(\widehat{B}_t\big)_{t\in\mathbb{R}_+}$ defined by

$$d\widehat{B}_t = (T-t)\sigma dt + dB_t$$

i.e.

$$\widehat{B}_T = \sigma \int_0^T (T-t)dt + B_T = \frac{\sigma^2}{2}T + B_T, \qquad T \geqslant 0,$$

is a standard Brownian motion under $\widehat{\mathbb{P}}$, and under $\widehat{\mathbb{P}}$ we have the dynamics

$$dr_r = \theta dt + \sigma dB_t = \big(\theta - (T-t)\sigma^2\big)dt + \sigma d\widehat{B}_t.$$

(7) We have

$$\mathbb{E}^*\left[e^{-\int_t^T r_s ds}(P(T,S) - K)^+ \,\Big|\, \mathcal{F}_t\right]$$

$$= P(t,T)\widehat{\mathbb{E}}\big[(P(T,S) - K)^+ \,\big|\, \mathcal{F}_t\big]$$

$$= P(t,T)$$

$$\times \widehat{\mathbb{E}}\left[\left(e^{-(S-T)r_0 - (S^2-T^2)\theta/2 + (S-T)^3\sigma^2/6 - \sigma(S-T)B_T} - K\right)^+ \,\Big|\, \mathcal{F}_t\right]$$

$$= P(t,T)$$

$$\times \widehat{\mathbb{E}}\left[\left(e^{-(S-T)r_0-(S^2-T^2)\theta/2+(S-T)^3\sigma^2/6-(S-T)\sigma(-\sigma T^2/2+\widehat{B}_T)}-K\right)^+\Big|\mathcal{F}_t\right]$$

$$= P(t,T)\,\mathbb{E}\left[(e^{m+X}-K)^+\mid\mathcal{F}_t\right],$$

where

$$m = -(S-T)r_0 - \frac{\theta}{2}(S^2-T^2) + \frac{\sigma^2}{6}(S-T)^3$$

$$-(S-T)\sigma\left(-\sigma\frac{T^2}{2}+\widehat{B}_t\right)$$

$$= -(S-T)r_0 - \frac{\theta}{2}(S^2-T^2) + \frac{\sigma^2}{6}(S-T)^3$$

$$+(S-T)\sigma\left(\sigma\frac{T^2}{2}+\sigma\frac{t^2}{2}-\sigma tT-B_t\right),$$

and X is centered Gaussian with conditional variance

$$v^2 = (S-T)^2(T-t)\sigma^2$$

given $\mathcal{F}_t$, $t \in [0,T]$. Hence we have

$$\mathbb{E}^*\left[e^{-\int_t^T r_s ds}(P(T,S)-K)^+\Big|\mathcal{F}_t\right]$$

$$= P(t,T)\,e^{m+v^2/2}\Phi\left(v+\frac{m-\log K}{v}\right) - KP(t,T)\Phi\left(\frac{m-\log K}{v}\right),$$

where

$$\Phi(z) = \int_{-\infty}^z e^{-y^2/2}\frac{dy}{\sqrt{2\pi}}, \qquad z \in \mathbb{R}.$$

Now,

$$t \mapsto \frac{P(t,S)}{P(t,T)}$$

is a martingale under $\widehat{\mathbb{P}}$ (see Proposition 8.3), hence

$$\widehat{\mathbb{E}}[P(T,S)\mid\mathcal{F}_t] = \widehat{\mathbb{E}}\left[\frac{P(T,S)}{P(T,T)}\Big|\mathcal{F}_t\right]$$

$$= \widehat{\mathbb{E}}\left[\frac{P(T,S)}{P(T,T)}\Big|\mathcal{F}_t\right]$$

$$= \frac{P(t,S)}{P(t,T)},$$

and

$$\frac{P(t,S)}{P(t,T)} = \widehat{\mathbb{E}}[P(T,S)\mid\mathcal{F}_t] = \mathbb{E}\left[e^{m+X}\mid\mathcal{F}_t\right],$$

where X is a centered Gaussian random variable with variance $v^2 > 0$, thus we have

$$m + \frac{v^2}{2} = \log \frac{P(t,S)}{P(t,T)}. \tag{S.12}$$

As a consequence, we have

$$\mathbb{E}^* \left[e^{-\int_t^T r_s ds} (P(T,S) - K)^+ \,\middle|\, \mathcal{F}_t \right]$$

$$= P(t,S)\Phi \left(\frac{v}{2} + \frac{1}{v} \log \frac{P(t,S)}{KP(t,T)} \right)$$

$$- KP(t,T)\Phi \left(-\frac{v}{2} + \frac{1}{v} \log \frac{P(t,S)}{KP(t,T)} \right).$$

Note that Relation (S.12) can be obtained through direct force calculations instead of using a martingale argument. We have

$$\log P(t,T) - \log P(t,S)$$

$$= -(T-t)r_0 - \frac{\theta}{2}(T^2 - t^2) + \frac{\sigma^2}{6}(T-t)^3 - \sigma(T-t)B_t$$

$$- \left(-(S-t)r_0 - \frac{\theta}{2}(S^2 - t^2) + \frac{\sigma^2}{6}(S-t)^3 - \sigma(S-t)B_t \right)$$

$$= -(T-S)r_0 - (T^2 - S^2)\theta/2 - (T-S)\sigma B_t$$

$$+ \frac{\sigma^2}{6}\left((T-t)^3 - (S-t)^3\right),$$

hence

$$m = \log P(t,S) - \log P(t,T)$$

$$+ \frac{\sigma^2}{6}\left((T-t)^3 - (S-t)^3\right) + \frac{\sigma^2}{6}(S-T)^3$$

$$+ (S-T)\frac{\sigma^2}{2}(T^2 + t^2) - \sigma^2 tT(S-T)$$

$$= \log P(t,S) - \log P(t,T)$$

$$+ \frac{\sigma^2}{6}\left(T^3 - 3tT^2 + 3Tt^2 - S^3 + 3tS^2 - 3St^2\right)$$

$$+ \frac{\sigma^2}{6}(S-T)^3 + (S-T)\frac{\sigma^2}{2}(T^2 + t^2) - \sigma^2 tT(S-T)$$

$$= \log P(t,S) - \log P(t,T) + \frac{\sigma^2}{6}\left(-3tT^2 + 3Tt^2 + 3tS^2 - 3St^2\right)$$

$$+ \frac{\sigma^2}{6}\left(-3TS^2 + 3ST^2\right) + \frac{\sigma^2}{2}(S-T)(T^2 + t^2) - \sigma^2 tT(S-T),$$

and

$$m + \frac{v^2}{2}$$

$$= \log P(t, S) - \log P(t, T) + \frac{\sigma^2}{6}\left(-3tT^2 + 3Tt^2 + 3tS^2 - 3St^2\right)$$

$$+ \frac{\sigma^2}{6}\left(-3TS^2 + 3ST^2\right) + (S - T)\frac{\sigma^2}{2}\left(T^2 + t^2\right)$$

$$+ \frac{\sigma^2}{2}\left(S^2 - 2ST + T^2\right)(T - t) - \sigma^2 tT(S - T)$$

$$= \log P(t, S) - \log P(t, T), \qquad 0 \leqslant t \leqslant T \leqslant S.$$

Exercise 8.3. Bond call options (2). (Exercise 7.2 continued).

(1) From the result of Problem 4.8-(9), we find

$$\mathbb{E}\left[\frac{d\widehat{\mathbb{P}}}{d\mathbb{P}^*}\,\bigg|\,\mathcal{F}_t\right] = e^{-\int_0^t r_s ds}\frac{P(t, T)}{P(0, T)}$$

$$= e^{-\int_0^t r_s ds}\exp\left(rt - \frac{\sigma^2}{2b^2}\int_0^t \left|1 - e^{-(T-s)b}\right|^2 ds - \frac{X_t}{b}\left(1 - e^{-(T-t)b}\right)\right)$$

$$= e^{-\int_0^t X_s ds}\exp\left(-\frac{\sigma^2}{2b^2}\int_0^t \left|1 - e^{-(T-s)b}\right|^2 ds - \frac{X_t}{b}\left(1 - e^{-(T-t)b}\right)\right)$$

$$= \exp\left(\frac{X_t - \sigma B_t}{b} - \frac{\sigma^2}{2b^2}\int_0^t \left|1 - e^{-(T-s)b}\right|^2 ds - \frac{X_t}{b}\left(1 - e^{-(T-t)b}\right)\right)$$

$$= \exp\left(-\frac{\sigma}{b}B_t - \frac{\sigma^2}{2b^2}\int_0^t \left|1 - e^{-(T-s)b}\right|^2 ds + \frac{X_t}{b}e^{-(T-t)b}\right)$$

$$= \exp\left(-\frac{\sigma}{b}B_t - \frac{\sigma^2}{2b^2}\int_0^t \left|1 - e^{-(T-s)b}\right|^2 ds + \frac{\sigma}{b}\int_0^t e^{-(T-s)b}dB_s\right)$$

$$= \exp\left(-\frac{\sigma}{b}\int_0^t \left(1 - e^{-(T-s)b}\right)dB_s - \frac{\sigma^2}{2b^2}\int_0^t \left|1 - e^{-(T-s)b}\right|^2 ds\right),$$

and in particular, for $t = T$ we get

$$\frac{d\widehat{\mathbb{P}}}{d\mathbb{P}^*} = \exp\left(-\frac{\sigma}{b}\int_0^T \left(1 - e^{-(T-s)b}\right)dB_s - \frac{\sigma^2}{2b^2}\int_0^T \left|1 - e^{-(T-s)b}\right|^2 ds\right).$$

(2) By the Girsanov Theorem 2.1, the process $\left(\widehat{B}_t\right)_{t \in \mathbb{R}_+}$ defined as

$$\widehat{B}_t := B_t + \frac{\sigma}{b}\int_0^t \left(1 - e^{-(T-s)b}\right)ds, \qquad 0 \leqslant t \leqslant T,$$

is a standard Brownian motion under the forward measure $\widehat{\mathbb{P}}$, and we have

$$dr_t = dX_t = -bX_t dt + \sigma dB_t = -bX_t dt - \frac{\sigma^2}{b}\left(1 - e^{-(T-t)b}\right)dt + \sigma d\widehat{B}_t.$$

(3) When $b = 0$, we have

$$P(t,T) = \exp\left(-(T-t)r - (T-t)X_t + \frac{\sigma^2}{2}\int_t^T (T-s)^2 ds\right)$$

and

$$X_T = -\sigma^2 \int_0^T (T-s)ds + \sigma\widehat{B}_T = -\frac{\sigma^2}{2}T^2 + \sigma\widehat{B}_T,$$

hence

$$\mathbb{E}^*\left[e^{-\int_0^T r_s ds}(P(T,S) - K)^+\right] = P(0,T)\widehat{\mathbb{E}}\left[(P(T,S) - K)^+\right]$$

$$= P(0,T)\widehat{\mathbb{E}}\left[\left(e^{-r(S-T)-(S-T)X_T+\sigma^2\int_T^S (S-s)^2 ds/2} - K\right)^+\right]$$

$$= P(0,T)\widehat{\mathbb{E}}\left[\left(e^{-r(S-T)-(S-T)(\sigma\widehat{B}_T-\sigma^2 T^2/2)+\sigma^2\int_T^S (S-s)^2 ds/2} - K\right)^+\right]$$

$$= P(0,T)\widehat{\mathbb{E}}\left[\left(e^{-r(S-T)+\sigma^2(S-T)T^2/2+\sigma^2(S-T)^3/6-(S-T)\sigma\widehat{B}_T} - K\right)^+\right].$$

From the relation

$$\mathbb{E}\left[(e^{m+X} - K)^+\right] = e^{m+v^2/2}\Phi\left(v + \frac{m - \log K}{v}\right) - K\Phi\left(\frac{m - \log K}{v}\right),$$

where

$$m = -r(S-T) + \frac{\sigma^2}{2}(S-T)T^2 + \frac{\sigma^2}{6}(S-T)^3$$

and X is a centered Gaussian random variable with variance $v^2 = \sigma^2 T(S-T)^2$, with

$$\Phi(z) = \int_{-\infty}^z e^{-y^2/2}\frac{dy}{\sqrt{2\pi}}, \qquad z \in \mathbb{R},$$

and

$$-\frac{v^2}{2} + \log\frac{P(0,S)}{P(0,T)} = -\frac{\sigma^2}{2}T(S-T)^2 - rS + \frac{\sigma^2}{6}S^3 - \left(-rT + \frac{\sigma^2}{6}T^3\right)$$

$$= m, \tag{S.13}$$

we finally obtain

$$\mathbb{E}^*\left[e^{-\int_0^T r_s ds}(P(T,S) - K)^+\right] = P(0,S)\Phi\left(\frac{1}{v}\log\frac{P(0,S)}{KP(0,T)} + \frac{v}{2}\right)$$

$$-KP(0,T)\Phi\left(\frac{1}{v}\log\frac{P(0,S)}{KP(0,T)}-\frac{v}{2}\right).$$

On the other hand, Relation (S.13) can be independently recovered from the fact that

$$t \mapsto \frac{P(t,S)}{P(t,T)}$$

is a martingale under $\widehat{\mathbb{P}}$, see Chapter 8. Hence, we have

$$\widehat{\mathbb{E}}[P(T,S) \mid \mathcal{F}_t] = \widehat{\mathbb{E}}\left[\frac{P(T,S)}{P(T,T)} \mid \mathcal{F}_t\right] = \frac{P(t,S)}{P(t,T)},$$

and

$$\frac{P(0,S)}{P(0,T)} = \widehat{\mathbb{E}}[P(T,S) \mid \mathcal{F}_0] = \mathbb{E}\left[(e^{m+X} - K)^+\right],$$

where X is a centered Gaussian random variable with variance $v^2 > 0$, which yields

$$m + \frac{1}{2}v^2 = \log\frac{P(0,S)}{P(0,T)}.$$

Exercise 8.4. Stochastic string model. (Exercise 7.6 continued). We have

$$\mathbb{E}^*\left[e^{-\int_0^T r_s ds}(P(T,S) - K)^+\right]$$

$$= e^{-rT}\,\mathbb{E}^*\left[\left(e^{-(S-T)r-\alpha T(S-T)^3/3+\sigma\int_0^{S-T}B(T,x)dx} - K\right)^+\right]$$

$$= e^{-rT}\,\mathbb{E}\left[(x\,e^{m+X} - K)^+\right],$$

where $x = e^{-(S-T)r}$, $m = -\alpha T(S-T)^3/3$, and

$$X = \sigma\int_0^{S-T}B(T,x)dx \simeq \mathcal{N}\left(0, \frac{\sigma^2}{3}t(T-t)^3\right).$$

Given the relation $\alpha = \sigma^2/2$, this yields

$$\mathbb{E}^*\left[e^{-\int_0^T r_s ds}(P(T,S) - K)^+\right]$$

$$= e^{-rS}\Phi\left(\sigma\sqrt{T(S-T)^3/12} + \frac{\log(e^{-(S-T)r}/K)}{\sigma\sqrt{T(S-T)^3/3}}\right)$$

$$-K e^{-rT}\Phi\left(-\sigma\sqrt{T(S-T)^3/12} + \frac{\log(e^{-(S-T)r}/K)}{\sigma\sqrt{T(S-T)^3/3}}\right)$$

$$= P(0,S)\Phi\left(\sigma\sqrt{T(S-T)^3/12} + \frac{\log(e^{-(S-T)r}/K)}{\sigma\sqrt{T(S-T)^3/3}}\right)$$

$$-KP(0,T)\Phi\left(-\sigma\sqrt{T(S-T)^3/12}+\frac{\log(\mathrm{e}^{-(S-T)r}/K)}{\sigma\sqrt{T(S-T)^3/3}}\right).$$

Exercise 8.5. (Exercise 5.5 continued).

(1) By Exercise 5.5-(7), we have

$$\mathbb{E}\left[\frac{\mathrm{d}\widehat{\mathbb{P}}_T}{\mathrm{d}\mathbb{P}^*}\,\bigg|\,\mathcal{F}_t\right]=\frac{P(t,T)}{P(0,T)}\,\mathrm{e}^{-\int_0^t r_s^{\mathrm{T}}ds}=\mathrm{e}^{\sigma B_t-\sigma^2 t/2},\quad 0\leqslant t\leqslant T.$$

(2) By the Girsanov Theorem 2.1, the process $\left(\widehat{B}_t\right)_{t\in\mathbb{R}_+}$ defined as

$$\widehat{B}_t:=B_t-\sigma t,\qquad t\geqslant 0,$$

is a standard Brownian motion under $\widehat{\mathbb{P}}_T$.

(3) We have

$$\begin{aligned}
\log P(T,S)&=-\mu(S-T)+\sigma\int_0^T\frac{S-T}{S-s}dB_s\\
&=-\mu(S-T)+\sigma\int_0^t\frac{S-T}{S-s}dB_s+\sigma\int_t^T\frac{S-T}{S-s}dB_s\\
&=\frac{S-T}{S-t}\log P(t,S)+\sigma\int_t^T\frac{S-T}{S-s}dB_s\\
&=\frac{S-T}{S-t}\log P(t,S)+\sigma\int_t^T\frac{S-T}{S-s}d\widehat{B}_s+\sigma^2\int_t^T\frac{S-T}{S-s}ds\\
&=\frac{S-T}{S-t}\log P(t,S)+\sigma\int_t^T\frac{S-T}{S-s}d\widehat{B}_s+(S-T)\sigma^2\log\frac{S-t}{S-T},
\end{aligned}$$

$0\leqslant t\leqslant T<S$.

(4) We have

$$\begin{aligned}
&P(t,T)\widehat{\mathbb{E}}\big[(P(T,S)-K)^+\,\big|\,\mathcal{F}_t\big]\\
&=P(t,T)\,\mathbb{E}\left[\left(\mathrm{e}^X-K\right)^+\big|\,\mathcal{F}_t\right]\\
&=P(t,T)\,\mathrm{e}^{m(t,T)+v^2(t,T)/2}\Phi\left(\frac{v(t,T)}{2}+\frac{m(t,T)+v^2(t,T)/2-\log K}{v(t,T)}\right)\\
&\quad-KP(t,T)\Phi\left(-\frac{v(t,T)}{2}+\frac{m(t,T)+v^2(t,T)/2-\log K}{v(t,T)}\right)\\
&=P(t,T)\,\mathrm{e}^{m(t,T)+v^2(t,T)/2}\Phi\left(v(t,T)+\frac{m(t,T)-\log K}{v(t,T)}\right)\\
&\quad-KP(t,T)\Phi\left(\frac{m(t,T)-\log K}{v(t,T)}\right),
\end{aligned}$$

with

$$m(t,T) := \frac{S-T}{S-t} \log P(t,S) + (S-T)\sigma^2 \log \frac{S-t}{S-T}$$

and

$$v^2(t,T) := \sigma^2 \int_t^T \frac{(S-T)^2}{(S-s)^2} ds$$

$$= (S-T)^2 \sigma^2 \left(\frac{1}{S-T} - \frac{1}{S-t} \right)$$

$$= (S-T)\sigma^2 \frac{T-t}{S-t}, \qquad 0 \leqslant t \leqslant T < S,$$

hence

$$P(t,T)\widehat{\mathbb{E}}\left[(P(T,S) - K)^+ \mid \mathcal{F}_t\right]$$

$$= P(t,T)(P(t,S))^{(S-T)(S-t)} \left(\frac{S-t}{S-T} \right)^{(S-T)\sigma^2} e^{v^2(t,T)/2}$$

$$\times \Phi \left(v(t,T) + \frac{1}{v(t,T)} \log \left(\frac{(P(t,S))^{(S-T)(S-t)}}{K} \left(\frac{S-t}{S-T} \right)^{(S-T)\sigma^2} \right) \right)$$

$$- K P(t,T)\Phi \left(\frac{1}{v(t,T)} \log \left(\frac{(P(t,S))^{(S-T)(S-t)}}{K} \left(\frac{S-t}{S-T} \right)^{(S-T)\sigma^2} \right) \right).$$

Exercise 8.6. Bond put options (1).

(1) We have

$$d\left(\frac{P(t,T_2)}{P(t,T_1)} \right)$$

$$= \frac{dP(t,T_2)}{P(t,T_1)} + P(t,T_2)d\left(\frac{1}{P(t,T_1)} \right) + dP(t,T_2) \cdot d\left(\frac{1}{P(t,T_1)} \right)$$

$$= \frac{dP(t,T_2)}{P(t,T_1)} + P(t,T_2)\left(-\frac{dP(t,T_1)}{(P(t,T_1))^2} + \frac{dP(t,T_1) \cdot dP(t,T_1)}{(P(t,T_1))^3} \right)$$

$$- \frac{dP(t,T_1) \cdot dP(t,T_2)}{(P(t,T_1))^2}$$

$$= \frac{1}{P(t,T_1)} \left(r_t P(t,T_2)dt + \zeta_2(t)P(t,T_2)dB_t \right)$$

$$- \frac{P(t,T_2)}{(P(t,T_1))^2} \left(r_t P(t,T_1)dt + \zeta_1(t)P(t,T_1)dB_t \right)$$

$$+ \frac{P(t,T_2)}{(P(t,T_1))^3} \left((r_t P(t,T_1)dt + \zeta_1(t)P(t,T_1)dB_t)^2 \right)$$

$$-\frac{(r_t P(t,T_1)dt + \zeta_1(t)P(t,T_1)dB_t) \cdot (r_t P(t,T_2)dt + \zeta_2(t)P(t,T_2)dB_t)}{(P(t,T_1))^2}$$

$$= \zeta_2(t)\frac{P(t,T_2)}{P(t,T_1)}dB_t - \zeta_1(t)\frac{P(t,T_2)}{P(t,T_1)}dB_t$$

$$+ (\zeta_1(t))^2 \frac{P(t,T_2)}{P(t,T_1)}dt - \zeta_1(t)\zeta_2(t)\frac{P(t,T_2)}{P(t,T_1)}dt$$

$$= -\frac{P(t,T_2)}{P(t,T_1)}\zeta_1(t)(\zeta_2(t) - \zeta_1(t))dt + \frac{P(t,T_2)}{P(t,T_1)}(\zeta_2(t) - \zeta_1(t))dB_t$$

$$= \frac{P(t,T_2)}{P(t,T_1)}(\zeta_2(t) - \zeta_1(t))(dB_t - \zeta_1(t)dt)$$

$$= (\zeta_2(t) - \zeta_1(t))\frac{P(t,T_2)}{P(t,T_1)}d\widehat{B}_t = (\zeta_2(t) - \zeta_1(t))\frac{P(t,T_2)}{P(t,T_1)}d\widehat{B}_t,$$

where $d\widehat{B}_t = dB_t - \zeta_1(t)dt$ is a *standard Brownian motion* under the T_1-forward measure $\widehat{\mathbb{P}}$.

(2) From Question (1) or Relation (8.16), we have

$$P(T_1,T_2) = \frac{P(T_1,T_2)}{P(T_1,T_1)}$$

$$= \frac{P(t,T_2)}{P(t,T_1)}\exp\left(\int_t^{T_1}(\zeta_2(s) - \zeta_1(s))d\widehat{B}_s - \frac{1}{2}\int_t^{T_1}(\zeta_2(s) - \zeta_1(s))^2 ds\right)$$

$$= \frac{P(t,T_2)}{P(t,T_1)}\exp\left(X - \frac{v^2}{2}\right),$$

where X is a centered Gaussian random variable with variance

$$v^2 = \int_t^{T_1}(\zeta_2(s) - \zeta_1(s))^2 ds,$$

independent of $\mathcal{F}_t$ under $\widehat{\mathbb{P}}$, $0 \leqslant t \leqslant T_1$. Hence, by the hint (8.36), we find

$$\mathbb{E}^*\left[e^{-\int_0^{T_1} r_s ds}(K - P(T_1,T_2))^+ \mid \mathcal{F}_t\right]$$

$$= P(t,T_1)\widehat{\mathbb{E}}\left[(K - P(T_1,T_2))^+ \mid \mathcal{F}_t\right]$$

$$= P(t,T_1)\left(K\Phi\left(\frac{v}{2} + \frac{1}{v}\log\frac{K}{x}\right) - \frac{P(t,T_2)}{P(t,T_1)}\Phi\left(-\frac{v}{2} + \frac{1}{v}\log\frac{K}{x}\right)\right)$$

$$= KP(t,T_1)\Phi\left(\frac{v}{2} + \frac{1}{v}\log\frac{K}{x}\right) - P(t,T_2)\Phi\left(-\frac{v}{2} + \frac{1}{v}\log\frac{K}{x}\right),$$

with $x = P(t,T_2)/P(t,T_1)$.

Exercise 8.7. We have

$$\frac{dP(t,S)}{P(t,S)} = r_t dt + \sigma(t, r_t) \frac{\partial}{\partial x} \log G(t, r_t) dB_t,$$

$$\frac{dP(t,T)}{P(t,T)} = r_t dt + \sigma(t, r_t) \frac{\partial}{\partial x} \log F(t, r_t) dB_t,$$

hence

$$d\widehat{B}_t = dB_t - \frac{dP(t,T)}{P(t,T)} \cdot dB_t = dB_t - \sigma(t, r_t) \frac{\partial}{\partial x} \log F(t, r_t) dt,$$

and

$$\frac{dP(t,S)}{P(t,S)}$$

$$= r_t dt + \sigma^2(t, r_t) \frac{\partial}{\partial x} \log F(t, r_t) \frac{\partial}{\partial x} \log G(t, r_t) dt + \sigma(t, r_t) \frac{\partial}{\partial x} \log G(t, r_t) d\widehat{B}_t.$$

When the bond price $P(t, S)$ takes the form $P(t, S) = G(t, r_t)$, we also have

$$dP(t,S)$$

$$= r_t P(t,S) dt + \sigma^2(t, r_t) \frac{\partial}{\partial x} \log F(t, r_t) \frac{\partial G}{\partial x}(t, r_t) dt + \sigma(t, r_t) \frac{\partial G}{\partial x}(t, r_t) d\widehat{B}_t.$$

Exercise 8.8. Forward contracts. We have

$$\mathbb{E}^* \left[e^{-\int_t^T r_s ds} (P(T, S) - K) \Big| \mathcal{F}_t \right] = P(t, T) \widehat{\mathbb{E}}[(P(T, S) - K) \mid \mathcal{F}_t]$$

$$= P(t, T) \widehat{\mathbb{E}} \left[\frac{P(T, S) - K}{P(T, T)} \Big| \mathcal{F}_t \right]$$

$$= P(t, T) \widehat{\mathbb{E}}[P(T, S) - K \mid \mathcal{F}_t]$$

$$= P(t, T) \widehat{\mathbb{E}}[P(T, S) \mid \mathcal{F}_t] - K P(t, T)$$

$$= P(t, T) \widehat{\mathbb{E}} \left[\frac{P(T, S)}{P(T, T)} \Big| \mathcal{F}_t \right] - K P(t, T)$$

$$= P(t, T) \frac{P(t, S)}{P(t, T)} - K P(t, T)$$

$$= P(t, S) - K P(t, T),$$

since

$$t \mapsto \frac{P(t, S)}{P(t, T)}$$

is a martingale under the forward measure $\widehat{\mathbb{P}}$. The corresponding (static) hedging strategy is given by buying one bond with maturity S and by short selling K units of the bond with maturity T.

Remark: The above result can also be obtained by a direct argument using the tower property (11.7), as follows:

$$
\mathbb{E}^*\left[e^{-\int_t^T r_s ds}(P(T,S) - K)\,\middle|\,\mathcal{F}_t\right]
$$
$$
= \mathbb{E}^*\left[e^{-\int_t^T r_s ds}\left(\mathbb{E}^*\left[e^{-\int_T^S r_s ds}\,\middle|\,\mathcal{F}_T\right] - K\right)\middle|\,\mathcal{F}_t\right]
$$
$$
= \mathbb{E}^*\left[e^{-\int_t^T r_s ds}\,\mathbb{E}^*\left[e^{-\int_T^S r_s ds} - K\,\middle|\,\mathcal{F}_T\right]\middle|\,\mathcal{F}_t\right]
$$
$$
= \mathbb{E}^*\left[\mathbb{E}^*\left[e^{-\int_t^S r_s ds} - K e^{-\int_t^T r_s ds}\,\middle|\,\mathcal{F}_T\right]\middle|\,\mathcal{F}_t\right]
$$
$$
= \mathbb{E}^*\left[e^{-\int_t^S r_s ds} - K e^{-\int_t^T r_s ds}\,\middle|\,\mathcal{F}_t\right]
$$
$$
= P(t,S) - K P(t,T), \qquad 0 \leqslant t \leqslant T.
$$

Exercise 8.9.

(1) The payoff of the convertible bond is given by $\mathrm{Max}(\alpha S_\tau, P(\tau,T))$.
(2) We have

$$
\mathrm{Max}(\alpha S_\tau, P(\tau,T)) = P(\tau,T)\mathbb{1}_{\{\alpha S_\tau \leqslant P(\tau,T)\}} + \alpha S_\tau \mathbb{1}_{\{\alpha S_\tau > P(\tau,T)\}}
$$
$$
= P(\tau,T) + (\alpha S_\tau - P(\tau,T))\mathbb{1}_{\{\alpha S_\tau > P(\tau,T)\}}
$$
$$
= P(\tau,T) + (\alpha S_\tau - P(\tau,T))^+
$$
$$
= P(\tau,T) + \alpha\left(S_\tau - \frac{P(\tau,T)}{\alpha}\right)^+,
$$

where the latter European call option payoff has the strike price $K := P(\tau,T)/\alpha$.
(3) From the Markov property applied at time $t \in [0,\tau]$, we will write the corporate bond price as a function $C(t, S_t, r_t)$ of the underlying asset price and interest rate, hence we have

$$
C(t, S_t, r_t) = \mathbb{E}^*\left[e^{-\int_t^\tau r_s ds}\,\mathrm{Max}(\alpha S_\tau, P(\tau,T))\,\middle|\,\mathcal{F}_t\right].
$$

The martingale property follows from the equalities

$$
e^{-\int_0^t r_s ds}C(t, S_t, r_t) = e^{-\int_0^t r_s ds}\,\mathbb{E}^*\left[e^{-\int_t^\tau r_s ds}\,\mathrm{Max}(\alpha S_\tau, P(\tau,T))\,\middle|\,\mathcal{F}_t\right]
$$
$$
= \mathbb{E}^*\left[e^{-\int_0^\tau r_s ds}\,\mathrm{Max}(\alpha S_\tau, P(\tau,T))\,\middle|\,\mathcal{F}_t\right].
$$

(4) We have

$$d\left(e^{-\int_0^t r_s ds} C(t, S_t, r_t)\right)$$

$$= -r_t e^{-\int_0^t r_s ds} C(t, S_t, r_t) dt + e^{-\int_0^t r_s ds} \frac{\partial C}{\partial t}(t, S_t, r_t) dt$$

$$+ e^{-\int_0^t r_s ds} \frac{\partial C}{\partial x}(t, S_t, r_t)\left(r S_t dt + \sigma S_t dB_t^{(1)}\right)$$

$$+ e^{-\int_0^t r_s ds} \frac{\partial C}{\partial y}(t, S_t, r_t)\left(\gamma(t, r_t) dt + \eta(t, S_t) dB_t^{(2)}\right)$$

$$+ \frac{\sigma^2}{2} e^{-\int_0^t r_s ds} S_t^2 \frac{\partial^2 C}{\partial x^2}(t, S_t, r_t) dt$$

$$+ \frac{1}{2} e^{-\int_0^t r_s ds} \eta^2(t, r_t) \frac{\partial^2 C}{\partial y^2}(t, S_t, r_t) dt$$

$$+ \rho \sigma S_t \eta(t, r_t) e^{-\int_0^t r_s ds} \frac{\partial^2 C}{\partial x \partial y}(t, S_t, r_t) dt. \qquad (S.14)$$

The martingale property of $\left(e^{-\int_0^t r_s ds} C(t, S_t, r_t)\right)_{t \in \mathbb{R}_+}$ shows that the sum of terms in factor of dt vanishes in the above Relation (S.14), which yields the PDE

$$0 = - yC(t, x, y) + \frac{\partial C}{\partial t}(t, x, y) dt + ry \frac{\partial C}{\partial x}(t, x, y) + \gamma(t, y) \frac{\partial C}{\partial y}(t, x, y)$$

$$+ \frac{\sigma^2}{2} x^2 \frac{\partial^2 C}{\partial x^2}(t, x, y) + \eta^2(t, y) \frac{1}{2} \frac{\partial^2 C}{\partial y^2}(t, x, y)$$

$$+ \rho \sigma x \eta(t, y) \frac{\partial^2 C}{\partial x \partial y}(t, x, y),$$

with the terminal condition

$$C(\tau, x, y) = \text{Max}(\alpha x, F(\tau, y)), \quad \text{where} \quad F(\tau, r_\tau) = P(\tau, T)$$

is the bond pricing function.

(5) Using the martingale property of discounted bond prices, see Proposition 4.1, the convertible bond can be priced as

$$\mathbb{E}^*\left[e^{-\int_t^\tau r_s ds} \text{Max}(\alpha S_\tau, P(\tau, T)) \mid \mathcal{F}_t\right] \qquad (S.15)$$

$$= \mathbb{E}^*\left[e^{-\int_t^\tau r_s ds} P(\tau, T) \mid \mathcal{F}_t\right]$$

$$+ \alpha \mathbb{E}^*\left[e^{-\int_t^\tau r_s ds}\left(S_\tau - \frac{P(\tau, T)}{\alpha}\right)^+ \mid \mathcal{F}_t\right]$$

$$= \mathbb{E}^*\left[e^{-\int_t^\tau r_s ds} P(\tau, T) \mid \mathcal{F}_t\right]$$

$$+ \alpha \, \mathbb{E}^* \left[\mathbb{E}^* \left[e^{- \int_t^\tau r_s ds} \left(S_\tau - \frac{P(\tau, T)}{\alpha} \right)^+ \Bigg| \mathcal{F}_\tau \right] \Bigg| \mathcal{F}_t \right]$$

$$= P(t, T)$$

$$+ \alpha \, \mathbb{E}^* \left[e^{- \int_t^\tau r_s ds} \, \mathbb{E}^* \left[e^{- \int_\tau^T r_s ds} P(\tau, T) \Big| \mathcal{F}_\tau \right] \left(S_\tau - \frac{P(\tau, T)}{\alpha} \right)^+ \Bigg| \mathcal{F}_t \right]$$

$$= P(t, T)$$

$$+ \alpha \, \mathbb{E}^* \left[\mathbb{E}^* \left[e^{- \int_t^\tau r_s ds} e^{- \int_\tau^T r_s ds} P(\tau, T) \left(S_\tau - \frac{P(\tau, T)}{\alpha} \right)^+ \Bigg| \mathcal{F}_\tau \right] \Bigg| \mathcal{F}_t \right]$$

$$= P(t, T) + \alpha \, \mathbb{E}^* \left[e^{- \int_t^\tau r_s ds} \, e^{- \int_\tau^T r_s ds} P(\tau, T) \left(S_\tau - \frac{P(\tau, T)}{\alpha} \right)^+ \Bigg| \mathcal{F}_t \right]$$

$$= P(t, T) + \alpha \, \mathbb{E}^* \left[e^{- \int_t^T r_s ds} \left(\frac{S_\tau}{P(\tau, T)} - \frac{1}{\alpha} \right)^+ \Bigg| \mathcal{F}_t \right]$$

$$= P(t, T) + \alpha P(t, T) \widehat{\mathbb{E}} \left[\left(\frac{S_\tau}{P(\tau, T)} - \frac{1}{\alpha} \right)^+ \Bigg| \mathcal{F}_t \right], \qquad 0 \leqslant t \leqslant \tau \leqslant T,$$

where we used the tower property (11.7).

(6) We find

$$dZ_t = (\sigma - \sigma_B(t)) Z_t d\widehat{B}_t,$$

where $\left(\widehat{B}_t \right)_{t \in \mathbb{R}_+}$ is a standard Brownian motion under the forward measure $\widehat{\mathbb{P}}$.

(7) By modeling $(Z_t)_{t \in \mathbb{R}_+}$ as the geometric Brownian motion

$$dZ_t = \sigma(t) Z_t d\widehat{B}_t,$$

Relation (S.15) shows that the convertible bond can be priced as

$$P(t, T) + \alpha S_t \Phi(d_+(t, T)) - P(t, T) \Phi(d_-(t, T)),$$

where

$$d_+(t, T) = \frac{1}{v(t, T)} \left(\frac{v^2(t, T)}{2} + \log \frac{S_t}{P(t, T)} \right),$$

$$d_-(t, T) = \frac{1}{v(t, T)} \left(-\frac{v^2(t, T)}{2} + \log \frac{S_t}{P(t, T)} \right),$$

and $v^2(t, T) = \int_t^T \sigma^2(s, T) ds$, $t \in [0, T]$.

Exercise 8.10. Bond option hedging (1).

(1) We have

$$P(t,T) = P(s,T) \exp\left(\int_s^t r_u du + \int_s^t \zeta^T(u) dB_u - \frac{1}{2} \int_s^t |\zeta^T(u)|^2 du \right),$$

$0 \leqslant s \leqslant t \leqslant T$.

(2) We have

$$d\left(e^{-\int_0^t r_s ds} P(t,T) \right) = e^{-\int_0^t r_s ds} \zeta^T(t) P(t,T) dB_t,$$

which gives a martingale under $\mathbb{P}^*$ after integration, from the properties of the Itô integral.

(3) By the martingale property of the previous question, we have

$$\mathbb{E}^* \left[e^{-\int_0^T r_s ds} \,\Big|\, \mathcal{F}_t \right] = \mathbb{E}^* \left[P(T,T) e^{-\int_0^T r_s ds} \,\Big|\, \mathcal{F}_t \right]$$

$$= P(t,T) e^{-\int_0^t r_s ds}, \qquad 0 \leqslant t \leqslant T.$$

We can also write

$$P(t,T) = e^{\int_0^t r_s ds} \mathbb{E}^* \left[e^{-\int_0^T r_s ds} \,\Big|\, \mathcal{F}_t \right]$$

$$= \mathbb{E}^* \left[e^{\int_0^t r_s ds} e^{-\int_0^T r_s ds} \,\Big|\, \mathcal{F}_t \right]$$

$$= \mathbb{E}^* \left[e^{-\int_t^T r_s ds} \,\Big|\, \mathcal{F}_t \right], \qquad 0 \leqslant t \leqslant T,$$

since $e^{-\int_0^t r_s ds}$ is an $\mathcal{F}_t$-measurable random variable.

(4) Itô's formula yields

$$d\left(\frac{P(t,S)}{P(t,T)} \right) = \frac{P(t,S)}{P(t,T)} \left(\zeta^S(t) - \zeta^T(t) \right) \left(dB_t - \zeta^T(t) dt \right)$$

$$= \frac{P(t,S)}{P(t,T)} \left(\zeta^S(t) - \zeta^T(t) \right) d\widehat{B}_t, \qquad (S.16)$$

where $\left(\widehat{B}_t \right)_{t \in \mathbb{R}_+}$ is a standard Brownian motion under $\widehat{\mathbb{P}}$ by the Girsanov theorem.

(5) From (S.16), we have

$$\frac{P(t,S)}{P(t,T)} = \frac{P(0,S)}{P(0,T)}$$

$$\times \exp\left(\int_0^t \left(\zeta^S(s) - \zeta^T(s) \right) dB_s - \frac{1}{2} \int_0^t \left(|\zeta^S(s)|^2 - |\zeta^T(s)|^2 \right) ds \right)$$

$$= \frac{P(0,S)}{P(0,T)} \exp\left(\int_0^t \left(\zeta^S(s) - \zeta^T(s) \right) d\widehat{B}_s - \frac{1}{2} \int_0^t |\zeta^S(s) - \zeta^T(s)|^2 ds \right),$$

hence

$$\frac{P(u,S)}{P(u,T)}$$

$$= \frac{P(t,S)}{P(t,T)} \exp\left(\int_t^u (\zeta^S(s) - \zeta^T(s)) d\widehat{B}_s - \frac{1}{2} \int_t^u |\zeta^S(s) - \zeta^T(s)|^2 ds \right),$$

$t \in [0, u]$, and for $u = T$ this yields

$$P(T,S)$$

$$= \frac{P(t,S)}{P(t,T)} \exp\left(\int_t^T (\zeta^S(s) - \zeta^T(s)) d\widehat{B}_s - \frac{1}{2} \int_t^T |\zeta^S(s) - \zeta^T(s)|^2 ds \right),$$

since $P(T,T) = 1$.

(6) Let $\widehat{\mathbb{P}}$ denote the forward measure associated to the bond price $P(t,T)$, $0 \leqslant t \leqslant T$. For all $0 < T < S$, we have

$$\mathbb{E}^*\left[e^{-\int_t^T r_s ds}(P(T,S) - K)^+ \,\Big|\, \mathcal{F}_t \right] = P(t,T)\widehat{\mathbb{E}}\left[(P(T,S) - K)^+\,\big|\, \mathcal{F}_t\right]$$

$$= P(t,T)$$

$$\times \widehat{\mathbb{E}}\left[\left(\frac{P(t,S)}{P(t,T)} e^{\int_t^T (\zeta^S(s)-\zeta^T(s)) d\widehat{B}_s - \int_t^T (\zeta^S(s)-\zeta^T(s))^2 ds/2} - K \right)^+ \,\Big|\, \mathcal{F}_t \right]$$

$$= P(t,T)$$

$$\times \widehat{\mathbb{E}}\left[\left(\frac{P(t,S)}{P(t,T)} \exp\left(X - \frac{1}{2}\int_t^T |\zeta^S(s) - \zeta^T(s)|^2 ds \right) - K \right)^+ \,\Big|\, \mathcal{F}_t \right]$$

$$= P(t,T)\widehat{\mathbb{E}}\left[\left(e^{X+m(t,T,S)} - K \right)^+ \,\big|\, \mathcal{F}_t \right],$$

where

$$m(t,T,S) := -\frac{1}{2}v^2(t,T,S) + \log\frac{P(t,S)}{P(t,T)},$$

and X is a centered Gaussian random variable with variance

$$v^2(t,T,S) := \int_t^T |\zeta^S(s) - \zeta^T(s)|^2 ds$$

given $\mathcal{F}_t$. Recall that when X is a centered Gaussian random variable with variance $v^2 > 0$, the expectation of $(e^{m+X} - K)^+$ is given, as in the standard Black-Scholes formula, by

$$\mathbb{E}\left[(e^{m+X} - K)^+\right] = e^{m+v^2/2}\Phi\left(v + \frac{m - \log K}{v}\right) - K\Phi\left(\frac{m - \log K}{v}\right),$$

where

$$\Phi(z) = \int_{-\infty}^z e^{-y^2/2}\frac{dy}{\sqrt{2\pi}}, \qquad z \in \mathbb{R},$$

denotes the standard normal cumulative distribution function and for simplicity of notation we dropped the indices t, T, S in $m(t, T, S)$ and $v^2(t, T, S)$. Consequently, we have

$$\mathbb{E}^* \left[e^{-\int_t^T r_s ds} (P(T, S) - K)^+ \,\middle|\, \mathcal{F}_t \right] = P(t, T) \widehat{\mathbb{E}} \left[(P(T, S) - K)^+ \right]$$

$$= P(t, S) \Phi \left(\frac{v}{2} + \frac{1}{v} \log \frac{P(t, S)}{KP(t, T)} \right)$$

$$- KP(t, T) \Phi \left(-\frac{v}{2} + \frac{1}{v} \log \frac{P(t, S)}{KP(t, T)} \right).$$

(7) The self-financing hedging strategy that hedges the bond call option is obtained by holding a (possibly fractional) quantity

$$\Phi \left(\frac{v}{2} + \frac{1}{v} \log \frac{P(t, S)}{KP(t, T)} \right)$$

of the bond with maturity S, and by shorting a quantity

$$K\Phi \left(-\frac{v}{2} + \frac{1}{v} \log \frac{P(t, S)}{KP(t, T)} \right)$$

of the bond with maturity T.

Exercise 8.11. Bond put options (2). (Exercise 6.3 continued).

(1) From Exercise 6.3-(7), we have

$$e^{\int_0^t r_s ds} P(t, T) = P(0, T) \exp \left(\int_0^t \zeta_s dB_s - \frac{1}{2} \int_0^t \zeta_s^2 ds \right),$$

hence

$$\frac{d\widehat{\mathbb{P}}}{d\mathbb{P}^*} = \mathbb{E} \left[\frac{d\widehat{\mathbb{P}}}{d\mathbb{P}^*} \,\middle|\, \mathcal{F}_T \right]$$

$$= \frac{1}{P(0, T)} e^{-\int_0^T r_s ds}$$

$$= \exp \left(\int_0^t \zeta_s dB_s - \frac{1}{2} \int_0^t \zeta_s^2 ds \right).$$

(2) We have

$$dr_t = (\theta(t) - ar_t) + \sigma dB_t = (\theta(t) - ar_t) + \sigma \left(\sigma C(t, T) dt + d\widehat{B}_t \right),$$

where $\left(\widehat{B}_t \right)_{t \in \mathbb{R}_+}$ is a standard Brownian motion under $\widehat{\mathbb{P}}$.

(3) We have

$$d\left(\frac{P(t,S)}{P(t,T)}\right) = \frac{P(t,S)}{P(t,T)}\left(\zeta_t^S - \zeta_t^T\right)\left(dB_t - \zeta_t^T dt\right) = \frac{P(t,S)}{P(t,T)}\left(\zeta_t^S - \zeta_t^T\right)d\widehat{B}_t,$$

$0 \leqslant t \leqslant T$.

(4) We have

$$\widehat{\mathbb{E}}[P(T,S) \mid \mathcal{F}_t] = \widehat{\mathbb{E}}\left[\frac{P(T,S)}{P(T,T)}\,\Big|\,\mathcal{F}_t\right] = \frac{P(t,S)}{P(t,T)}, \qquad 0 \leqslant t \leqslant T \leqslant S,$$

hence

$$\begin{aligned}
\frac{P(t,S)}{P(t,T)} &= \widehat{\mathbb{E}}[P(T,S) \mid \mathcal{F}_T] \\
&= \widehat{\mathbb{E}}\left[e^{A(T,S)+X_T C(T,S)} \mid \mathcal{F}_T\right] \\
&= \exp\left(A(T,S) + C(T,S)\,\mathbb{E}[X_T \mid \mathcal{F}_t] + \frac{1}{2}C^2(T,S)\,\mathrm{Var}[X_T \mid \mathcal{F}_t]\right),
\end{aligned}$$

hence

$$A(T,S) + C(T,S)\,\mathbb{E}[X_T \mid \mathcal{F}_t] + \frac{1}{2}C^2(T,S)\,\mathrm{Var}[X_T \mid \mathcal{F}_t] = \log\frac{P(t,S)}{P(t,T)}.$$

(5) We have, using the call-put duality

$$\begin{aligned}
P(t,T)&\widehat{\mathbb{E}}\left[(K - P(T,S))^+ \mid \mathcal{F}_t\right] \\
&= P(t,T)\widehat{\mathbb{E}}[K - P(T,S) \mid \mathcal{F}_t] + P(t,T)\widehat{\mathbb{E}}\left[(P(T,S) - K)^+ \mid \mathcal{F}_t\right] \\
&= KP(t,T) - P(t,T)\widehat{\mathbb{E}}[P(T,S) \mid \mathcal{F}_t] + P(t,T)\widehat{\mathbb{E}}\left[(P(T,S) - K)^+ \mid \mathcal{F}_t\right] \\
&= KP(t,T) - P(t,S) + P(t,T)\widehat{\mathbb{E}}\left[\left(e^X - K\right)^+ \mid \mathcal{F}_t\right],
\end{aligned}$$

where

$$m_t = A(T,S) + C(T,S)\,\mathbb{E}[X_T \mid \mathcal{F}_t],$$

and X is a centered Gaussian random variable with variance

$$v^2(t,T) = C^2(T,S)\,\mathrm{Var}[X_T \mid \mathcal{F}_t]$$

given $\mathcal{F}_t$, hence

$$\begin{aligned}
P(t,T)&\widehat{\mathbb{E}}\left[(K - P(T,S))^+ \mid \mathcal{F}_t\right] \\
&= KP(t,T) - P(t,S) + P(t,S)\Phi\left(\frac{v(t,T)}{2} + \frac{m_t + v^2(t,T)/2 - \log K}{v(t,T)}\right) \\
&\quad - KP(t,T)\Phi\left(-\frac{v(t,T)}{2} + \frac{m_t + v^2(t,T)/2 - \log K}{v(t,T)}\right)
\end{aligned}$$

$$= KP(t,T) - P(t,S) + P(t,S)\Phi\left(\frac{v(t,T)}{2} + \frac{1}{v(t,T)}\log\frac{P(t,S)}{KP(t,T)}\right)$$

$$-KP(t,T)\Phi\left(-\frac{v(t,T)}{2} + \frac{1}{v(t,T)}\log\frac{P(t,S)}{KP(t,T)}\right)$$

$$= KP(t,T)\Phi\left(\frac{v(t,T)}{2} + \frac{1}{v(t,T)}\log\frac{KP(t,T)}{P(t,S)}\right)$$

$$-P(t,S)\Phi\left(-\frac{v(t,T)}{2} + \frac{1}{v(t,T)}\log\frac{KP(t,T)}{P(t,S)}\right).$$

Exercise 8.12.

(1) It suffices to check that the definition of $\left(B_t^R\right)_{t\in\mathbb{R}_+}$ implies the covariation identity

$$
\begin{aligned}
dB_t^S \cdot dB_t^R &= dB_t^S \cdot \left(\rho dB_t^S + \sqrt{1-\rho^2}dB_t\right)\\
&= \rho dB_t^S \cdot dB_t^S + \sqrt{1-\rho^2}dB_t^S \cdot dB_t\\
&= \rho dB_t^S \cdot dB_t^S\\
&= \rho dt
\end{aligned}
$$

by Itô's calculus, which shows that $\left(B_t^R\right)_{t\in\mathbb{R}_+}$ is a standard Brownian motion by the Lévy characterization of Brownian motion, see Section 1.4.

(2) By the Itô formula in two variables (1.12), we have

$$
\begin{aligned}
dX_t &= d\left(\frac{S_t}{R_t}\right)\\
&= \frac{dS_t}{R_t} - S_t\frac{dR_t}{R_t^2} + \frac{S_t}{R_t^3}dR_t \cdot dR_t - \frac{1}{R_t^2}dS_t \cdot dR_t\\
&= \frac{1}{R_t}\left(rS_tdt + \sigma^S S_t dB_t^S\right) - \frac{S_t}{R_t^2}\left(r^R R_t dt + \sigma^R R_t dB_t^R\right)\\
&\quad + \frac{S_t}{R_t^3}\left(r^R R_t dt + \sigma^R R_t dB_t^R\right) \cdot \left(r^R R_t dt + \sigma^R R_t dB_t^R\right)\\
&\quad - \frac{1}{R_t^2}\left(rS_tdt + \sigma^S S_t dB_t^S\right) \cdot \left(r^R R_t dt + \sigma^R R_t dB_t^R\right)\\
&= \frac{1}{R_t}\left(rS_tdt + \sigma^S S_t dB_t^S\right) - \frac{S_t}{R_t^2}\left(r^R R_t dt + \sigma^R R_t dB_t^R\right)\\
&\quad + \frac{S_t}{R_t}\left(\sigma^R\right)^2 dt - \rho\frac{S_t}{R_t}\sigma^S\sigma^R dt\\
&= \left(r - r^R + \left(\sigma^R\right)^2 - \rho\sigma^R\sigma^S\right)X_tdt + X_t\left(\sigma^S dB_t^S - \sigma^R dB_t^R\right)
\end{aligned}
$$

$$= \left(r - r^R + \left(\sigma^R\right)^2 - \rho\sigma^R\sigma^S\right)X_t dt + \widehat{\sigma}X_t dB_t^X,$$

where

$$\widehat{\sigma} := \sqrt{\left(\sigma^S\right)^2 - 2\rho\sigma^R\sigma^S + \left(\sigma^R\right)^2}$$

and $\left(B_t^X\right)_{t\in\mathbb{R}_+}$ defined by

$$\begin{aligned}
dB_t^X &:= \frac{1}{\widehat{\sigma}}\left(\sigma^S dB_t^S - \sigma^R dB_t^R\right) \\
&= \frac{1}{\widehat{\sigma}}\left(\sigma^S dB_t^S - \sigma^R\left(\rho dB_t^S + \sqrt{1-\rho^2}dB_t\right)\right) \\
&= \frac{\sigma^S - \rho\sigma^R}{\widehat{\sigma}}dB_t^S - \sqrt{1-\rho^2}\frac{\sigma^R}{\widehat{\sigma}}dB_t, \qquad t \geqslant 0,
\end{aligned}$$

is a standard Brownian motion under $\mathbb{P}^*$ by the Lévy characterization, as

$$\begin{aligned}
&dB_t^X \bullet dB_t^X \\
&= \left(\frac{\sigma^S - \rho\sigma^R}{\widehat{\sigma}}dB_t^S - \sqrt{1-\rho^2}\frac{\sigma^R}{\widehat{\sigma}}dB_t\right) \\
&\quad \bullet \left(\frac{\sigma^S - \rho\sigma^R}{\widehat{\sigma}}dB_t^S - \sqrt{1-\rho^2}\frac{\sigma^R}{\widehat{\sigma}}dB_t\right) \\
&= \left(\frac{\sigma^S - \rho\sigma^R}{\widehat{\sigma}}\right)^2 dB_t^S \bullet dB_t^S + 2\sqrt{1-\rho^2}\frac{\left(\sigma^S - \rho\sigma^R\right)\sigma^R}{\left(\widehat{\sigma}\right)^2}dB_t^S \bullet dB_t \\
&\quad + \left(1-\rho^2\right)\frac{\left(\sigma^R\right)^2}{\left(\widehat{\sigma}\right)^2}dB_t \bullet dB_t \\
&= \left(\frac{\sigma^S - \rho\sigma^R}{\widehat{\sigma}}\right)^2 dt + \left(1-\rho^2\right)\frac{\left(\sigma^R\right)^2}{\left(\widehat{\sigma}\right)^2}dt \\
&= \frac{\left(\sigma^S - \rho\sigma^R\right)^2 + \left(1-\rho^2\right)\left(\sigma^R\right)^2}{\left(\widehat{\sigma}\right)^2}dt \\
&= \frac{\left(\sigma^S - \rho\sigma^R\right)^2 + \left(1-\rho^2\right)\left(\sigma^R\right)^2}{\left(\sigma^S\right)^2 - 2\rho\sigma^R\sigma^S + \left(\sigma^R\right)^2}dt \\
&= dt,
\end{aligned}$$

due to the definition of $\widehat{\sigma}$.

(3) By a standard calculation, we have

$$\begin{aligned}
dY_t &= d\left(e^{at}X_t\right) \\
&= a\,e^{at}X_t dt + e^{at}dX_t
\end{aligned}$$

$$= r \, \mathrm{e}^{at} X_t dt + \mathrm{e}^{at} \widehat{\sigma} X_t dB_t^X$$

$$= r Y_t dt + \widehat{\sigma} Y_t dB_t^X.$$

(4) Letting $\widetilde{Y}_t = \mathrm{e}^{-rt} Y_t = \mathrm{e}^{(a-r)t} S_t / R_t$, $t \in \mathbb{R}_+$, we find, using the Black-Scholes formula,

$$\mathbb{E}^* \left[\left(\frac{S_T}{R_T} - K \right)^+ \Big| \mathcal{F}_t \right] = \mathrm{e}^{-aT} \mathbb{E}^* \left[(Y_T - \mathrm{e}^{aT} K)^+ \Big| \mathcal{F}_t \right]$$

$$= \mathrm{e}^{-(a-r)T} \mathbb{E}^* \left[(\widetilde{Y}_T - \mathrm{e}^{(a-r)T} K)^+ \Big| \mathcal{F}_t \right]$$

$$= \mathrm{e}^{-(a-r)T} \left(\widetilde{Y}_t \Phi \left(\frac{(r-a+\widehat{\sigma}^2/2)(T-t)}{\widehat{\sigma}\sqrt{T-t}} + \frac{1}{\widehat{\sigma}\sqrt{T-t}} \log \frac{S_t}{KR_t} \right) \right.$$

$$\left. - K \mathrm{e}^{(a-r)T} \Phi \left(\frac{(r-a-\widehat{\sigma}^2/2)(T-t)}{\widehat{\sigma}\sqrt{T-t}} + \frac{1}{\widehat{\sigma}\sqrt{T-t}} \log \frac{S_t}{KR_t} \right) \right)$$

$$= \frac{S_t}{R_t} \mathrm{e}^{(r-a)(T-t)} \Phi \left(\frac{(r-a+\widehat{\sigma}^2/2)(T-t)}{\widehat{\sigma}\sqrt{T-t}} + \frac{1}{\widehat{\sigma}\sqrt{T-t}} \log \frac{S_t}{KR_t} \right)$$

$$- K \Phi \left(\frac{(r-a-\widehat{\sigma}^2/2)(T-t)}{\widehat{\sigma}\sqrt{T-t}} + \frac{1}{\widehat{\sigma}\sqrt{T-t}} \log \frac{S_t}{KR_t} \right),$$

hence the price of the quanto option is given by

$$\mathrm{e}^{-(T-t)r} \mathbb{E}^* \left[\left(\frac{S_T}{R_T} - K \right)^+ \Big| \mathcal{F}_t \right]$$

$$= \frac{S_t}{R_t} \mathrm{e}^{-(T-t)a} \Phi \left(\frac{(r-a+\widehat{\sigma}^2/2)(T-t)}{\widehat{\sigma}\sqrt{T-t}} + \frac{1}{\widehat{\sigma}\sqrt{T-t}} \log \frac{S_t}{KR_t} \right)$$

$$- K \mathrm{e}^{-(T-t)r} \Phi \left(\frac{(r-a-\widehat{\sigma}^2/2)(T-t)}{\widehat{\sigma}\sqrt{T-t}} + \frac{1}{\widehat{\sigma}\sqrt{T-t}} \log \frac{S_t}{KR_t} \right),$$

where $\Phi(\cdot)$ denotes the standard normal cumulative distribution function.

Problem 8.13. Bond option hedging (2).

(1) We have

$$\mathbb{E}^* \left[\mathrm{e}^{-\int_t^T r_s ds} (P(T,S) - K)^+ \Big| \mathcal{F}_t \right] = V_T = V_0 + \int_0^T dV_t$$

$$= P(0,T) \widehat{\mathbb{E}} [(P(T,S) - K)^+] + \int_0^t \xi_s^T dP(s,T) + \int_0^t \xi_s^S dP(s,S).$$

(2) We have

$$d\widetilde{V}_t = d \left(\mathrm{e}^{-\int_0^t r_s ds} V_t \right)$$

$$= -r_t e^{-\int_0^t r_s ds} V_t dt + e^{-\int_0^t r_s ds} dV_t$$

$$= -r_t e^{-\int_0^t r_s ds} \left(\xi_t^T P(t,T) + \xi_t^S P(t,S) \right) dt$$

$$+ e^{-\int_0^t r_s ds} \xi_t^T dP(t,T) + e^{-\int_0^t r_s ds} \xi_t^S dP(t,S)$$

$$= \xi_t^T d\widetilde{P}(t,T) + \xi_t^S d\widetilde{P}(t,S).$$

(3) As in Section 8.3 and the solution of Exercise 8.10-(6), we have

$$\mathbb{E}^* \left[e^{-\int_t^T r_s ds} (P(T,S) - K)^+ \,\middle|\, \mathcal{F}_t \right] = P(t,T) \widehat{\mathbb{E}} \left[(P(T,S) - K)^+ \,\middle|\, \mathcal{F}_t \right]$$

$$= P(t,T)$$

$$\times \widehat{\mathbb{E}} \left[\left(\frac{P(t,S)}{P(t,T)} e^{\int_t^T (\zeta^S(s) - \zeta^T(s)) d\widehat{B}_s - \int_t^T |\zeta^S(s) - \zeta^T(s)|^2 ds/2} - K \right)^+ \,\middle|\, \mathcal{F}_t \right]$$

$$= P(t,T) \, \mathbb{E} \left[(e^X - K)^+ \,\middle|\, \mathcal{F}_t \right]$$

$$= P(t,T) e^{m(t,T) + v^2(t,T)/2}$$

$$\times \Phi \left(\frac{v(t,T)}{2} + \frac{1}{v(t,T)} \left(m(t,T) + \frac{1}{2} v^2(t,T) - \log K \right) \right)$$

$$- K P(t,T) \Phi \left(-\frac{v(t,T)}{2} + \frac{1}{v(t,T)} \left(m(t,T) + \frac{1}{2} v^2(t,T) - \log K \right) \right),$$

where X is a centered Gaussian random variable under $\widehat{\mathbb{P}}$, with

$$m(t,T) = -\frac{1}{2} \int_t^T |\zeta^S(s) - \zeta^T(s)|^2 ds + \log \frac{P(t,S)}{P(t,T)}$$

and variance

$$v^2(t,T) := \int_t^T |\zeta^S(s) - \zeta^T(s)|^2 ds,$$

given $\mathcal{F}_t$, *i.e.*

$$P(t,T) \widehat{\mathbb{E}} \left[(P(T,S) - K)^+ \,\middle|\, \mathcal{F}_t \right]$$

$$= P(t,S) \Phi \left(\frac{v(t,T)}{2} + \frac{1}{v(t,T)} \log \frac{P(t,S)}{K P(t,T)} \right)$$

$$- K P(t,T) \Phi \left(-\frac{v(t,T)}{2} + \frac{1}{v(t,T)} \log \frac{P(t,S)}{K P(t,T)} \right)$$

$$= P(t,T) C(X_t, v(t,T)),$$

with $X_t := P(t,S)/P(t,T)$, $t \in [0,T]$.

(4) By Itô's formula and the martingale property of the processes $(X_t)_{t \in [0,T]}$ and

$$\widehat{V}_t = C(X_t, v(t,T)) = \widehat{\mathbb{E}} \left[(P(T,S) - K)^+ \,\middle|\, \mathcal{F}_t \right], \quad 0 \leqslant t \leqslant T,$$

under $\widehat{\mathbb{P}}$ by Proposition 8.3, we have

$$
\begin{aligned}
\widehat{V}_t &= \widehat{\mathbb{E}}\big[(P(T,S)-K)^+ \mid \mathcal{F}_t\big] \\
&= C(X_t, v(t,T)) \\
&= C(X_0, v(0,T)) + \int_0^t \frac{\partial C}{\partial x}(X_s, v(s,T))dX_s \\
&= \widehat{\mathbb{E}}\big[(P(T,S)-K)^+\big] + \int_0^t \frac{\partial C}{\partial x}(X_s, v(s,T))dX_s,
\end{aligned}
$$

since the process $\big(\widehat{\mathbb{E}}\big[(P(T,S)-K)^+ \mid \mathcal{F}_t\big]\big)_{t\in[0,T]}$ is a martingale under $\widehat{\mathbb{P}}$, and this yields (8.39), see also Privault and Teng (2012) and references therein.

(5) By (S.16), we have

$$
\begin{aligned}
d\widehat{V}_t &= d\left(\frac{V_t}{P(t,T)}\right) \\
&= d\widehat{\mathbb{E}}\big[(P(T,S)-K)^+ \mid \mathcal{F}_t\big] \\
&= dC(X_t, v(t,T)) \\
&= \frac{\partial C}{\partial x}(X_t, v(t,T))dX_t \\
&= \big(\zeta^S(t) - \zeta^T(t)\big)\frac{P(t,S)}{P(t,T)}\frac{\partial C}{\partial x}(X_t, v(t,T))d\widehat{B}_t.
\end{aligned}
$$

(6) We have

$$
\begin{aligned}
dV_t &= d\big(P(t,T)\widehat{V}_t\big) \\
&= P(t,T)d\widehat{V}_t + \widehat{V}_t dP(t,T) + d\widehat{V}_t \cdot dP(t,T) \\
&= P(t,S)\frac{\partial C}{\partial x}(X_t, v(t,T))\big(\zeta^S(t) - \zeta^T(t)\big)d\widehat{B}_t + \widehat{V}_t dP(t,T) \\
&\quad + P(t,S)\frac{\partial C}{\partial x}(X_t, v(t,T))\big(\zeta^S(t) - \zeta^T(t)\big)\zeta^T(t)dt \\
&= P(t,S)\frac{\partial C}{\partial x}(X_t, v(t,T))\big(\zeta^S(t) - \zeta^T(t)\big)dB_t + \widehat{V}_t dP(t,T).
\end{aligned}
$$

(7) We have

$$
\begin{aligned}
d\widetilde{V}_t &= d\big(e^{-\int_0^t r_s ds}V_t\big) \\
&= -r_t e^{-\int_0^t r_s ds}V_t dt + e^{-\int_0^t r_s ds}dV_t \\
&= \widetilde{P}(t,S)\frac{\partial C}{\partial x}(X_t, v(t,T))\big(\zeta^S(t) - \zeta^T(t)\big)dB_t + \widehat{V}_t d\widetilde{P}(t,T).
\end{aligned}
$$

(8) By numeraire invariance, see *e.g.* Huang (1985), Jamshidian (1996), or page 184 of Protter (2001), we have

$$dV_t = d\big(P(t,T)\widehat{V}_t\big)$$

$$= \widehat{V}_t dP(t,T) + P(t,T)d\widehat{V}_t + dP(t,T) \cdot d\widehat{V}_t$$

$$= \widehat{V}_t dP(t,T) + P(t,T)\frac{\partial C}{\partial x}(X_t, v(t,T))dX_t$$

$$+ \frac{\partial C}{\partial x}(X_t, v(t,T))dP(t,T) \cdot dX_t$$

$$= \frac{\partial C}{\partial x}(X_t, v(t,T))X_t dP(t,T) + \frac{\partial C}{\partial x}(X_t, v(t,T))P(t,T)dX_t$$

$$+ \frac{\partial C}{\partial x}(X_t, v(t,T))dP(t,T) \cdot dX_t$$

$$+ \left(\widehat{V}_t - \frac{P(t,S)}{P(t,T)}\frac{\partial C}{\partial x}(X_t, v(t,T))\right) dP(t,T)$$

$$= \left(\widehat{V}_t - \frac{P(t,S)}{P(t,T)}\frac{\partial C}{\partial x}(X_t, v(t,T))\right) dP(t,T) + \frac{\partial C}{\partial x}(X_t, v(t,T))dP(t,S),$$

since

$$dP(t,S) = d(X_t P(t,T))$$

$$= X_t dP(t,T) + P(t,T)dX_t + dP(t,T) \cdot dX_t.$$

(9) We have

$$d\widetilde{V}_t = \widetilde{P}(t,S)\frac{\partial C}{\partial x}(X_t, v(t,T))(\zeta^S(t) - \zeta^T(t))dB_t + \widehat{V}_t d\widetilde{P}(t,T)$$

$$= \frac{\partial C}{\partial x}(X_t, v(t,T))d\widetilde{P}(t,S)$$

$$- \frac{P(t,S)}{P(t,T)}\frac{\partial C}{\partial x}(X_t, v(t,T))d\widetilde{P}(t,T) + \widehat{V}_t d\widetilde{P}(t,T)$$

$$= \left(\widehat{V}_t - \frac{P(t,S)}{P(t,T)}\frac{\partial C}{\partial x}(X_t, v(t,T))\right) d\widetilde{P}(t,T)$$

$$+ \frac{\partial C}{\partial x}(X_t, v(t,T))d\widetilde{P}(t,S),$$

hence, by identification with (8.38), the hedging strategy $(\xi_t^T, \xi_t^S)_{t\in[0,T]}$ of the bond call option is given by

$$\xi_t^T = \widehat{V}_t - \frac{P(t,S)}{P(t,T)}\frac{\partial C}{\partial x}(X_t, v(t,T))$$

$$= C(X_t, v(t,T)) - \frac{P(t,S)}{P(t,T)}\frac{\partial C}{\partial x}(X_t, v(t,T)),$$

and

$$\xi_t^S = \frac{\partial C}{\partial x}(X_t, v(t,T)), \qquad 0 \leqslant t \leqslant T.$$

(10) We have

$$
\frac{\partial C}{\partial x}(x,v) = \frac{\partial}{\partial x}\left(x\Phi\left(\frac{v}{2} + \frac{1}{v}\log\frac{x}{K}\right) - K\Phi\left(-\frac{v}{2} + \frac{1}{v}\log\frac{x}{K}\right)\right)
$$

$$
= x\frac{\partial}{\partial x}\Phi\left(\frac{v}{2} + \frac{1}{v}\log\frac{x}{K}\right) - K\frac{\partial}{\partial x}\Phi\left(-\frac{v}{2} + \frac{1}{v}\log\frac{x}{K}\right)
$$

$$
+ \Phi\left(\frac{v}{2} + \frac{1}{v}\log\frac{x}{K}\right)
$$

$$
= \frac{1}{v\sqrt{2\pi}}\exp\left(-\frac{1}{2}\left(\frac{v}{2} + \frac{1}{v}\log\frac{x}{K}\right)^2\right)
$$

$$
- \frac{K}{xv\sqrt{2\pi}}\exp\left(-\frac{1}{2}\left(-\frac{v}{2} + \frac{1}{v}\log\frac{x}{K}\right)^2\right) + \Phi\left(\frac{v}{2} + \frac{1}{v}\log\frac{x}{K}\right)
$$

$$
= \Phi\left(\frac{\log(x/K) + v^2/2}{v}\right),
$$

see also (2.22). As a consequence, we find

$$
\xi_t^T = C(X_t, v(t,T)) - \frac{P(t,S)}{P(t,T)}\frac{\partial C}{\partial x}(X_t, v(t,T))
$$

$$
= \frac{P(t,S)}{P(t,T)}\Phi\left(\frac{v^2(t,T)/2 + \log X_t}{v(t,T)}\right)
$$

$$
- K\Phi\left(-\frac{v(t,T)}{2} + \frac{1}{v(t,T)}\log\frac{P(t,S)}{KP(t,T)}\right)
$$

$$
- \frac{P(t,S)}{P(t,T)}\Phi\left(\frac{\log(X_t/K) + v^2(t,T)/2}{v(t,T)}\right)
$$

$$
= -K\Phi\left(\frac{\log(X_t/K) - v^2(t,T)/2}{v(t,T)}\right),
$$

and

$$
\xi_t^S = \frac{\partial C}{\partial x}(X_t, v(t,T)) = \Phi\left(\frac{\log(X_t/K) + v^2(t,T)/2}{v(t,T)}\right),
$$

$t \in [0,T)$, and the hedging strategy is given by

$$
V_T = \mathbb{E}^*\left[e^{-\int_t^T r_s ds}(P(T,S) - K)^+ \,\Big|\, \mathcal{F}_t\right]
$$

$$
= V_0 + \int_0^t \xi_s^T dP(s,T) + \int_0^t \xi_s^S dP(s,S)
$$

$$
= V_0 - K\int_0^t \Phi\left(\frac{\log(X_s/K) - v^2(s,T)/2}{v(s,T)}\right) dP(s,T)
$$

$$+ \int_0^t \Phi \left(\frac{\log(X_s/K) + v^2(s,T)/2}{v(s,T)} \right) dP(s,S).$$

Consequently, the bond call option can be hedged by short selling a quantity

$$-\xi_t^T = K\Phi \left(\frac{\log(X_t/K) - v^2(t,T)/2}{v(t,T)} \right)$$

$$= K\Phi \left(-\frac{v(t,T)}{2} + \frac{1}{v(t,T)} \log \frac{P(t,S)}{KP(t,T)} \right),$$

of a bond with maturity T, and by holding a quantity

$$\xi_t^S = \Phi \left(\frac{\log(X_t/K) + v^2(t,T)/2}{v(t,T)} \right)$$

$$= \Phi \left(\frac{v(t,T)}{2} + \frac{1}{v(t,T)} \log \frac{P(t,S)}{KP(t,T)} \right)$$

of the bond with maturity S.

Chapter 9

Exercise 9.1.

(1) We price the swaption at $t = 0$, with $T_1 = 4$ years, $T_2 = 5$ years, $T_3 = 6$ years, $T_4 = 7$ years, $\kappa = 5\%$, and the swap rate $(S(t, T_1, T_4))_{t \in [0, T_1]}$ is modeled as a driftless geometric Brownian motion with volatility coefficient $\hat{\sigma} = \sigma_{1,4}(t) = 0.2$ under the forward swap measure $\widehat{\mathbb{P}}_{1.4}$. The discount factors are given by $P(0, T_1) = e^{-4r}$, $P(0, T_2) = e^{-5r}$, $P(0, T_3) = e^{-6r}$, $P(0, T_4) = e^{-7r}$, where $r = 5\%$.

(2) By Proposition 9.6 the price of the swaption is

$$(P(0, T_1) - P(0, T_4))\Phi(d_+(t, T_1))$$
$$-\kappa\Phi(d_-(t, T_1))(P(0, T_2) + P(0, T_3) + P(0, T_4)),$$

where $d_+(t, T_1)$ and $d_-(t, T_1)$ are given in Proposition 9.6, and the LIBOR swap rate $S(0, T_1, T_4)$ is given by

$$S(0, T_1, T_4) = \frac{P(0, T_1) - P(0, T_4)}{P(0, T_1, T_4)}$$

$$= \frac{P(0, T_1) - P(0, T_4)}{P(0, T_2) + P(0, T_3) + P(0, T_4)}$$

$$
\begin{aligned}
&= \frac{e^{-4r} - e^{-7r}}{e^{-5r} + e^{-6r} + e^{-7r}} \\
&= \frac{e^{3r} - 1}{e^{2r} + e^{r} + 1} \\
&= \frac{e^{0.15} - 1}{e^{0.1} + e^{0.05} + 1} \\
&= 0.051271096.
\end{aligned}
$$

By Proposition 9.6 we also have

$$
d_+(t, T_1) = \frac{\log(0.051271096/0.05) + (0.2)^2 \times 4/2}{0.2\sqrt{4}} = 0.526161682,
$$

and

$$
d_-(t, T_1) = \frac{\log(0.051271096/0.05) - (0.2)^2 2 \times 4/2}{0.2\sqrt{4}} = 0.005214714,
$$

Hence, the price of the swaption is given by

$$
\begin{aligned}
\left(e^{-4r} - e^{-7r}\right) &\Phi(0.526161682) \tag{S.17} \\
&- \kappa \Phi(0.005214714)\left(e^{-5r} + e^{-6r} + e^{-7r}\right) \\
&= (0.818730753 - 0.70468809) \times 0.700612 \\
&\quad - 0.05 \times 0.50208 \times (0.818730753 + 0.740818221 + 0.70468809) \\
&= 2.3058251\%.
\end{aligned}
$$

Finally, we need to multiply (S.17) by the notional principal amount of $10 million, *i.e.* $100,000 by interest percentage point, or $1,000 by basis point, which yields $230,582.51.

Exercise 9.2.

(1) We price the floorlet at $t = 0$, with $T_1 = 9$ months, $T_2 = 1$ year, $\kappa = 4.5\%$. The LIBOR rate $(L(t, T_1, T_2))_{t \in [0, T_1]}$ is modeled as a driftless geometric Brownian motion with volatility coefficient $\widehat{\sigma} = \sigma_{1,2}(t) = 0.1$ under the forward measure $\widehat{\mathbb{P}}_2$. The discount factors are given by

$$
P(0, T_1) = e^{-9r/12} \simeq 0.970809519
$$

and

$$
P(0, T_2) = e^{-r} \simeq 0.961269954,
$$

with $r = 3.95\%$.

(2) By (9.8), the price of the floorlet is

$$\mathbb{E}^* \left[e^{-\int_0^{T_2} r_s ds} (\kappa - L(T_1, T_1, T_2))^+ \right]$$
$$= P(0, T_2) \big(\kappa \Phi(-d_-(t, T_1)) - L(0, T_1, T_2) \Phi(-d_+(t, T_1)) \big), \quad \text{(S.18)}$$

where

$$d_+(t, T_1) = \frac{\log(L(0, T_1, T_2)/\kappa) + \sigma^2 T_1/2}{\sigma_1 \sqrt{T_1}},$$

and

$$d_-(t, T_1) = \frac{\log(L(0, T_1, T_2)/\kappa) - \sigma^2 T_1/2}{\sigma \sqrt{T_1}},$$

are given in Proposition 9.2, and the LIBOR rate $L(0, T_1, T_2)$ is given by

$$L(0, T_1, T_2) = \frac{P(0, T_1) - P(0, T_2)}{(T_2 - T_1)P(0, T_2)}$$
$$= \frac{e^{-3r/4} - e^{-r}}{0.25 e^{-r}}$$
$$= 4\big(e^{r/4} - 1 \big)$$
$$\simeq 3.9695675\%.$$

Hence, we have

$$d_+(t, T_1) = \frac{\log(0.039695675/0.045) + (0.1)^2 \times 0.75/2}{0.1 \times \sqrt{0.75}} \simeq -1.404927033,$$

and

$$d_-(t, T_1) = \frac{\log(0.039695675/0.045) - (0.1)^2 \times 0.75/2}{0.1 \times \sqrt{0.75}} \simeq -1.491529573,$$

hence

$$\mathbb{E}^* \left[e^{-\int_0^{T_2} r_s ds} (\kappa - L(T_1, T_1, T_2))^+ \right]$$
$$= 0.961269954 \times \big(\kappa \Phi(1.491529573) - L(0, T_1, T_2) \times \Phi(1.404927033) \big)$$
$$= 0.961269954 \times (0.045 \times 0.932089 - 0.039695675 \times 0.919979)$$
$$\simeq 0.52147141\%.$$

Finally, we need to multiply (S.18) by the notional principal amount of $1 million per interest rate percentage point, *i.e.* $10,000 per percentage point or $100 per basis point, which yields $5214.71.

Exercise 9.3.

(1) We have

$$L(t, T_1, T_2) = L(0, T_1, T_2) \exp\left(\int_0^t \gamma_1(s) dB_s^{(2)} - \frac{1}{2} \int_0^t |\gamma_1(s)|^2 ds \right),$$

$0 \leqslant t \leqslant T_1$, and $L(t, T_2, T_3) = b$, hence

$$\frac{P(t, T_2)}{P(t, T_3)} = 1 + \delta b,$$

and $\widehat{\mathbb{P}}_2 = \widehat{\mathbb{P}}_3$ up to time T_2, while $\widehat{\mathbb{P}}_2 = \widehat{\mathbb{P}}_{1,2}$ up to time T_1 by (8.35).

(2) We have

$$\mathbb{E}^* \left[e^{-\int_t^{T_2} r_s ds} (L(T_1, T_1, T_2) - \kappa)^+ \,\Big|\, \mathcal{F}_t \right]$$

$$= P(t, T_2) \widehat{\mathbb{E}}_2 \left[(L(T_1, T_1, T_2) - \kappa)^+ \mid \mathcal{F}_t \right]$$

$$= P(t, T_2) \widehat{\mathbb{E}}_2 \left[\left(L(t, T_1, T_2) e^{\int_t^{T_1} \gamma_1(s) dB_s^{(2)} - \frac{1}{2} \int_t^{T_1} |\gamma_1(s)|^2 ds} - \kappa \right)^+ \,\Big|\, \mathcal{F}_t \right]$$

$$= P(t, T_2) \mathrm{Bl}(L(t, T_1, T_2), \kappa, \sigma_1(t), 0, T_1 - t),$$

where

$$\sigma_1^2(t) = \frac{1}{T_1 - t} \int_t^{T_1} |\gamma_1(s)|^2 ds,$$

and

$$\mathbb{E}^* \left[e^{-\int_t^{T_3} r_s ds} (L(T_2, T_2, T_3) - \kappa)^+ \,\Big|\, \mathcal{F}_t \right] = P(t, T_3) \widehat{\mathbb{E}}_3 \left[(b - \kappa)^+ \mid \mathcal{F}_t \right]$$

$$= P(t, T_3)(b - \kappa)^+.$$

(3) We have

$$\frac{P(t, T_1)}{P(t, T_1, T_3)} = \frac{P(t, T_1)}{\delta P(t, T_2) + \delta P(t, T_3)}$$

$$= \frac{P(t, T_1)}{\delta P(t, T_2)} \frac{1}{1 + P(t, T_3)/P(t, T_2)}$$

$$= \frac{1 + \delta b}{\delta(\delta b + 2)} (1 + \delta L(t, T_1, T_2)), \qquad 0 \leqslant t \leqslant T_1,$$

and

$$\frac{P(t, T_3)}{P(t, T_1, T_3)} = \frac{P(t, T_3)}{P(t, T_2) + P(t, T_3)}$$

$$= \frac{1}{1 + P(t, T_2)/P(t, T_3)}$$

$$= \frac{1}{(2 + \delta b)\delta}, \qquad 0 \leqslant t \leqslant T_2. \qquad \text{(S.19)}$$

(4) We have

$$S(t, T_1, T_3) = \frac{P(t, T_1)}{P(t, T_1, T_3)} - \frac{P(t, T_3)}{P(t, T_1, T_3)}$$

$$= \frac{1 + \delta b}{\delta(2 + \delta b)}(1 + \delta L(t, T_1, T_2)) - \frac{1}{\delta(2 + \delta b)}$$

$$= \frac{b}{2 + \delta b} + \frac{1 + \delta b}{2 + \delta b}L(t, T_1, T_2), \qquad 0 \leqslant t \leqslant T_2,$$

and

$$dS(t, T_1, T_3) = \gamma_1(t)\frac{1 + \delta b}{2 + \delta b}L(t, T_1, T_2)dB_t^{(2)}$$

$$= \gamma_1(t)\left(S(t, T_1, T_3) - \frac{b}{2 + \delta b}\right)dB_t^{(2)}$$

$$= \sigma_{1,3}(t)S(t, T_1, T_3)dB_t^{(2)},$$

$0 \leqslant t \leqslant T_2$, with

$$\sigma_{1,3}(t) = \left(1 - \frac{b}{S(t, T_1, T_3)(2 + \delta b)}\right)\gamma_1(t)$$

$$= \left(1 - \frac{b}{b + (1 + \delta b)L(t, T_1, T_2)}\right)\gamma_1(t)$$

$$= \frac{(1 + \delta b)L(t, T_1, T_2)}{b + (1 + \delta b)L(t, T_1, T_2)}\gamma_1(t)$$

$$= \frac{(1 + \delta b)L(t, T_1, T_2)}{(2 + \delta b)S(t, T_1, T_2)}\gamma_1(t).$$

Exercise 9.4. Swaption hedging (1).

(1) We have

$$S(T_i, T_i, T_j) = S(t, T_i, T_j)\exp\left(\int_t^{T_i} \sigma_{i,j}(s)dB_s^{i,j} - \frac{1}{2}\int_t^{T_i} |\sigma_{i,j}(s)|^2 ds\right),$$

and

$$V_t = P(t, T_i, T_j)\widehat{\mathbb{E}}_{i,j}\left[(S(T_i, T_i, T_j) - \kappa)^+ \mid \mathcal{F}_t\right]$$

$$= P(t, T_i, T_j)$$

$$\times \widehat{\mathbb{E}}_{i,j}\left[\left(S(t, T_i, T_j)e^{\int_t^{T_i} \sigma_{i,j}(s)dB_s^{i,j} - \int_t^{T_i} |\sigma_{i,j}(s)|^2 ds/2} - \kappa\right)^+ \mid \mathcal{F}_t\right]$$

$$= P(t, T_i, T_j)Bl\big(\kappa, v(t, T_i)/\sqrt{T_i - t}, 0, T_i - t\big)$$

$$= P(t, T_i, T_j)C(S(t, T_i, T_j), \kappa, v(t, T_i)),$$

by Lemma 2.3, where

$$v^2(t, T_i) = \int_t^{T_i} |\sigma_{i,j}(s)|^2 ds,$$

and

$$C(x, \kappa, v) = x\Phi\left(\frac{v}{2} + \frac{1}{v}\log\frac{x}{K}\right) - \kappa\Phi\left(-\frac{v}{2} + \frac{1}{v}\log\frac{x}{K}\right). \qquad (S.20)$$

(2) We apply the Itô formula to the forward portfolio value process

$$\widehat{V}_t = C(S_t, \kappa, v(t, T_i)), \qquad 0 \leqslant t \leqslant T_i, \qquad (S.21)$$

and use the fact that both $\widehat{V}_t$ and $(S(t, T_i, T_j))_{t\in[0,T_i]}$ are martingales under $\widehat{\mathbb{P}}_{i,j}$.

(3) By numeraire invariance, letting $S_t := S(t, T_i, T_j)$, by (S.20) and (S.21) we have, as in Problem 8.13-(8),

$$
\begin{aligned}
dV_t &= d\big(P(t, T_i, T_j)\widehat{V}_t\big) \\
&= \widehat{V}_t dP(t, T_i, T_j) + P(t, T_i, T_j)d\widehat{V}_t + dP(t, T_i, T_j)\boldsymbol{\cdot}\ d\widehat{V}_t \\
&= \widehat{V}_t dP(t, T_i, T_j) + P(t, T_i, T_j)\frac{\partial C}{\partial x}(S_t, \kappa, v(t, T_i))dS_t \\
&\quad + \frac{\partial C}{\partial x}(S_t, \kappa, v(t, T_i))dP(t, T_i, T_j)\boldsymbol{\cdot}\ dS_t \\
&= S_t\frac{\partial C}{\partial x}(S_t, \kappa, v(t, T_i))dP(t, T_i, T_j) + P(t, T_i, T_j)\frac{\partial C}{\partial x}(S_t, \kappa, v(t, T_i))dS_t \\
&\quad + \frac{\partial C}{\partial x}(S_t, \kappa, v(t, T_i))dP(t, T_i, T_j)\boldsymbol{\cdot}\ dS_t \\
&\quad + \left(\widehat{V}_t - S_t\frac{\partial C}{\partial x}(S_t, \kappa, v(t, T_i))\right)dP(t, T_i, T_j) \\
&= \left(\widehat{V}_t - S_t\frac{\partial C}{\partial x}(S_t, \kappa, v(t, T_i))\right)dP(t, T_i, T_j) \\
&\quad + \frac{\partial C}{\partial x}(S_t, \kappa, v(t, T_i))d(P(t, T_i) - P(t, T_j)), \qquad (S.22)
\end{aligned}
$$

since

$$
\begin{aligned}
d(P(t, T_i) - P(t, T_j)) &= d(S_t P(t, T_i, T_j)) \\
&= S_t dP(t, T_i, T_j) + P(t, T_i, T_j)dS_t + dP(t, T_i, T_j)\boldsymbol{\cdot}\ dS_t,
\end{aligned}
$$

see also Privault and Teng (2012) and references therein.

(4) By (2.22) and (S.22), we have

$$dV_t = \Phi\left(\frac{\log(S_t/K)}{v(t, T_i)} + \frac{v(t, T_i)}{2}\right)d(P(t, T_i) - P(t, T_j)) \qquad (S.23)$$

$$-\kappa\Phi\left(\frac{\log(S_t/K)}{v(t,T_i)} - \frac{v(t,T_i)}{2}\right)dP(t,T_i,T_j)$$

$$= \Phi\left(\frac{\log(S_t/K)}{v(t,T_i)} + \frac{v(t,T_i)}{2}\right)d(P(t,T_i) - P(t,T_j))$$

$$-\kappa\Phi\left(\frac{\log(S_t/K)}{v(t,T_i)} - \frac{v(t,T_i)}{2}\right)\sum_{k=i}^{j-1}(T_{k+1}-T_k)dP(t,T_{k+1})$$

$$= \Phi\left(\frac{\log(S_t/K)}{v(t,T_i)} + \frac{v(t,T_i)}{2}\right)dP(t,T_i)$$

$$-\kappa\Phi\left(\frac{\log(S_t/K)}{v(t,T_i)} - \frac{v(t,T_i)}{2}\right)\sum_{k=i}^{j-2}(T_{k+1}-T_k)dP(t,T_{k+1})$$

$$-\Phi\left(\frac{\log(S_t/K)}{v(t,T_i)} + \frac{v(t,T_i)}{2}\right)dP(t,T_j)$$

$$-(T_j - T_{j-1})\kappa\Phi\left(\frac{\log(S_t/K)}{v(t,T_i)} - \frac{v(t,T_i)}{2}\right)dP(t,T_j).$$

Consequently, by matching (9.28) to (S.23), the hedging strategy of the swaption is given by

$$\xi_t^i = \Phi\left(\frac{\log(S_t/K)}{v(t,T_i)} + \frac{v(t,T_i)}{2}\right),$$

$$\xi_t^k = -(T_k - T_{k-1})\kappa\Phi\left(\frac{\log(S_t/K)}{v(t,T_i)} - \frac{v(t,T_i)}{2}\right), \quad i+1 \leqslant k \leqslant j-1,$$

and

$$\xi_t^j = -\Phi\left(\frac{v(t,T_i)}{2} + \frac{\log(S_t/K)}{v(t,T_i)}\right)$$

$$-(T_j - T_{j-1})\kappa\Phi\left(-\frac{v(t,T_i)}{2} + \frac{\log(S_t/K)}{v(t,T_i)}\right).$$

Exercise 9.5. Floorlet pricing.

(1) The forward measure $\widehat{\mathbb{P}}_S$ is defined from the bond price $P(t,S)$, and this gives

$$F_t = P(t,S)\widehat{\mathbb{E}}[(\kappa - L(T,T,S))^+ \mid \mathcal{F}_t].$$

(2) The LIBOR rate $L(t,T,S)$ is a driftless geometric Brownian motion with volatility σ under the forward measure $\widehat{\mathbb{P}}_S$. Indeed, the LIBOR rate $L(t,T,S)$ can be written as the forward price $L(t,T,S) = \widehat{X}_t =$

$X_t/P(t, S)$, where $X_t = (P(t, T) - P(t, S))/(S - T)$. Since both discounted bond prices $e^{-\int_0^t r_s ds} P(t, T)$ and $e^{-\int_0^t r_s ds} P(t, S)$ are martingales under $\mathbb{P}^*$, the same is true of X_t. Hence, $L(t, T, S) = X_t/P(t, S)$ becomes a martingale under the forward measure $\widehat{\mathbb{P}}_S$ by Proposition 2.1, and computing its dynamics under $\widehat{\mathbb{P}}_S$ amounts to removing any "dt" term in (9.29), *i.e.*

$$dL(t, T, S) = \sigma L(t, T, S) d\widehat{B}_t, \qquad 0 \leqslant t \leqslant T,$$

hence $L(t, T, S) = L(0, T, S) e^{\sigma \widehat{B}_t - \sigma^2 t/2}$, where $(\widehat{B}_t)_{t \in \mathbb{R}_+}$ is a standard Brownian motion under $\widehat{\mathbb{P}}_S$.

(3) We find

$$
\begin{aligned}
F_t &= P(t, S) \widehat{\mathbb{E}}\big[(\kappa - L(T, T, S))^+ \,\big|\, \mathcal{F}_t \big] \\
&= P(t, S) \widehat{\mathbb{E}}\big[(\kappa - L(t, T, S) e^{-(T-t)\sigma^2/2 + \sigma(\widehat{B}_T - \widehat{B}_t)})^+ \,\big|\, \mathcal{F}_t \big] \\
&= P(t, S)\big(\kappa \Phi(-d_-(t, T)) - \widehat{X}_t \Phi(-d_+(t, T)) \big) \\
&= \kappa P(t, S) \Phi(-d_-(t, T)) - P(t, S) L(t, T, S) \Phi(-d_+(t, T)) \\
&= \kappa P(t, S) \Phi(-d_-(t, T)) - \frac{P(t, T) - P(t, S)}{S - T} \Phi(-d_+(t, T)),
\end{aligned}
$$

where $e^m = L(t, T, S) e^{-(T-t)\sigma^2/2}$, $v^2 = (T - t)\sigma^2$, and

$$d_+(t, T) = \frac{\log(L(t, T, S)/\kappa)}{\sigma\sqrt{T - t}} + \frac{\sigma\sqrt{T - t}}{2},$$

and

$$d_-(t, T) = \frac{\log(L(t, T, S)/\kappa)}{\sigma\sqrt{T - t}} - \frac{\sigma\sqrt{T - t}}{2},$$

because $L(t, T, S)$ is a driftless geometric Brownian motion with volatility σ under the forward measure $\widehat{\mathbb{P}}_S$.

Exercise 9.6. Jamshidian's trick.

(1) We have

$$P(T_i, T_i, T_j) = \sum_{k=i}^{j-1} c_{k+1} P(T_i, T_{k+1}).$$

(2) We let $\tilde{c}_k = 1$, $k = i+1, \ldots, j-1$.

(3) The swaption can be priced as

$$\mathbb{E}^* \left[e^{- \int_t^{T_i} r_s ds} \left(P(T_i, T_i) - P(T_i, T_j) - \kappa P(T_i, T_i, T_j) \right)^+ \,\Big|\, \mathcal{F}_t \right]$$

$$= \mathbb{E}^* \left[e^{- \int_t^{T_i} r_s ds} \left(1 - \kappa \sum_{k=i}^{j-1} c_{k+1} P(T_i, T_{k+1}) \right)^+ \,\Big|\, \mathcal{F}_t \right]$$

$$= \kappa \, \mathbb{E}^* \left[e^{- \int_t^{T_i} r_s ds} \left(\sum_{k=i}^{j-1} c_{k+1} F_{k+1}(T_i, \gamma_\kappa) - \sum_{k=i}^{j-1} c_{k+1} F_{k+1}(T_i, r_{T_i}) \right)^+ \,\Big|\, \mathcal{F}_t \right]$$

$$= \kappa \, \mathbb{E}^* \left[e^{- \int_t^{T_i} r_s ds} \left(\sum_{k=i}^{j-1} c_{k+1} \left(F_{k+1}(T_i, \gamma_\kappa) - F_{k+1}(T_i, r_{T_i}) \right) \right)^+ \,\Big|\, \mathcal{F}_t \right]$$

$$= \kappa \, \mathbb{E}^* \left[e^{- \int_t^{T_i} r_s ds} \sum_{k=i}^{j-1} c_{k+1} \mathbb{1}_{\{ r_{T_i} \leqslant \gamma_k \}} \left(F_{k+1}(T_i, \gamma_\kappa) - F_{k+1}(T_i, r_{T_i}) \right) \,\Big|\, \mathcal{F}_t \right]$$

$$= \kappa \, \mathbb{E}^* \left[e^{- \int_t^{T_i} r_s ds} \sum_{k=i}^{j-1} c_{k+1} \left(F_{k+1}(T_i, \gamma_\kappa) - F_{k+1}(T_i, r_{T_i}) \right)^+ \,\Big|\, \mathcal{F}_t \right]$$

$$= \kappa \sum_{k=i}^{j-1} c_{k+1} \, \mathbb{E}^* \left[e^{- \int_t^{T_i} r_s ds} \left(F_{k+1}(T_i, \gamma_\kappa) - P(T_i, T_{k+1}) \right)^+ \,\Big|\, \mathcal{F}_t \right]$$

$$= \kappa \sum_{k=i}^{j-1} c_{k+1} P(t, T_i) \widehat{\mathbb{E}}_i \left[\left(F_{k+1}(T_i, \gamma_\kappa) - P(T_i, T_{k+1}) \right)^+ \,\Big|\, \mathcal{F}_t \right],$$

which is a weighted sum of bond put option prices with strike prices $F_{k+1}(T_i, \gamma_\kappa)$, $k = i, \dots, j-1$, computable from Proposition 8.6.

Exercise 9.7. Vasicek caplet pricing. Recall that the forward measure is defined by

$$\frac{d\widehat{\mathbb{P}}_{i|\mathcal{F}_t}}{d\mathbb{P}_{|\mathcal{F}_t}} = \frac{e^{- \int_t^{T_i} r_s ds}}{P(t, T_i)}, \qquad 0 \leqslant t \leqslant T_i, \quad i = 1, 2,$$

with

$$\mathbb{E} \left[\frac{d\widehat{\mathbb{P}}_i}{d\mathbb{P}^*} \,\Big|\, \mathcal{F}_t \right] = \frac{P(t, T_i)}{P(0, T_i)} e^{- \int_0^t r_s ds}, \qquad 0 \leqslant t \leqslant T_i, \quad i = 1, 2.$$

The forward swap measure is defined by

$$\frac{d\widehat{\mathbb{P}}_{1,2|\mathcal{F}_t}}{d\mathbb{P}_{|\mathcal{F}_t}} = \frac{P(T_1, T_2)}{P(t, T_2)} e^{- \int_t^{T_1} r_s ds}, \qquad 0 \leqslant t \leqslant T_1,$$

with

$$\mathbb{E}\left[\frac{d\widehat{\mathbb{P}}_{1,2}}{d\mathbb{P}^*}\,\bigg|\,\mathcal{F}_t\right] = \frac{P(t,T_2)}{P(0,T_2)}\,e^{-\int_0^t r_s ds}, \qquad 0 \leqslant t \leqslant T_1.$$

In particular,

$$\mathbb{E}\left[\frac{d\widehat{\mathbb{P}}_{1,2}}{d\mathbb{P}^*}\,\bigg|\,\mathcal{F}_{T_1}\right] = \mathbb{E}\left[\frac{d\widehat{\mathbb{P}}_2}{d\mathbb{P}^*}\,\bigg|\,\mathcal{F}_{T_1}\right] = \frac{P(T_1,T_2)}{P(0,T_2)}\,e^{-\int_0^{T_1} r_s ds},$$

which means that

$$\mathbb{E}^*\left[e^{-\int_0^{T_2} r_s ds}F\right] = P(0,T_2)\,\mathbb{E}^*\left[\frac{P(T_1,T_2)}{P(0,T_2)}\,e^{-\int_0^{T_1} r_s ds}F\right]$$

$$= P(0,T_2)\widehat{\mathbb{E}}_2[F]$$

$$= P(0,T_2)\widehat{\mathbb{E}}_{1,2}[F],$$

for all F integrable and $\mathcal{F}_{T_1}$-measurable. Moreover,

$$dB_t^{(i)} := dB_t - \zeta_i(t)dt$$

is a standard Brownian motion under $\widehat{\mathbb{P}}_i$, $i = 1, 2$. In addition, $\left(B_t^{(2)}\right)_{t\in[0,T_1]}$ is also a standard Brownian motion until time T_1 under $\widehat{\mathbb{P}}_{1,2}$. We have

$$\widehat{\mathbb{E}}_{1,2}\left[\frac{d\widehat{\mathbb{P}}_k}{d\widehat{\mathbb{P}}_{1,2}}\,\bigg|\,\mathcal{F}_t\right] = \frac{P(0,T_2)}{P(0,T_k)}\frac{P(t,T_k)}{P(t,T_2)} \qquad 0 \leqslant t \leqslant T_1, \quad k = 1, 2,$$

and the process

$$t \mapsto \frac{P(t,T_1)}{P(t,T_2)}, \qquad 0 \leqslant t \leqslant T_1,$$

is an $(\mathcal{F}_t)_{t\in[0,T_1]}$-martingale under both $\widehat{\mathbb{P}}_2$ and $\widehat{\mathbb{P}}_{1,2}$, while

$$t \mapsto \frac{P(t,T_2)}{P(t,T_1)}, \qquad 0 \leqslant t \leqslant T_1,$$

is an $(\mathcal{F}_t)_{t\in[0,T_1]}$-martingale under $\widehat{\mathbb{P}}_1$.

(1) We have

$$\frac{dP(t,T_i)}{P(t,T_i)} = r_t dt + \zeta_i(t)dB_t, \qquad i = 1, 2,$$

and

$$P(T,T_i) = P(t,T_i)\exp\left(\int_t^T r_s ds + \int_t^T \zeta_i(s)dB_s - \frac{1}{2}\int_t^T |\zeta_i(s)|^2 ds\right),$$

$0 \leqslant t \leqslant T \leqslant T_i$, $i = 1, 2$, hence

$$\log P(T, T_i) = \log P(t, T_i) + \int_t^T r_s ds + \int_t^T \zeta_i(s) dB_s - \frac{1}{2} \int_t^T |\zeta_i(s)|^2 ds,$$

$0 \leqslant t \leqslant T \leqslant T_i$, $i = 1, 2$, and

$$d \log P(t, T_i) = r_t dt + \zeta_i(t) dB_t - \frac{1}{2} |\zeta_i(t)|^2 dt, \quad i = 1, 2.$$

In the Vasicek model

$$dr_t = -br_t dt + \sigma dB_t,$$

where $(B_t)_{t \in \mathbb{R}_+}$ is a standard Brownian motion under $\mathbb{P}$, we have

$$\zeta_i(t) = -\frac{\sigma}{b} \left(1 - e^{-(T_i - t)b}\right), \qquad 0 \leqslant t \leqslant T_i, \quad i = 1, 2.$$

(2) Consider the standard Brownian motion

$$B_t^{(i)} = B_t - \int_0^t \zeta_i(s) ds$$

under $\widehat{\mathbb{P}}_i$, $i = 1, 2$. We have

$$\frac{P(T, T_1)}{P(T, T_2)}$$

$$= \frac{P(t, T_1)}{P(t, T_2)} \exp \left(\int_t^T (\zeta_1(s) - \zeta_2(s)) dB_s - \frac{1}{2} \int_t^T (|\zeta_1(s)|^2 - |\zeta_2(s)|^2) ds \right)$$

$$= \frac{P(t, T_1)}{P(t, T_2)} \exp \left(\int_t^T (\zeta_1(s) - \zeta_2(s)) dB_s^{(2)} - \frac{1}{2} \int_t^T |\zeta_1(s) - \zeta_2(s)|^2 ds \right),$$

which is an $(\mathcal{F}_t)_{t \in [0, \min(T_1, T_2)]}$-martingale under $\widehat{\mathbb{P}}_2$ and under $\widehat{\mathbb{P}}_{1,2}$, and

$$\frac{P(T, T_2)}{P(T, T_1)}$$

$$= \frac{P(t, T_2)}{P(t, T_1)} \exp \left(-\int_t^T (\zeta_1(s) - \zeta_2(s)) dB_s^{(1)} - \frac{1}{2} \int_t^T |\zeta_1(s) - \zeta_2(s)|^2 ds \right),$$

which is an $(\mathcal{F}_t)_{t \in [0, \min(T_1, T_2)]}$-martingale under $\widehat{\mathbb{P}}_1$.

(3) We have

$$f(t, T_1, T_2) = -\frac{1}{T_2 - T_1} (\log P(t, T_2) - \log P(t, T_1))$$

and in the Vasicek case,

$$f(t, T_1, T_2) = -\frac{\sigma^2}{2b} - \frac{1}{T_2 - T_1} \left(\frac{r_t}{b} + \frac{\sigma^2}{b^3} \right) \left(e^{-(T_2 - t)b} - e^{-(T_1 - t)b} \right)$$

$$+ \frac{\sigma^2}{4(T_2 - T_1)b^3} \left(e^{-2(T_2 - t)b} - e^{-2(T_1 - t)b} \right).$$

(4) We have

$$df(t, T_1, T_2) = -\frac{1}{T_2 - T_1} d\log \frac{P(t, T_2)}{P(t, T_1)}$$

$$= -\frac{1}{T_2 - T_1} \left((\zeta_2(t) - \zeta_1(t))dB_t - \frac{1}{2}(|\zeta_2(t)|^2 - |\zeta_1(t)|^2)dt \right)$$

$$= -\frac{1}{T_2 - T_1}$$
$$\times \left((\zeta_2(t) - \zeta_1(t))(dB_t^{(2)} + \zeta_2(t)dt) - \frac{1}{2}(|\zeta_2(t)|^2 - |\zeta_1(t)|^2)dt \right)$$

$$= -\frac{1}{T_2 - T_1} \left((\zeta_2(t) - \zeta_1(t))dB_t^{(2)} + \frac{1}{2}|\zeta_2(t) - \zeta_1(t)|^2 dt \right).$$

(5) We have

$$f(T, T_1, T_2) = -\frac{1}{T_2 - T_1} \log \frac{P(T, T_2)}{P(T, T_1)}$$

$$= f(t, T_1, T_2)$$
$$- \frac{1}{T_2 - T_1} \left(\int_t^T (\zeta_2(s) - \zeta_1(s))dB_s - \frac{1}{2}(|\zeta_2(s)|^2 - |\zeta_1(s)|^2)ds \right)$$

$$= f(t, T_1, T_2)$$
$$- \frac{1}{T_2 - T_1} \left(\int_t^T (\zeta_2(s) - \zeta_1(s))dB_s^{(2)} - \frac{1}{2}\int_t^T |\zeta_2(s) - \zeta_1(s)|^2 ds \right)$$

$$= f(t, T_1, T_2)$$
$$- \frac{1}{T_2 - T_1} \left(\int_t^T (\zeta_2(s) - \zeta_1(s))dB_s^{(1)} + \frac{1}{2}\int_t^T |\zeta_2(s) - \zeta_1(s)|^2 ds \right).$$

Hence, $f(T, T_1, T_2)$ has a Gaussian distribution given $\mathcal{F}_t$ with conditional mean

$$m_2 := f(t, T_1, T_2) + \frac{1}{2(T_2 - T_1)} \int_t^T |\zeta_2(s) - \zeta_1(s)|^2 ds$$

under $\widehat{\mathbb{P}}_2$, resp.

$$m_1 := f(t, T_1, T_2) - \frac{1}{2(T_2 - T_1)} \int_t^T |\zeta_2(s) - \zeta_1(s)|^2 ds$$

under $\widehat{\mathbb{P}}_1$, and variance

$$v^2 := \frac{1}{(T_2 - T_1)^2} \int_t^T |\zeta_2(s) - \zeta_1(s)|^2 ds.$$

Hence

$$(T_2 - T_1) \mathbb{E}^* \left[e^{-\int_t^{T_2} r_s ds} (f(T_1, T_1, T_2) - \kappa)^+ \Big| \mathcal{F}_t \right]$$

$$= (T_2 - T_1)P(t, T_2)\widehat{\mathbb{E}}_2\big[(f(T_1, T_1, T_2) - \kappa)^+ \mid \mathcal{F}_t\big]$$

$$= (T_2 - T_1)P(t, T_2)\widehat{\mathbb{E}}_2\big[(m_2 + X - \kappa)^+ \mid \mathcal{F}_t\big]$$

$$= (T_2 - T_1)P(t, T_2)\left(\frac{v}{\sqrt{2\pi}}e^{-(\kappa - m_2)^2/(2v^2)} + (m_2 - \kappa)\Phi\left(\frac{m_2 - \kappa}{v}\right)\right).$$

(6) We have

$$f(t, T_1, T_2) = -\frac{1}{T_2 - T_1}(\log P(t, T_2) - \log P(t, T_1))$$

$$= r_t + \frac{1}{T_2 - T_1}\frac{\sigma^2}{6}\left((T_1 - t)^3 - (T_2 - t)^3\right).$$

Exercise 9.8. Caplet pricing on the LIBOR. (Exercise 9.7 continued).

(1) We have

$$L(T, T_1, T_2) = \frac{1}{T_2 - T_1}\left(\frac{P(T, T_1)}{P(T, T_2)} - 1\right)$$

$$= \frac{1}{T_2 - T_1}\left(\frac{P(t, T_1)}{P(t, T_2)}e^{\int_t^T(\zeta_1(s) - \zeta_2(s))dB_s - \int_t^T(|\zeta_1(s)|^2 - |\zeta_2(s)|^2)ds/2} - 1\right)$$

$$= \frac{1}{T_2 - T_1}\left(\frac{P(t, T_1)}{P(t, T_2)}e^{\int_t^T(\zeta_1(s) - \zeta_2(s))dB_s^{(2)} - \int_t^T|\zeta_1(s) - \zeta_2(s)|^2 ds/2} - 1\right)$$

$$= \frac{1}{T_2 - T_1}\left(\frac{P(t, T_1)}{P(t, T_2)}e^{\int_t^T(\zeta_1(s) - \zeta_2(s))dB_s^{(1)} + \int_t^T|\zeta_1(s) - \zeta_2(s)|^2 ds/2} - 1\right),$$

and by Itô calculus,

$$dL(t, T_1, T_2) = \frac{1}{T_2 - T_1}d\left(\frac{P(t, T_1)}{P(t, T_2)}\right)$$

$$= \frac{1}{T_2 - T_1}\frac{P(t, T_1)}{P(t, T_2)}$$

$$\times \left((\zeta_1(t) - \zeta_2(t))dB_t + \frac{1}{2}|\zeta_1(t) - \zeta_2(t)|^2 dt - \frac{|\zeta_1(t)|^2 - |\zeta_2(t)|^2}{2}dt\right)$$

$$= \left(\frac{1}{T_2 - T_1} + L(t, T_1, T_2)\right)$$

$$\times ((\zeta_1(t) - \zeta_2(t))dB_t + \zeta_2(t)(\zeta_2(t) - \zeta_1(t))dt)$$

$$= \left(\frac{1}{T_2 - T_1} + L(t, T_1, T_2)\right)$$

$$\times \left((\zeta_1(t) - \zeta_2(t))dB_t^{(1)} + (|\zeta_2(t)|^2 - |\zeta_1(t)|^2)dt\right)$$

$$= \left(\frac{1}{T_2 - T_1} + L(t, T_1, T_2) \right) (\zeta_1(t) - \zeta_2(t)) dB_t^{(2)}, \quad 0 \leqslant t \leqslant T_1,$$

$$(S.24)$$

hence $(T_2 - T_1)^{-1} + L(t, T_1, T_2)$ is a geometric Brownian motion, with

$$\frac{1}{T_2 - T_1} + L(T, T_1, T_2)$$

$$= \left(\frac{1}{T_2 - T_1} + L(t, T_1, T_2) \right)$$

$$\times \exp \left(\int_t^T (\zeta_1(s) - \zeta_2(s)) dB_s^{(2)} - \frac{1}{2} \int_t^T |\zeta_1(s) - \zeta_2(s)|^2 ds \right),$$

$0 \leqslant t \leqslant T \leqslant T_1$.

(2) We have

$$(T_2 - T_1) \mathbb{E}^* \left[e^{- \int_t^{T_2} r_s ds} (L(T_1, T_1, T_2) - \kappa)^+ \Big| \mathcal{F}_t \right]$$

$$= (T_2 - T_1) \mathbb{E}^* \left[e^{- \int_t^{T_1} r_s ds} P(T_1, T_2) (L(T_1, T_1, T_2) - \kappa)^+ \Big| \mathcal{F}_t \right]$$

$$= P(t, T_1, T_2) \widehat{\mathbb{E}}_{1,2} \big[(L(T_1, T_1, T_2) - \kappa)^+ \big| \mathcal{F}_t \big].$$

Since $\left(B_t^{(2)} \right)_{t \in [0, T_1]}$ is a standard Brownian motion until time T_1 under $\widehat{\mathbb{P}}_{1,2}$, by (S.24) the forward rate

$$L(T_1, T_1, T_2) = -\frac{1}{T_2 - T_1}$$

$$+ \left(\frac{1}{T_2 - T_1} + L(t, T_1, T_2) \right)$$

$$\times \exp \left(\int_t^{T_1} (\zeta_1(s) - \zeta_2(s)) dB_s^{(2)} - \frac{1}{2} \int_t^{T_1} |\zeta_1(s) - \zeta_2(s)|^2 ds \right)$$

has same distribution as

$$\left(\frac{1}{T_2 - T_1} + L(t, T_1, T_2) \right) e^{X - \text{Var}[X]/2} - \frac{1}{T_2 - T_1}$$

$$= \frac{1}{T_2 - T_1} \left(\frac{P(t, T_1)}{P(t, T_2)} e^{X - \text{Var}[X]/2} - 1 \right),$$

where X is a centered Gaussian random variable under $\widehat{\mathbb{P}}_2$, with variance

$$\int_t^{T_1} |\zeta_1(s) - \zeta_2(s)|^2 ds$$

given $\mathcal{F}_t$. Hence we have

$$(T_2 - T_1)\, \mathbb{E}^* \left[e^{-\int_t^{T_2} r_s ds} (L(T_1, T_1, T_2) - \kappa)^+ \,\Big|\, \mathcal{F}_t \right]$$

$$= P(t, T_1, T_2)$$

$$\times \mathrm{Bl}\left(\frac{1}{T_2 - T_1} + L(t, T_1, T_2), \frac{\int_t^{T_1} |\zeta_1(s) - \zeta_2(s)|^2 ds}{T_1 - t}, \kappa + \frac{1}{T_2 - T_1}, T_1 - t \right).$$

Exercise 9.9. Swaption pricing.

(1) We have

$$L(t, T_1, T_2) = L(0, T_1, T_2) \exp\left(\int_0^t \gamma_1(s) dW_s^{(2)} - \frac{1}{2} \int_0^t |\gamma_1(s)|^2 ds \right),$$

$0 \leqslant t \leqslant T_1$, and $L(t, T_2, T_3) = b$, $0 \leqslant t \leqslant T_2$. Note that we have $P(t, T_2)/P(t, T_3) = 1 + \delta b$, hence $\widehat{\mathbb{P}}_2 = \widehat{\mathbb{P}}_3 = \widehat{\mathbb{P}}_{1,2}$ up to time T_1.

(2) We apply the change of numeraire Proposition 8.1 under the forward measure $\mathbb{P}_2$.

(3) We have

$$\mathbb{E}^* \left[e^{-\int_t^{T_2} r_s ds} (L(T_1, T_1, T_2) - \kappa)^+ \,\Big|\, \mathcal{F}_t \right]$$

$$= P(t, T_2)\widehat{\mathbb{E}}_2 \left[(L(T_1, T_1, T_2) - \kappa)^+ \,\big|\, \mathcal{F}_t \right]$$

$$= P(t, T_2)\widehat{\mathbb{E}}_2 \left[\left(L(t, T_1, T_2)\, e^{\int_t^{T_1} \gamma_1(s) dW_s^{(2)} - \int_t^{T_1} |\gamma_1(s)|^2 ds/2} - \kappa \right)^+ \,\Big|\, \mathcal{F}_t \right]$$

$$= P(t, T_2)\mathrm{Bl}(L(t, T_1, T_2), \kappa, \sigma_1(t), 0, T_1 - t),$$

where

$$\sigma_1^2(t) = \frac{1}{T_1 - t} \int_t^{T_1} |\gamma_1(s)|^2 ds, \qquad 0 \leqslant t < T_1.$$

(4) We have

$$\frac{P(t, T_1)}{P(t, T_1, T_3)} = \frac{P(t, T_1)}{\delta P(t, T_2) + \delta P(t, T_3)}$$

$$= \frac{P(t, T_1)}{\delta P(t, T_2)} \frac{1}{1 + P(t, T_3)/P(t, T_2)}$$

$$= \frac{1 + \delta b}{\delta(\delta b + 2)} (1 + \delta L(t, T_1, T_2)), \qquad 0 \leqslant t \leqslant T_1,$$

and

$$\frac{P(t, T_3)}{P(t, T_1, T_3)} = \frac{P(t, T_3)}{P(t, T_2) + P(t, T_3)}$$

$$= \frac{1}{1 + P(t,T_2)/P(t,T_3)}$$

$$= \frac{1}{\delta}\frac{1}{2 + \delta b}, \qquad 0 \leqslant t \leqslant T_2.$$

(5) We have

$$S(t,T_1,T_3) = \frac{P(t,T_1)}{P(t,T_1,T_3)} - \frac{P(t,T_3)}{P(t,T_1,T_3)}$$

$$= \frac{1 + \delta b}{\delta(2 + \delta b)}(1 + \delta L(t,T_1,T_2)) - \frac{1}{\delta(2 + \delta b)}$$

$$= \frac{1}{2 + \delta b}(b + (1 + \delta b)L(t,T_1,T_2)), \qquad 0 \leqslant t \leqslant T_2.$$

We have

$$dS(t,T_1,T_3) = \frac{1 + \delta b}{2 + \delta b}L(t,T_1,T_2)\gamma_1(t)dW_t^{(2)}$$

$$= \left(S(t,T_1,T_3) - \frac{b}{2 + \delta b}\right)\gamma_1(t)dW_t^{(2)}$$

$$= S(t,T_1,T_3)\sigma_{1,3}(t)dW_t^{(2)}, \qquad 0 \leqslant t \leqslant T_2,$$

with

$$\sigma_{1,3}(t) = \left(1 - \frac{b}{S(t,T_1,T_3)(2 + \delta b)}\right)\gamma_1(t)$$

$$= \left(1 - \frac{b}{b + (1 + \delta b)L(t,T_1,T_2)}\right)\gamma_1(t)$$

$$= \frac{(1 + \delta b)L(t,T_1,T_2)}{b + (1 + \delta b)L(t,T_1,T_2)}\gamma_1(t)$$

$$= \frac{(1 + \delta b)L(t,T_1,T_2)}{(2 + \delta b)S(t,T_1,T_3)}\gamma_1(t).$$

(6) The process $(W_t^{(2)})_{t \in \mathbb{R}_+}$ is a standard Brownian motion under $\widehat{\mathbb{P}}_2 = \widehat{\mathbb{P}}_{1,3}$ and

$$P(t,T_1,T_3)\widehat{\mathbb{E}}_{1,3}\left[(S(T_1,T_1,T_3) - \kappa)^+ \mid \mathcal{F}_t\right]$$
$$= P(t,T_2)\mathrm{Bl}(S(t,T_1,T_2),\kappa,\tilde{\sigma}_{1,3}(t),0,T_1 - t),$$

where $|\tilde{\sigma}_{1,3}(t)|^2$ is the approximation of the volatility

$$\frac{1}{T_1 - t}\int_t^{T_1}|\sigma_{1,3}(s)|^2 ds = \frac{1}{T_1 - t}\int_t^{T_1}\left(\frac{(1 + \delta b)L(s,T_1,T_2)}{(2 + \delta b)S(s,T_1,T_3)}\right)^2\gamma_1(s)ds$$

obtained by freezing the random component of $\sigma_{1,3}(s)$ at time t, *i.e.*

$$\tilde{\sigma}_{1,3}^2(t) = \frac{1}{T_1 - t}\left(\frac{(1 + \delta b)L(t,T_1,T_2)}{(2 + \delta b)S(t,T_1,T_3)}\right)^2\int_t^{T_1}|\gamma_1(s)|^2 ds.$$

Exercise 9.10. Caplet pricing.

(1) The LIBOR rate $L(t, T, S)$ is a driftless geometric Brownian motion with deterministic volatility function $\sigma(t)$ under the forward measure $\widehat{\mathbb{P}}_S$.

Explanation: The LIBOR rate $L(t, T, S)$ can be written as the forward price $L(t, T, S) = \widehat{X}_t = X_t / P(t, S)$ where $X_t = (P(t, T) - P(t, S))/(S - T)$. Since both discounted bond prices $e^{-\int_0^t r_s ds} P(t, T)$ and $e^{-\int_0^t r_s ds} P(t, S)$ are martingales under $\mathbb{P}^*$, the same is true of X_t. Hence, $L(t, T, S) = X_t / P(t, S)$ becomes a martingale under the forward measure $\widehat{\mathbb{P}}_S$ by Proposition 2.1, and computing its dynamics under $\widehat{\mathbb{P}}_S$ amounts to removing any "dt" term in the original stochastic differential equation defining $L(t, T, S)$, *i.e.* we find

$$dL(t, T, S) = \sigma(t) L(t, T, S) d\widehat{B}_t, \qquad 0 \leqslant t \leqslant T,$$

hence

$$L(t, T, S) = L(0, T, S) \exp\left(\int_0^t \sigma(s) d\widehat{B}_s - \frac{1}{2} \int_0^t \sigma^2(s) ds \right),$$

where $\left(\widehat{B}_t \right)_{t \in \mathbb{R}_+}$ is a standard Brownian motion under $\widehat{\mathbb{P}}_S$.

(2) Choosing $P(t, S)$ as a numeraire, we have

$$\mathbb{E}^* \left[e^{-\int_t^S r_s ds} \phi(L(T, T, S)) \,\Big|\, \mathcal{F}_t \right] = \mathbb{E}^* \left[e^{-\int_t^S r_s ds} \phi(L(T, T, S)) \,\Big|\, \mathcal{F}_t \right]$$

$$= P(t, S) \widehat{\mathbb{E}}[\phi(L(T, T, S)) \mid \mathcal{F}_t].$$

(3) Given the solution

$$L(T, T, S) = L(0, T, S) \exp\left(\int_0^T \sigma(s) d\widehat{B}_s - \frac{1}{2} \int_0^T \sigma^2(s) ds \right)$$

$$= L(t, T, S) \exp\left(\int_t^T \sigma(s) d\widehat{B}_s - \frac{1}{2} \int_t^T \sigma^2(s) ds \right),$$

we find

$$P(t, S) \widehat{\mathbb{E}} \left[\phi(L(T, T, S)) \mid \mathcal{F}_t \right]$$

$$= P(t, S) \widehat{\mathbb{E}} \left[\phi\left(L(t, T, S) e^{\int_t^T \sigma(s) d\widehat{B}_s - \int_t^T \sigma^2(s) ds / 2} \right) \,\Big|\, \mathcal{F}_t \right]$$

$$= P(t, S) \int_{-\infty}^{\infty} \phi\left(L(t, T, S) e^{x - \eta^2/2} \right) e^{-x^2/(2\eta^2)} \frac{dx}{\sqrt{2\pi\eta^2}},$$

because $\displaystyle\int_t^T \sigma(s) d\widehat{B}_s$ is a centered Gaussian variable with variance $\eta^2 := \displaystyle\int_t^T \sigma^2(s) ds$, independent of $\mathcal{F}_t$ under the forward measure $\widehat{\mathbb{P}}$.

Exercise 9.11. Receiver swaption pricing.

(1) Using the annuity numeraire $P(t, T_i, T_j)$, we have

$$
\mathbb{E}^* \left[e^{-\int_t^{T_i} r_s ds} P(T_i, T_i, T_j) \phi(S(T_i, T_i, T_j)) \,\Big|\, \mathcal{F}_t \right]
$$

$$
= P(t, T_i, T_j) \widehat{\mathbb{E}}_{i,j} \left[\frac{P(T_i, T_i, T_j)}{P(T_i, T_i, T_j)} \phi(S(T_i, T_i, T_j)) \,\Big|\, \mathcal{F}_t \right]
$$

$$
= P(t, T_i, T_j) \widehat{\mathbb{E}}_{i,j} [\phi(S(T_i, T_i, T_j)) \mid \mathcal{F}_t].
$$

(2) Since $S(t, T_i, T_j)$ is a forward price for the numeraire $P(t, T_i, T_j)$, it is a martingale under the forward swap measure $\widehat{\mathbb{P}}_{i,j}$ and we have

$$
S(t, T_i, T_j) = \sigma S(t, T_i, T_j) d\widehat{B}_t^{i,j}, \qquad 0 \leqslant t \leqslant T_i,
$$

where $\left(\widehat{B}_t^{i,j}\right)_{t \in \mathbb{R}_+}$ is a standard Brownian motion under the forward swap measure $\widehat{\mathbb{P}}_{i,j}$.

(3) Given the solution

$$
S(T_i, T_i, T_j) = S(0, T_i, T_j) e^{\sigma \widehat{B}_{T_i} - \sigma^2 T_i / 2}
$$

$$
= S(t, T_i, T_j) e^{\sigma(\widehat{B}_{T_i} - \widehat{B}_t) - (T_i - t)\sigma^2 / 2}
$$

of (9.32), we find

$$
P(t, T_i, T_j) \widehat{\mathbb{E}}_{i,j} \left[\phi(S(T_i, T_i, T_j)) \mid \mathcal{F}_t \right]
$$

$$
= P(t, T_i, T_j) \widehat{\mathbb{E}}_{i,j} \left[\phi\big(S(t, T_i, T_j) e^{\sigma(\widehat{B}_{T_i} - \widehat{B}_t) - (T_i - t)\sigma^2 / 2}\big) \mid \mathcal{F}_t \right]
$$

$$
= P(t, T_i, T_j)
$$

$$
\times \int_{-\infty}^{\infty} \phi\big(S(t, T_i, T_j) e^{\sigma x - (T_i - t)\sigma^2 / 2}\big) e^{-x^2 / (2(T_i - t))} \frac{dx}{\sqrt{2\pi(T_i - t)}},
$$

because $\widehat{B}_{T_i} - \widehat{B}_t$ is a centered Gaussian variable with variance $T_i - t$, independent of $\mathcal{F}_t$ under the forward measure $\widehat{\mathbb{P}}_{i,j}$.

(4) We find

$$
P(t, T_i, T_j) \widehat{\mathbb{E}}_{i,j} \left[(\kappa - S(T_i, T_i, T_j))^+ \mid \mathcal{F}_t \right]
$$

$$
= P(t, T_i, T_j) \widehat{\mathbb{E}}_{i,j} \left[\big(\kappa - S(t, T_i, T_j) e^{-(T_i - t)\sigma^2 / 2 + \sigma(\widehat{B}_{T_i} - \widehat{B}_t)}\big)^+ \mid \mathcal{F}_t \right]
$$

$$
= P(t, T_i, T_j) \big(\kappa \Phi(-d_-(t, T_i)) - \widehat{X}_t \Phi(-d_+(t, T_i))\big)
$$

$$
= P(t, T_i, T_j) \kappa \Phi(-d_-(t, T_i)) - P(t, T_i, T_j) S(t, T_i, T_j) \Phi(-d_+(t, T_i))
$$

$$
= P(t, T_i, T_j) \kappa \Phi(-d_-(t, T_i)) - (P(t, T_i) - P(t, T_j)) \Phi(-d_+(t, T_i)),
$$

where

$$
d_+(t, T_i) = \frac{\log(S(t, T_i, T_j) / \kappa)}{\sigma \sqrt{T_i - t}} + \frac{\sigma}{2} \sqrt{T_i - t},
$$

and

$$d_-(t,T_i) = \frac{\log(S(t,T_i,T_j)/\kappa)}{\sigma\sqrt{T_i-t}} - \frac{\sigma}{2}\sqrt{T_i-t},$$

because $S(t,T_i,T_j)$ is a driftless geometric Brownian motion with volatility σ under the forward measure $\widehat{\mathbb{P}}_{i,j}$.

Exercise 9.12. Caplet hedging.

(1) Apply the Itô formula to $d(P(t,T_1)/P(t,T_2))$.
(2) We have

$$L(T_1,T_1,T_2) = L(t,T_1,T_2)\exp\left(\int_t^{T_1}\sigma(s)d\widehat{B}_s - \frac{1}{2}\int_t^{T_1}|\sigma(s)|^2 ds\right).$$

(3) We have

$$P(t,T_2)\widehat{\mathbb{E}}\big[(L(T_1,T_1,T_2)-\kappa)^+\,|\,\mathcal{F}_t\big]$$

$$= P(t,T_2)\widehat{\mathbb{E}}\left[\left(L(t,T_1,T_2)\,e^{\int_t^{T_1}\sigma(s)d\widehat{B}_s - \int_t^{T_1}|\sigma(s)|^2 ds/2} - \kappa\right)^+\,\Big|\,\mathcal{F}_t\right]$$

$$= P(t,T_2)\text{Bl}\left(L(t,T_1,T_2),\kappa,\frac{v(t,T_1)}{\sqrt{T_1-t}},0,T_1-t\right)$$

$$= P(t,T_2)L(t,T_1,T_2)\Phi\left(\frac{v(t,T_1)}{2} + \frac{\log(x/K)}{v(t,T_1)}\right)$$

$$-\kappa P(t,T_2)\Phi\left(-\frac{v(t,T_1)}{2} + \frac{\log(x/K)}{v(t,T_1)}\right).$$

(4) Integrate the self-financing condition (9.36) between 0 and t.
(5) We have

$$d\widetilde{V}_t = d\big(e^{-\int_0^t r_s ds}V_t\big)$$

$$= -r_t e^{-\int_0^t r_s ds}V_t dt + e^{-\int_0^t r_s ds}dV_t$$

$$= -r_t e^{-\int_0^t r_s ds}\xi_t^{(1)}P(t,T_1) - r_t e^{-\int_0^t r_s ds}\xi_t^{(2)}P(t,T_2)dt$$

$$+ e^{-\int_0^t r_s ds}\xi_t^{(1)}dP(t,T_1) + e^{-\int_0^t r_s ds}\xi_t^{(2)}dP(t,T_2)$$

$$= \xi_t^{(1)}d\widetilde{P}(t,T_1) + \xi_t^{(2)}d\widetilde{P}(t,T_2), \qquad 0 \leqslant t \leqslant T_1,$$

since

$$\frac{d\widetilde{P}(t,T_1)}{\widetilde{P}(t,T_1)} = \zeta_1(t)dt, \qquad \frac{d\widetilde{P}(t,T_2)}{\widetilde{P}(t,T_2)} = \zeta_2(t)dt.$$

(6) We apply the Itô formula and the fact that

$$t \mapsto \widehat{\mathbb{E}}\big[(L(T_1, T_1, T_2) - \kappa)^+ \,|\, \mathcal{F}_t\big]$$

and $(L(t, T_1, T_2))_{t \in \mathbb{R}_+}$ are both martingales under $\widehat{\mathbb{P}}$. Next, we use the fact that

$$\widehat{V}_t = \widehat{\mathbb{E}}\big[(L(T_1, T_1, T_2) - \kappa)^+ \,|\, \mathcal{F}_t\big],$$

and apply the result of Question (6).

(7) Apply the Itô rule to $V_t = P(t, T_2)\widehat{V}_t$ using Relation (9.34) and the result of Question (6).

(8) We have

$$
\begin{aligned}
dV_t &= \frac{\partial C}{\partial x}(L(t, T_1, T_2), v(t, T_1))P(t, T_1)(\zeta_1(t) - \zeta_2(t))dB_t + \widehat{V}_t dP(t, T_2) \\
&= \frac{\partial C}{\partial x}(L(t, T_1, T_2), v(t, T_1))P(t, T_1)(\zeta_1(t) - \zeta_2(t))dB_t \\
&\quad + \zeta_2(t)\widehat{V}_t P(t, T_2)dB_t \\
&= (1 + L(t, T_1, T_2))\frac{\partial C}{\partial x}(L(t, T_1, T_2), v(t, T_1))P(t, T_2)(\zeta_1(t) - \zeta_2(t))dB_t \\
&\quad + \zeta_2(t)\widehat{V}_t P(t, T_2)dB_t \\
&= \zeta_1(t)L(t, T_1, T_2)\frac{\partial C}{\partial x}(L(t, T_1, T_2), v(t, T_1))P(t, T_2)dB_t \\
&\quad - \zeta_2(t)L(t, T_1, T_2)\frac{\partial C}{\partial x}(L(t, T_1, T_2), v(t, T_1))P(t, T_2)dB_t \\
&\quad + \frac{\partial C}{\partial x}(L(t, T_1, T_2), v(t, T_1))P(t, T_2)(\zeta_1(t) - \zeta_2(t))dB_t \\
&\quad + \zeta_2(t)\widehat{V}_t P(t, T_2)dB_t \\
&= \zeta_1(t)L(t, T_1, T_2)\frac{\partial C}{\partial x}(L(t, T_1, T_2), v(t, T_1))P(t, T_2)dB_t \\
&\quad + \zeta_2(t)\left(\widehat{V}_t - L(t, T_1, T_2)\frac{\partial C}{\partial x}(L(t, T_1, T_2), v(t, T_1))\right)P(t, T_2)dB_t \\
&\quad + \frac{\partial C}{\partial x}(L(t, T_1, T_2), v(t, T_1))P(t, T_2)(\zeta_1(t) - \zeta_2(t))dB_t,
\end{aligned}
$$

hence

$$
\begin{aligned}
d\widetilde{V}_t &= \zeta_1(t)L(t, T_1, T_2)\frac{\partial C}{\partial x}(L(t, T_1, T_2), v(t, T_1))\widetilde{P}(t, T_2)dB_t \\
&\quad + \left(\widehat{V}_t - \zeta_2(t)L(t, T_1, T_2)\frac{\partial C}{\partial x}(L(t, T_1, T_2), v(t, T_1))\right)\widetilde{P}(t, T_2)dB_t \\
&\quad + (\zeta_1(t) - \zeta_2(t))\frac{\partial C}{\partial x}(L(t, T_1, T_2), v(t, T_1))\widetilde{P}(t, T_2)dB_t
\end{aligned}
$$

$$= \zeta_1(t)\big(\widetilde{P}(t,T_1) - \widetilde{P}(t,T_2)\big)\frac{\partial C}{\partial x}(L(t,T_1,T_2), v(t,T_1))dB_t$$

$$+ \left(\widehat{V}_t - L(t,T_1,T_2)\frac{\partial C}{\partial x}(L(t,T_1,T_2), v(t,T_1))\right) d\widetilde{P}(t,T_2)$$

$$+ (\zeta_1(t) - \zeta_2(t))\frac{\partial C}{\partial x}(L(t,T_1,T_2), v(t,T_1))\widetilde{P}(t,T_2)dB_t$$

$$= \big(\zeta_1(t)\widetilde{P}(t,T_1) - \zeta_2(t)\widetilde{P}(t,T_2)\big)\frac{\partial C}{\partial x}(L(t,T_1,T_2), v(t,T_1))dB_t$$

$$+ \left(\widehat{V}_t - L(t,T_1,T_2)\frac{\partial C}{\partial x}(L(t,T_1,T_2), v(t,T_1))\right) d\widetilde{P}(t,T_2)$$

$$= \frac{\partial C}{\partial x}(L(t,T_1,T_2), v(t,T_1))d\big(\widetilde{P}(t,T_1) - \widetilde{P}(t,T_2)\big)$$

$$+ \left(\widehat{V}_t - L(t,T_1,T_2)\frac{\partial C}{\partial x}(L(t,T_1,T_2), v(t,T_1))\right) d\widetilde{P}(t,T_2),$$

hence the hedging portfolio is given by

$$\xi_t^1 = \frac{\partial C}{\partial x}(L(t,T_1,T_2), v(t,T_1))$$

and

$$\xi_t^2 = \widehat{V}_t - L(t,T_1,T_2)\frac{\partial C}{\partial x}(L(t,T_1,T_2), v(t,T_1))$$

$$- \frac{\partial C}{\partial x}(L(t,T_1,T_2), v(t,T_1)),$$

$0 \leqslant t \leqslant T_1$.

Exercise 9.13.

(1) We have

$$L(t,T_1,T_2) = L(0,T_1,T_2)\,e^{\gamma B_t^{(2)} - \gamma^2 t/2}, \qquad 0 \leqslant t \leqslant T_1.$$

(2) We have

$$P(t,T_2)\widehat{\mathbb{E}}_2\left[(L(T_1,T_1,T_2) - \kappa)^+ \mid \mathcal{F}_t\right]$$
$$= P(t,T_2)\mathrm{Bl}(L(t,T_1,T_2), \kappa, \gamma, 0, T_1 - t), \qquad 0 \leqslant t \leqslant T_1.$$

Exercise 9.14. (Exercise 9.3 continued). From Relations (8.31) and (S.19) above, we have $\widehat{\mathbb{P}}_3 = \widehat{\mathbb{P}}_{1,3}$ up to time T_1, hence $\left(B^{(3)}\right)_{t\in[0,T_1]}$ is a standard Brownian motion under $\widehat{\mathbb{P}}_{1,3}$, and

$$\mathbb{E}^*\left[e^{-\int_t^{T_1} r_s ds} P(T_1,T_2,T_3)(S(T_1,T_1,T_3) - \kappa)^+ \,\Big|\, \mathcal{F}_t\right]$$

$$= P(t, T_1, T_3)\widehat{\mathbb{E}}_{1,3}\left[(S(T_1, T_1, T_3) - \kappa)^+ \mid \mathcal{F}_t\right]$$

$$= P(t, T_1, T_3)\mathrm{Bl}(S(t, T_1, T_3), \kappa, \tilde{\sigma}_{1,3}(t), 0, T_1 - t), \quad 0 \leqslant t \leqslant T_1,$$

where $\tilde{\sigma}_{1,3}(t)$ is the approximate volatility obtained by freezing the random component of $\sigma_{1,3}(s)$ at time t as in Proposition 9.9, *i.e.* by (9.25), we have

$$|\tilde{\sigma}_{1,3}(t)|^2$$

$$= \frac{1}{T_1 - t}$$

$$\times \sum_{k,l=1}^{2} \delta_k \delta_l v_{l+1}^{1,3}(t) v_{k+1}^{1,3}(t) \frac{L(t, T_l, T_{l+1}) L(t, T_k, T_{k+1})}{|S(t, T_1, T_3)|^2} \int_t^{T_1} \gamma_l(s)\gamma_k(s)ds.$$

Exercise 9.15. (Exercise 9.13 continued).

(1) We have

$$P(t, T_1) = P(t, T_2)(1 + \delta L(t, T_1, T_2)), \quad 0 \leqslant t \leqslant T_1,$$

hence

$$dP(t, T_1) = P(t, T_2)\delta dL(t, T_1, T_2) + (1 + \delta L(t, T_1, T_2))dP(t, T_2)$$

$$+ \delta dP(t, T_2) \cdot dL(t, T_1, T_2)$$

$$= P(t, T_2)\delta\gamma L(t, T_1, T_2)dB_t^{(2)}$$

$$+ (1 + \delta L(t, T_1, T_2))P(t, T_2)(r_t dt + \zeta_2(t)dB_t)$$

$$+ \zeta_2(t)\delta\gamma L(t, T_1, T_2)P(t, T_2)dt$$

$$= P(t, T_2)\delta\gamma L(t, T_1, T_2)dB_t$$

$$+ (1 + \delta L(t, T_1, T_2))P(t, T_2)(r_t dt + \zeta_2(t)dB_t)$$

$$= P(t, T_2)(\delta\gamma L(t, T_1, T_2) + \zeta_2(t)(1 + \delta L(t, T_1, T_2)))dB_t$$

$$+ P(t, T_2)(1 + \delta L(t, T_1, T_2))r_t dt$$

$$= \frac{P(t, T_1)}{1 + \delta L(t, T_1, T_2)}(\delta\gamma L(t, T_1, T_2) + \zeta_2(t)(1 + \delta L(t, T_1, T_2)))dB_t$$

$$+ P(t, T_2)(1 + \delta L(t, T_1, T_2))r_t dt$$

$$= P(t, T_1)\left(\frac{\gamma\delta L(t, T_1, T_2)}{1 + \delta L(t, T_1, T_2)} + \zeta_2(t)\right)dB_t$$

$$+ P(t, T_2)(1 + \delta L(t, T_1, T_2))r_t dt$$

$$= P(t, T_1)\left(\frac{\gamma\delta L(t, T_1, T_2)}{1 + \delta L(t, T_1, T_2)} + \zeta_2(t)\right)dB_t + r_t P(t, T_1)dt,$$

hence we have

$$\zeta_1(t) = \frac{\delta\gamma L(t, T_1, T_2)}{1 + \delta L(t, T_1, T_2)} + \zeta_2(t), \quad 0 \leqslant t \leqslant T_1.$$

(2) We have

$$\frac{dL(t, T_1, T_2)}{L(t, T_1, T_2)} = \gamma dB_t^{(2)} = \gamma dB_t - \gamma\zeta_2(t)dt, \qquad 0 \leqslant t \leqslant T_1,$$

(3) Assuming that

$$\frac{dL(s, T_1, T_2)}{L(s, T_1, T_2)} = \gamma dB_s - \gamma\zeta_2(t)ds, \qquad t \leqslant s \leqslant T_1,$$

we get

$$L(s, T_1, T_2) = L(t, T_1, T_2) e^{\gamma(B_s - B_t) - \gamma^2(s-t)/2 - (s-t)\gamma\zeta_2(t)}, \qquad 0 \leqslant t \leqslant s.$$

On the other hand, we have $\widehat{\mathbb{P}}_1 = \mathbb{P}$ since $\zeta_1 = 0$, hence $(B_t)_{t \in \mathbb{R}_+}$ is a standard Brownian motion under $\widehat{\mathbb{P}}_1$, and

$$P(t, T_1)\widehat{\mathbb{E}}_1\left[(P(T_1, T_2) - K)^+ \,\middle|\, \mathcal{F}_t\right]$$
$$= P(t, T_1)\,\mathbb{E}^*\left[(P(T_1, T_2) - K)^+ \,\middle|\, \mathcal{F}_t\right]$$
$$= P(t, T_1)$$
$$\times \mathbb{E}^*\left[\left((1 + \delta L(t, T_1, T_2) e^{\gamma(B_{T_1} - B_t) - (T_1 - t)\gamma^2/2 - (T_1 - t)\gamma\zeta_2(t)})^{-1} - K\right)^+ \,\middle|\, \mathcal{F}_t\right]$$
$$= P(t, T_1) \int_{-\infty}^{\infty}\left((1 + \delta L(t, T_1, T_2) e^{\gamma x - (T_1 - t)\gamma^2/2 - (T_1 - t)\gamma\zeta_2(t)})^{-1} - K\right)^+$$
$$\times \frac{e^{-x^2/(2(T_1 - t))}}{\sqrt{2(T_1 - t)\pi}} dx.$$

Exercise 9.16.

(1) We have

$$dX_t = \sum_{i=2}^{n} c_i d\widehat{P}(t, T_i) = \sum_{i=2}^{n} c_i \sigma_i(t) \widehat{P}(t, T_i) d\widehat{B}_t = \sigma_t X_t d\widehat{B}_t,$$

from which we obtain

$$\sigma_t = \frac{1}{X_t}\sum_{i=2}^{n} c_i \sigma_i(t) \widehat{P}(t, T_i) = \frac{\displaystyle\sum_{i=2}^{n} c_i \sigma_i(t) \widehat{P}(t, T_i)}{\displaystyle\sum_{i=2}^{n} c_i \widehat{P}(t, T_i)} = \frac{\displaystyle\sum_{i=2}^{n} c_i \sigma_i(t) P(t, T_i)}{\displaystyle\sum_{i=2}^{n} c_i P(t, T_i)}.$$

(2) Approximating the random process σ_t by the deterministic function of time

$$\widehat{\sigma}(t) := \frac{1}{X_0}\sum_{i=2}^{n} c_i \sigma_i(t) \widehat{P}(0, T_i)$$

$$= \frac{\sum_{i=2}^{n} c_i \sigma_i(t) \widehat{P}(0, T_i)}{\sum_{i=2}^{n} c_i \widehat{P}(0, T_i)} = \frac{\sum_{i=2}^{n} c_i \sigma_i(t) P(0, T_i)}{\sum_{i=2}^{n} c_i P(0, T_i)},$$

we find

$$P(0, T_1) \widehat{\mathbb{E}} \left[(X_{T_0} - \kappa)^+ \right]$$

$$\simeq P(0, T_1) \widehat{\mathbb{E}} \left[\left(X_0 \, e^{\int_0^{T_0} |\widehat{\sigma}(t)|^2 dW_t^{(1)} - \int_0^{T_0} |\widehat{\sigma}(t)|^2 dt/2} - \kappa \right)^+ \right]$$

$$= P(0, T_1) \mathrm{Bl}_{\mathrm{call}} \left(X_0, \kappa, \sqrt{\frac{1}{T_0} \int_0^{T_0} |\widehat{\sigma}(t)|^2 dt}, 0, T \right)$$

$$= X_0 P(0, T_1) \Phi \left(\frac{v(t, T_0)}{2} + \frac{1}{v(t, T_0)} \log \frac{X_0}{\kappa} \right)$$

$$\quad - \kappa P(0, T_1) \Phi \left(-\frac{v(t, T_0)}{2} + \frac{1}{v(t, T_0)} \log \frac{X_0}{\kappa} \right)$$

$$= \sum_{j=2}^{n} c_j P(0, T_j) \Phi \left(\frac{v(t, T_0)}{2} - \frac{1}{v(t, T_0)} \log \left(\sum_{i=2}^{n} \frac{c_i P(0, T_i)}{\kappa P(0, T_1)} \right) \right)$$

$$\quad - \kappa P(0, T_1) \Phi \left(-\frac{v(t, T_0)}{2} - \frac{1}{v(t, T_0)} \log \left(\sum_{i=2}^{n} \frac{c_i P(0, T_i)}{\kappa P(0, T_1)} \right) \right),$$

with $v^2(t, T_0) := \int_0^{T_0} |\widehat{\sigma}(t)|^2 dt$.

Problem 9.17. Swaption hedging (2).

(1) We have

$$S(T_i, T_i, T_j) = S(t, T_i, T_j) \exp \left(\int_t^{T_i} \sigma_{i,j}(s) dB_s^{i,j} - \frac{1}{2} \int_t^{T_i} |\sigma_{i,j}(s)|^2 ds \right).$$

(2) We have

$$P(t, T_i, T_j) \widehat{\mathbb{E}}_{i,j} \left[(S(T_i, T_i, T_j) - \kappa)^+ \mid \mathcal{F}_t \right]$$

$$= P(t, T_i, T_j)$$

$$\quad \times \widehat{\mathbb{E}}_{i,j} \left[\left(S(t, T_i, T_j) \, e^{\int_t^{T_i} \sigma_{i,j}(s) dB_s^{i,j} - \int_t^{T_i} |\sigma_{i,j}(s)|^2 ds/2} - \kappa \right)^+ \mid \mathcal{F}_t \right]$$

$$= P(t, T_i, T_j) \mathrm{Bl} \left(S(t, T_i, T_j), \kappa, \frac{v(t, T_i)}{\sqrt{T_i - t}}, 0, T_i - t \right)$$

$$= S(t, T_i, T_j) P(t, T_i, T_j) \Phi \left(\frac{\log(x/K)}{v(t, T_i)} + \frac{v(t, T_i)}{2} \right)$$

$$- \kappa P(t, T_i, T_j) \Phi \left(\frac{\log(x/K)}{v(t, T_i)} - \frac{v(t, T_i)}{2} \right),$$

where

$$v^2(t, T_i) = \int_t^{T_i} |\sigma_{i,j}(s)|^2 ds.$$

(3) Integrate the self-financing condition (9.42) between 0 and t.

(4) We have

$$d\tilde{V}_t = d\left(e^{-\int_0^t r_s ds} V_t \right)$$

$$= -r_t e^{-\int_0^t r_s ds} V_t \, dt + e^{-\int_0^t r_s ds} dV_t$$

$$= -r_t e^{-\int_0^t r_s ds} \sum_{k=i}^j \xi_t^{(k)} P(t, T_k) dt + e^{-\int_0^t r_s ds} \sum_{k=i}^j \xi_t^{(k)} dP(t, T_k)$$

$$= \sum_{k=i}^j \xi_t^{(k)} d\tilde{P}(t, T_k), \qquad 0 \leqslant t \leqslant T_i,$$

since

$$\frac{d\tilde{P}(t, T_k)}{\tilde{P}(t, T_k)} = \zeta_k(t) dt, \qquad k = i, \ldots, j.$$

(5) We apply the Itô formula and the fact that

$$t \mapsto \widehat{\mathbb{E}}_{i,j} \left[\left(S(T_i, T_i, T_j) - \kappa \right)^+ \mid \mathcal{F}_t \right]$$

and $(S_t)_{t \in \mathbb{R}_+}$ are both martingales under $\widehat{\mathbb{P}}_{i,j}$.

(6) Use the fact that

$$\widehat{V}_t = \widehat{\mathbb{E}}_{i,j} \left[\left(S(T_i, T_i, T_j) - \kappa \right)^+ \mid \mathcal{F}_t \right],$$

and apply the result of Question (5).

(7) Apply the Itô rule to $V_t = P(t, T_i, T_j) \widehat{V}_t$ using Relation (9.39) and the result of Question (6).

(8) From the expression (9.41) of the swap rate volatilities, we have

$$dV_t = S_t \frac{\partial C}{\partial x} (S_t, v(t, T_i))$$

$$\times \left(\sum_{k=i}^{j-1} (T_{k+1} - T_k) P(t, T_{k+1})(\zeta_i(t) - \zeta_{k+1}(t)) + P(t, T_j)(\zeta_i(t) - \zeta_j(t)) \right) dB_t$$

$$+ \widehat{V}_t dP(t, T_i, T_j)$$

$$= S_t \frac{\partial C}{\partial x}(S_t, v(t, T_i))$$

$$\times \left(\sum_{k=i}^{j-1} (T_{k+1} - T_k) P(t, T_{k+1})(\zeta_i(t) - \zeta_{k+1}(t)) + P(t, T_j)(\zeta_i(t) - \zeta_j(t)) \right) dB_t$$

$$+ \widehat{V}_t \sum_{k=i}^{j-1} (T_{k+1} - T_k)\zeta_{k+1}(t)P(t, T_{k+1})dB_t$$

$$= S_t \frac{\partial C}{\partial x}(S_t, v(t, T_i)) \sum_{k=i}^{j-1} (T_{k+1} - T_k) P(t, T_{k+1})(\zeta_i(t) - \zeta_{k+1}(t))dB_t$$

$$+ \frac{\partial C}{\partial x}(S_t, v(t, T_i))P(t, T_j)(\zeta_i(t) - \zeta_j(t))dB_t$$

$$+ \widehat{V}_t \sum_{k=i}^{j-1} (T_{k+1} - T_k)\zeta_{k+1}(t)P(t, T_{k+1})dB_t$$

$$= S_t\zeta_i(t) \frac{\partial C}{\partial x}(S_t, v(t, T_i)) \sum_{k=i}^{j-1} (T_{k+1} - T_k) P(t, T_{k+1})dB_t$$

$$- S_t \frac{\partial C}{\partial x}(S_t, v(t, T_i)) \sum_{k=i}^{j-1} (T_{k+1} - T_k) P(t, T_{k+1})\zeta_{k+1}(t)dB_t$$

$$+ \frac{\partial C}{\partial x}(S_t, v(t, T_i))P(t, T_j)(\zeta_i(t) - \zeta_j(t))dB_t$$

$$+ \widehat{V}_t \sum_{k=i}^{j-1} (T_{k+1} - T_k)\zeta_{k+1}(t)P(t, T_{k+1})dB_t$$

$$= S_t\zeta_i(t) \frac{\partial C}{\partial x}(S_t, v(t, T_i)) \sum_{k=i}^{j-1} (T_{k+1} - T_k) P(t, T_{k+1})dB_t$$

$$+ \left(\widehat{V}_t - S_t \frac{\partial C}{\partial x}(S_t, v(t, T_i)) \right) \sum_{k=i}^{j-1} (T_{k+1} - T_k) P(t, T_{k+1})\zeta_{k+1}(t)dB_t$$

$$+ \frac{\partial C}{\partial x}(S_t, v(t, T_i))P(t, T_j)(\zeta_i(t) - \zeta_j(t))dB_t.$$

(9) We have

$$d\widetilde{V}_t = S_t\zeta_i(t) \frac{\partial C}{\partial x}(S_t, v(t, T_i)) \sum_{k=i}^{j-1} (T_{k+1} - T_k) \widetilde{P}(t, T_{k+1})dB_t$$

$$+ \left(\widehat{V}_t - S_t \frac{\partial C}{\partial x}(S_t, v(t, T_i)) \right) \sum_{k=i}^{j-1} (T_{k+1} - T_k) \widetilde{P}(t, T_{k+1})\zeta_{k+1}(t)dB_t$$

$$+\frac{\partial C}{\partial x}(S_t, v(t, T_i))\widetilde{P}(t, T_j)(\zeta_i(t) - \zeta_j(t))dB_t$$

$$= (\widetilde{P}(t, T_i) - \widetilde{P}(t, T_j))\zeta_i(t)\frac{\partial C}{\partial x}(S_t, v(t, T_i))dB_t$$

$$+ \left(\widehat{V}_t - S_t\frac{\partial C}{\partial x}(S_t, v(t, T_i))\right)d\widetilde{P}(t, T_i, T_j)$$

$$+\frac{\partial C}{\partial x}(S_t, v(t, T_i))\widetilde{P}(t, T_j)(\zeta_i(t) - \zeta_j(t))dB_t$$

$$= (\zeta_i(t)\widetilde{P}(t, T_i) - \zeta_j(t)\widetilde{P}(t, T_j))\frac{\partial C}{\partial x}(S_t, v(t, T_i))dB_t$$

$$+ \left(\widehat{V}_t - S_t\frac{\partial C}{\partial x}(S_t, v(t, T_i))\right)d\widetilde{P}(t, T_i, T_j)$$

$$= \frac{\partial C}{\partial x}(S_t, v(t, T_i))d(\widetilde{P}(t, T_i) - \widetilde{P}(t, T_j))$$

$$+ \left(\widehat{V}_t - S_t\frac{\partial C}{\partial x}(S_t, v(t, T_i))\right)d\widetilde{P}(t, T_i, T_j).$$

(10) Taking $v := v(t, T_i)$, we have

$$\frac{\partial C}{\partial x}(x, v) = \frac{\partial}{\partial x}\left(x\Phi\left(\frac{v}{2} + \frac{1}{v}\log\frac{x}{\kappa}\right) - \kappa\Phi\left(-\frac{v}{2} + \frac{1}{v}\log\frac{x}{\kappa}\right)\right)$$

$$= x\frac{\partial}{\partial x}\Phi\left(\frac{v}{2} + \frac{1}{v}\log\frac{x}{\kappa}\right) - \kappa\frac{\partial}{\partial x}\Phi\left(-\frac{v}{2} + \frac{1}{v}\log\frac{x}{\kappa}\right) + \Phi\left(\frac{v}{2} + \frac{1}{v}\log\frac{x}{\kappa}\right)$$

$$= \frac{1}{v\sqrt{2\pi}}e^{-(v/2 + v^{-1}\log(x/\kappa))^2/2} - \frac{\kappa}{vx\sqrt{2\pi}}e^{-(-v/2 + v^{-1}\log(x/\kappa))^2/2}$$

$$+ \Phi\left(\frac{v}{2} + \frac{1}{v}\log\frac{x}{\kappa}\right) = \Phi\left(\frac{v}{2} + \frac{1}{v}\log\frac{x}{\kappa}\right).$$

(11) We have

$$d\widetilde{V}_t = \frac{\partial C}{\partial x}(S_t, v(t, T_i))d(\widetilde{P}(t, T_i) - \widetilde{P}(t, T_j))$$

$$+ \left(\widehat{V}_t - S_t\frac{\partial C}{\partial x}(S_t, v(t, T_i))\right)d\widetilde{P}(t, T_i, T_j)$$

$$= \Phi\left(\frac{v(t, T_i)}{2} + \frac{1}{v(t, T_i)}\log\frac{S_t}{K}\right)d(\widetilde{P}(t, T_i) - \widetilde{P}(t, T_j))$$

$$-\kappa\Phi\left(-\frac{v(t, T_i)}{2} + \frac{1}{v(t, T_i)}\log\frac{S_t}{K}\right)d\widetilde{P}(t, T_i, T_j).$$

(12) The hedging strategy can be obtained by comparing the results of Questions (4) and (11).

Chapter 10

Exercise 10.1. By absence of arbitrage we have $(1-\alpha)\,\mathrm{e}^{-r_d T} = \mathrm{e}^{-rT}$, hence $\alpha = 1 - \mathrm{e}^{-(r-r_d)T}$.

Exercise 10.2.

(1) The bond payoff $\mathbb{1}_{\{\tau > T-t\}}$ is discounted according to the risk-free rate, before taking expectation.
(2) We have $\mathbb{E}^*\left[\mathbb{1}_{\{\tau > T-t\}}\right] = \mathrm{e}^{-\lambda(T-t)}$, hence $P_d(t,T) = \mathrm{e}^{-(\lambda+r)(T-t)}$.
(3) We have $P_M(t,T) = \mathrm{e}^{-(\lambda+r)(T-t)}$, hence $\lambda = -r + (T-t)^{-1}\log P_M(t,T)$.

Exercise 10.3. From the terminal data of Figure S.2 on McDonald's Corp, we infer $S_t^{i,j} = 0.10790\%$, with $T_i = $ Mar 20, 2015, $t = $ Apr 12, 2015, and $\xi = 40\%$.

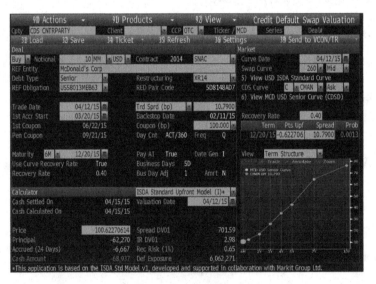

Fig. S.2: CDS market data.

Next, from the discount factors of Figure S.3 we solve Equation (10.10) numerically in Table S.1 below to find the default rate $\lambda_1 = 0.0017987468$ and default probability 0.0012460256, which is consistent with the value of 0.0013 in Figure S.2, see also Castellacci (2008).

Date	Delta	Discount Factor	Premium Leg	Protection Leg
Jun 22, 2015	0.2611111	0.99952277	0.0002814722	0.0002814708
Sep 21, 2015	0.2527778	0.99827639	0.0002721533	0.000272154
Dec 21, 2015	0.2527778	0.99607821	0.0002715541	0.0002715548
		Total	0.0008251796	0.0008251796

Table S.1: CDS market data.

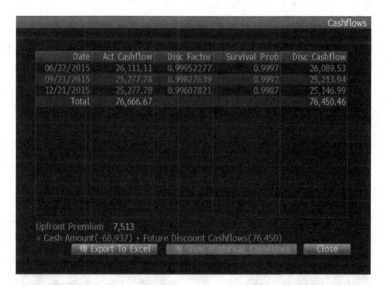

Fig. S.3: CDS discount factors and survival probabilities.

Exercise 10.4.

(1) By equating the premium and protection legs, we find

$$(1 - \xi) \sum_{k=i}^{j-1} P(t, T_{k+1}) \left(Q(t, T_k) - Q(t, T_{k+1})\right)$$

$$= S_t^{i,j} \sum_{k=i}^{j-1} \delta_k P(t, T_{k+1}) Q(t, T_{k+1}).$$

For $j = i + 1$ this yields

$$(1 - \xi) P(t, T_{i+1}) \left(Q(t, T_i) - Q(t, T_{i+1})\right) = S_t^{i,i+1} \delta_i P(t, T_{i+1}) Q(t, T_{i+1}),$$

hence

$$Q(t, T_{i+1}) = \frac{1 - \xi}{S_t^{i,i+1} \delta_i + 1 - \xi},$$

with $Q(t, T_i) = 1$, and the recurrence relation

$$(1 - \xi) P(t, T_{j+1}) \left(Q(t, T_j) - Q(t, T_{j+1}) \right)$$

$$+ (1 - \xi) \sum_{k=i}^{j-1} P(t, T_{k+1}) \left(Q(t, T_k) - Q(t, T_{k+1}) \right)$$

$$= S_t^{i,j} \delta_j P(t, T_{j+1}) Q(t, T_{j+1}) + S_t^{i,j} \sum_{k=i}^{j-1} \delta_k P(t, T_{k+1}) Q(t, T_{k+1}),$$

i.e.

$$Q(t, T_{j+1}) = \frac{(1 - \xi) Q(t, T_j)}{1 - \xi + S_t^{i,j} \delta_j}$$

$$+ \sum_{k=i}^{j-1} \frac{P(t, T_{k+1}) \left((1 - \xi) Q(t, T_k) - Q(t, T_{k+1}) \left((1 - \xi) + \delta_k S_t^{i,j} \right) \right)}{P(t, T_{j+1})(1 - \xi + S_t^{i,j} \delta_j)}.$$

(2) From the terminal data on the Coca-Cola Company, see Figure S.4,

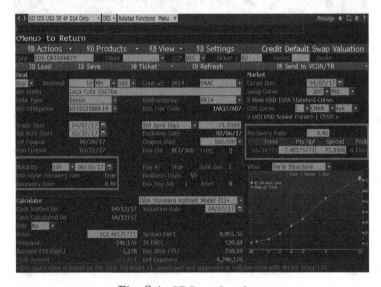

Fig. S.4: CDS market data.

we obtain the spread values $S_t^{1,1}, \ldots, S_t^{1,8}$ in base points (bp) and compute the corresponding the corresponding survival probabilities in Table S.2, where $1 - Q(t, T_8) = 1 - 0.880602 = 0.119398$ is consistent with the default probability of 0.1163 in Figure S.4.

k	Maturity	T_k	$S_t^{1,k}$ (bp)	$Q(t, T_k)$
1	6M	0.5	10.97	0.999087
2	1Y	1	12.25	0.997961
3	2Y	2	14.32	0.995235
4	3Y	3	19.91	0.990037
5	4Y	4	26.48	0.982293
6	5Y	5	33.29	0.972122
7	7Y	7	52.91	0.937632
8	10Y	10	71.91	0.880602

Table S.2: Spread and survival probabilities.

Problem 10.5. Defaultable bonds.

(1) Use the fact that $(r_t, \lambda_t)_{t \in [0,T]}$ is a Markov process.

(2) Use the tower property (11.7) of the conditional expectation given $\mathcal{F}_t$.

(3) We have

$$d\left(e^{-\int_0^t (r_s + \lambda_s)ds} P(t, T) \right)$$

$$= -(r_t + \lambda_t) e^{-\int_0^t (r_s + \lambda_s)ds} P(t, T)dt + e^{-\int_0^t (r_s + \lambda_s)ds} dP(t, T)$$

$$= -(r_t + \lambda_t) e^{-\int_0^t (r_s + \lambda_s)ds} P(t, T)dt + e^{-\int_0^t (r_s + \lambda_s)ds} dF(t, r_t, \lambda_t)$$

$$= -(r_t + \lambda_t) e^{-\int_0^t (r_s + \lambda_s)ds} P(t, T)dt + e^{-\int_0^t (r_s + \lambda_s)ds} \frac{\partial F}{\partial x}(t, r_t, \lambda_t)dr_t$$

$$+ e^{-\int_0^t (r_s + \lambda_s)ds} \frac{\partial F}{\partial y}(t, r_t, \lambda_t)d\lambda_t + \frac{\sigma^2}{2} e^{-\int_0^t (r_s + \lambda_s)ds} \frac{\partial^2 F}{\partial x^2}(t, r_t, \lambda_t)dt$$

$$+ \frac{\eta^2}{2} e^{-\int_0^t (r_s + \lambda_s)ds} \frac{\partial^2 F}{\partial y^2}(t, r_t, \lambda_t)dt$$

$$+ \rho\sigma\eta e^{-\int_0^t (r_s + \lambda_s)ds} \frac{\partial^2 F}{\partial x \partial y}(t, r_t, \lambda_t)dt + e^{-\int_0^t (r_s + \lambda_s)ds} \frac{\partial F}{\partial t}(t, r_t, \lambda_t)dt$$

$$= \sigma e^{-\int_0^t (r_s + \lambda_s)ds} \frac{\partial F}{\partial x}(t, r_t, \lambda_t)dB_t^{(1)}$$

$$+ \eta e^{-\int_0^t (r_s + \lambda_s)ds} \frac{\partial F}{\partial y}(t, r_t, \lambda_t)dB_t^{(2)} + e^{-\int_0^t (r_s + \lambda_s)ds}$$

$$\times \left(-(r_t + \lambda_t)P(t,T) - ar_t \frac{\partial F}{\partial x}(t, r_t, \lambda_t) - b\lambda_t \frac{\partial F}{\partial y}(t, r_t, \lambda_t) \right.$$

$$+ \frac{\sigma^2}{2} \frac{\partial^2 F}{\partial x^2}(t, r_t, \lambda_t) + \frac{\eta^2}{2} \frac{\partial^2 F}{\partial y^2}(t, r_t, \lambda_t)$$

$$\left. + \rho\sigma\eta \frac{\partial^2 F}{\partial x \partial y}(t, r_t, \lambda_t) + \frac{\partial F}{\partial t}(t, r_t, \lambda_t) \right) dt,$$

hence the bond pricing PDE is given by

$$-ax\frac{\partial F}{\partial x}(t,x,y) - by\frac{\partial F}{\partial y}(t,x,y) + \frac{\sigma^2}{2}\frac{\partial^2 F}{\partial x^2}(t,x,y) + \frac{\eta^2}{2}\frac{\partial^2 F}{\partial y^2}(t,x,y)$$

$$+ \rho\sigma\eta \frac{\partial^2 F}{\partial x \partial y}(t,x,y) + \frac{\partial F}{\partial t}(t, r_t, \lambda_t) = (x+y)F(t,x,y).$$

(4) We have

$$\int_0^t r_s ds = \frac{1}{a}\left(\sigma B_t^{(1)} + r_0 - r_t\right)$$

$$= \frac{\sigma}{a}\left(B_t^{(1)} r_0 - r_0 e^{-at} - \int_0^t e^{-(t-s)a} dB_s^{(1)}\right)$$

$$= \frac{\sigma}{a}\int_0^t \left(1 - e^{-(t-s)a}\right) dB_s^{(1)} + r_0 \frac{\sigma}{a}\left(1 - e^{-at}\right),$$

hence

$$\int_t^T r_s ds = \int_0^T r_s ds - \int_0^t r_s ds$$

$$= \frac{\sigma}{a}\int_0^T \left(1 - e^{-(T-s)a}\right) dB_s^{(1)} - \frac{\sigma}{a}\int_0^t \left(1 - e^{-(t-s)a}\right) dB_s^{(1)}$$

$$+ \frac{\sigma r_0}{a}\left(e^{-at} - e^{-aT}\right)$$

$$= -\frac{\sigma}{a}\int_0^t \left(e^{-(T-s)a} - e^{-(t-s)a}\right) dB_s^{(1)} - \frac{\sigma}{a}\int_t^T \left(e^{-(T-s)a} - 1\right) dB_s^{(1)}$$

$$+ \frac{\sigma}{a}r_0\left(e^{-at} - e^{-aT}\right)$$

$$= \frac{\sigma}{a}\left(1 - e^{-(T-t)a}\right)\left(r_0 e^{-at} + \int_0^t e^{-(t-s)a} dB_s^{(1)}\right)$$

$$+ \frac{\sigma}{a}\int_t^T \left(1 - e^{-(T-s)a}\right) dB_s^{(1)}$$

$$= \frac{1}{a}\left(1 - e^{-(T-t)a}\right)r_t + \frac{\sigma}{a}\int_t^T \left(1 - e^{-(T-s)a}\right) dB_s^{(1)}.$$

The answer for λ_t is similar.

(5) As a consequence of the previous question, we have

$$
\mathbb{E}\left[\int_t^T r_s ds + \int_t^T \lambda_s ds \,\Big|\, \mathcal{F}_t\right] = C(a,t,T)r_t + C(b,t,T)\lambda_t,
$$

and

$$
\mathrm{Var}\left[\int_t^T r_s ds + \int_t^T \lambda_s ds \,\Big|\, \mathcal{F}_t\right]
$$

$$
= \mathrm{Var}\left[\int_t^T r_s ds \,\Big|\, \mathcal{F}_t\right] + \mathrm{Var}\left[\int_t^T \lambda_s ds \,\Big|\, \mathcal{F}_t\right]
$$

$$
+ 2\,\mathrm{Cov}\left(\int_t^T X_s ds, \int_t^T Y_s ds \,\Big|\, \mathcal{F}_t\right)
$$

$$
= \frac{\sigma^2}{a^2}\int_t^T \left|1 - e^{-(T-s)a}\right|^2 ds
$$

$$
+ 2\rho\frac{\sigma\eta}{ab}\int_t^T \left(1 - e^{-(T-s)a}\right)\left(1 - e^{-(T-s)b}\right) ds
$$

$$
+ \frac{\eta^2}{b^2}\int_t^T \left|1 - e^{-(T-s)b}\right|^2 ds
$$

$$
= \sigma^2 \int_t^T C^2(a,s,T) ds + 2\rho\sigma\eta \int_t^T C(a,s,T)C(b,s,T) ds
$$

$$
+ \eta^2 \int_t^T C^2(b,s,T) ds,
$$

from the Itô isometry.

(6) We have

$$
P(t,T) = \mathbb{1}_{\{\tau>t\}}\, \mathbb{E}\left[\exp\left(-\int_t^T r_s ds - \int_t^T \lambda_s ds\right) \,\Big|\, \mathcal{F}_t\right]
$$

$$
= \mathbb{1}_{\{\tau>t\}} \exp\left(-\mathbb{E}\left[\int_t^T r_s ds \,\Big|\, \mathcal{F}_t\right] - \mathbb{E}\left[\int_t^T \lambda_s ds \,\Big|\, \mathcal{F}_t\right]\right)
$$

$$
\times \exp\left(\frac{1}{2}\mathrm{Var}\left[\int_t^T r_s ds + \int_t^T \lambda_s ds \,\Big|\, \mathcal{F}_t\right]\right)
$$

$$
= \mathbb{1}_{\{\tau>t\}} \exp\left(-C(a,t,T)r_t - C(b,t,T)\lambda_t\right)
$$

$$
\times \exp\left(\frac{\sigma^2}{2}\int_t^T C^2(a,s,T) ds + \frac{\eta^2}{2}\int_t^T C^2(b,s,T) e^{-(T-s)b} ds\right)
$$

$$
\times \exp\left(\rho\sigma\eta \int_t^T C(a,s,T)C(b,s,T) ds\right).
$$

(7) This is a direct consequence of the answers to Questions (3) and (6).

(8) The above analysis shows that

$$\mathbb{P}(\tau > T \mid \mathcal{G}_t) = \mathbb{1}_{\{\tau > t\}} \, \mathbb{E}\left[e^{-\int_t^T \lambda_s ds} \,\middle|\, \mathcal{F}_t \right]$$

$$= \mathbb{1}_{\{\tau > t\}} \exp\left(-C(b, t, T)\lambda_t + \frac{\eta^2}{2} \int_t^T C^2(b, s, T) ds \right),$$

for $a = 0$ and

$$\mathbb{E}\left[e^{-\int_t^T r_s ds} \,\middle|\, \mathcal{F}_t \right] = \exp\left(-C(a, t, T)r_t + \frac{\sigma^2}{2} \int_t^T C^2(a, s, T) ds \right),$$

for $b = 0$, and this implies

$$U(t, T) = \exp\left(\rho \sigma \eta \int_t^T C(a, s, T) C(b, s, T) ds \right)$$

$$= \exp\left(\rho \frac{\sigma \eta}{ab} \left(T - t - C(a, t, T) - C(b, t, T) + C(a+b, t, T) \right) \right).$$

(9) We have

$$f(t, T) = -\mathbb{1}_{\{\tau > t\}} \frac{\partial}{\partial T} \log P(t, T)$$

$$= \mathbb{1}_{\{\tau > t\}} \left(r_t \, e^{-(T-t)a} - \frac{\sigma^2}{2} C^2(a, t, T) + \lambda_t \, e^{-(T-t)b} - \frac{\eta^2}{2} C^2(b, t, T) \right)$$

$$- \mathbb{1}_{\{\tau > t\}} \rho \sigma \eta C(a, t, T) C(b, t, T).$$

(10) We use the relation

$$\mathbb{P}(\tau > T \mid \mathcal{G}_t) = \mathbb{1}_{\{\tau > t\}} \, \mathbb{E}\left[e^{-\int_t^T \lambda_s ds} \,\middle|\, \mathcal{F}_t \right]$$

$$= \mathbb{1}_{\{\tau > t\}} \exp\left(-C(b, t, T)\lambda_t + \frac{\eta^2}{2} \int_t^T C^2(b, s, T) ds \right)$$

$$= \mathbb{1}_{\{\tau > t\}} \, e^{-\int_t^T f_2(t, u) du},$$

where $f_2(t, T)$ is the Vasicek forward rate corresponding to λ_t, *i.e.*

$$f_2(s, T) = \lambda_t \, e^{-(T-s)b} - \frac{\eta^2}{2} C^2(b, s, T).$$

(11) In this case, we have $\rho = 0$ and

$$P(t, T) = \mathbb{P}(\tau > T \mid \mathcal{G}_t) \mathbb{E}\left[e^{-\int_t^T r_s ds} \,\middle|\, \mathcal{F}_t \right],$$

since $U(t, T) = 0$.

Bibliography

Albanese, C. and Lawi, S. (2005). Laplace transforms for integrals of Markov processes, *Markov Process. Related Fields* **11**, 4, pp. 677–724. 42

Bielecki, T.-R. and Rutkowski, M. (2002). *Credit risk: modelling, valuation and hedging*, Springer Finance (Springer-Verlag, Berlin). ix

Björk, T. (2004). On the geometry of interest rate models, in *Paris-Princeton Lectures on Mathematical Finance 2003*, Lecture Notes in Math., Vol. 1847 (Springer, Berlin), pp. 133–215. ix, 102

Black, F. (1976). The pricing of commodity contracts, *J. of Financial Economics* **3**, pp. 167–179. 178

Brace, A., Gatarek, D. and Musiela, M. (1997). The market model of interest rate dynamics, *Math. Finance* **7**, 2, pp. 127–155. 119, 132

Brigo, D. and Mercurio, F. (2006). *Interest rate models—theory and practice*, 2nd edn., Springer Finance (Springer-Verlag, Berlin). ix, 261

Cairns, A. (2004). *Interest rate models* (Princeton University Press, Princeton, NJ). ix

Carmona, R. A. and Tehranchi, M. R. (2006). *Interest rate models: an infinite dimensional stochastic analysis perspective*, Springer Finance (Springer-Verlag, Berlin). ix

Castellacci, G. (2008). Bootstrapping credit curves from CDS spread curves, Available at https://dx.doi.org/10.2139/ssrn.2042177. 225, 336

Chan, K., Karolyi, G., Longstaff, F. and Sanders, A. (1992). An empirical comparison of alternative models of the short-term interest rate, *The Journal of Finance* **47**, 3, pp. 1209–1227, papers and Proceedings of the Fifty-Second Annual Meeting of the American Finance Association, New Orleans, Louisiana. 50, 52

Charpentier, A. (ed.) (2014). *Computational Actuarial Science with R*, The R Series (Chapman & Hall/CRC, USA). 90, 103

Chen, R.-R., Cheng, X., Fabozzi, F. and Liu, B. (2008). An explicit, multi-factor credit default swap pricing model with correlated factors, *Journal of Financial and Quantitative Analysis* **43**, 1, pp. 123–160. 220

Chen, R.-R. and Huang, J.-Z. (2001). Credit spread bounds and their implications for credit risk modeling, Working paper, Rutgers University and Penn State University. 220

Chung, K. (2002). *Green, Brown, and probability & Brownian motion on the line* (World Scientific Publishing Co. Inc., River Edge, NJ). 235

Courtadon, G. (1982). The pricing of options on default-free bonds, *The Journal of Financial and Quantitative Analysis* **17**, 1, pp. 75–100. 50

Cox, J., Ingersoll, J. and Ross, S. (1985). A theory of the term structure of interest rates, *Econometrica* **53**, pp. 385–407. 42, 44, 63

Dellacherie, C., Maisonneuve, B. and Meyer, P. (1992). *Probabilités et Potentiel*, Vol. 5 (Hermann). 217

Dothan, L. (1978). On the term structure of interest rates, *Jour. of Fin. Ec.* **6**, pp. 59–69. 44

Duffie, D. and Singleton, K. (2003). *Credit risk*, Princeton Series in Finance (Princeton University Press, Princeton, NJ). 219

Elliott, R. and Jeanblanc, M. (1999). Incomplete markets with jumps and informed agents, *Math. Methods Oper. Res.* **50**, 3, pp. 475–492. 218

Elliott, R., Jeanblanc, M. and Yor, M. (2000). On models of default risk, *Math. Finance* **10**, 2, pp. 179–195. 217

Faff, R. and Gray, P. (2006). On the estimation and comparison of short-rate models using the generalised method of moments, *Journal of Banking and Finance* **30**, pp. 3131–3146. 255

Feller, W. (1951). Two singular diffusion problems, *Ann. of Math. (2)* **54**, pp. 173–182. 42

Filipović, D. (2009). *Term-structure models. A graduate course*, Springer Finance (Springer-Verlag, Berlin). ix

Guirreri, S. (2015). *YieldCurve: Modelling and Estimation of the Yield Curve*, R package version 4.1. 90

Guo, X., Jarrow, R. and Menn, C. (2007). A note on Lando's formula and conditional independence, Preprint. 216, 219

Harrison, J. and Pliska, S. (1981). Martingales and stochastic integrals in the theory of continuous trading, *Stochastic Process. Appl.* **11**, pp. 215–260. 122

Heath, D., Jarrow, R. and Morton, A. (1992). Bond pricing and the term structure of interest rates: a new methodology, *Econometrica* **60**, pp. 77–105. 119, 122

Ho, S. and Lee, S. (1986). Term structure movements and pricing interest rate contingent claims, *Journal of Finance* **41**, pp. 1011–1029. 45

Huang, C.-F. (1985). Information structures and viable price systems, *Journal of Mathematical Economics* **14**, pp. 215–240. 307

Hull, J. and White, A. (1990). Pricing interest rate derivative securities, *The Review of Financial Studies* **3**, pp. 537–592. 45, 131

Ikeda, N. and Watanabe, S. (1989). *Stochastic Differential Equations and Diffusion Processes* (North-Holland). 5, 9

Jacod, J. (1985). Grossissement initial, hypothèse (H') et théorème de Girsanov, in T. Jeulin and M. Yor (eds.), *Grossissements de filtrations: exemples et applications*, Lecture Notes in Mathematics, Vol. 1118 (Springer-Verlag, Berlin), pp. 6–14. 218

Jacod, J. and Protter, P. (2000). *Probability essentials* (Springer-Verlag, Berlin). 231, 233

James, J. and Webber, N. (2001). *Interest rate modelling, Wileys Series in Financial Engineering*, Vol. XVIII (Cambridge University Press). ix, 44

Jamshidian, F. (1989). An exact bond option formula, *The Journal of Finance* **XLIV**, 1, pp. 205–209. 153, 199

Jamshidian, F. (1996). Sorting out swaptions, *Risk Magazine* **9**, 3, pp. 59–60. 307

Jeulin, T. (1980). *Semi-martingales et grossissement d'une filtration, Lecture Notes in Mathematics*, Vol. 833 (Springer Verlag). 218

Kim, Y.-J. (2002). Option pricing under stochastic interest rates: An empirical investigation, *Asia-Pacific Financial Markets* **9**, pp. 23–44. 155

Lando, D. (1998). On Cox processes and credit risky securities, *Review of Derivative Research* **2**, pp. 99–120. 216, 219

Lim, N. and Privault, N. (2016). Analytic bond pricing for short rate dynamics evolving on matrix lie groups, *Quant. Finance* **16**, pp. 119–129. 44

Lindström, E. (2007). Estimating parameters in diffusion processes using an approximate maximum likelihood approach, *Annals of Operations Research* **151**, pp. 269–288. 255

Mamon, R. (2004). Three ways to solve for bond prices in the Vasicek model, *Journal of Applied Mathematics and Decision Sciences* **8**, 1, pp. 1–14. 283

Marsh, T. and Rosenfeld, E. (1983). Stochastic processes for interest rates and equilibrium bond prices, *The Journal of Finance* **38**, 2, pp. 635–646, papers and Proceedings Forty-First Annual Meeting American Finance Association New York, N.Y. 44, 71

Mikosch, T. (1998). *Elementary stochastic calculus—with finance in view*, Advanced Series on Statistical Science & Applied Probability, Vol. 6 (World Scientific Publishing Co. Inc., River Edge, NJ). 25

Nelson, C. and Siegel, A. (1987). Parsimonious modeling of yield curves, *Journal of Business* **60**, pp. 473–489. 97

Øksendal, B. (2003). *Stochastic differential equations*, sixth edn., Universitext (Springer-Verlag, Berlin). 25

Privault, N. (2009). *Stochastic analysis in discrete and continuous settings with normal martingales*, Lecture Notes in Mathematics, Vol. 1982 (Springer-Verlag, Berlin). 5, 9, 209

Privault, N. (2014). *Stochastic Finance: An Introduction with Market Examples*, Financial Mathematics Series (Chapman & Hall/CRC). 24, 25, 152, 157, 158, 178, 186, 219, 222

Privault, N. and Teng, T.-R. (2012). Risk-neutral hedging in bond markets, *Risk and Decision Analysis* **3**, pp. 201–209. 306, 314

Privault, N. and Wei, X. (2009). Calibration of the LIBOR market model - implementation in PREMIA, Bankers, Markets & Investors **99**, pp. 20–28. 193, 195

Protter, P. (2001). A partial introduction to financial asset pricing theory, *Stochastic Process. Appl.* **91**, 2, pp. 169–203. 33, 307

Protter, P. (2004). *Stochastic integration and differential equations*, Stochastic Modelling and Applied Probability, Vol. 21, 2nd edn. (Springer-Verlag, Berlin). 1, 12, 16, 29, 34, 59, 61, 148, 217, 231

Rebonato, R. (1996). *Interest-Rate Option Models* (John Wiley & Sons). 194

Revuz, D. and Yor, M. (1994). *Continuous Martingales and Brownian Motion* (Springer-Verlag). 12

Santa-Clara, P. and Sornette, D. (2001). The dynamics of the forward interest rate curve with stochastic string shocks, *The Review of Financial Studies* **14**, 1, pp. 149–185. 136

Schoenmakers, J. (2002). Calibration of LIBOR models to caps and swaptions: a way around intrinsic instabilities via parsimonious structures and a collateral market criterion, WIAS Preprint No 740, Berlin. 194

Schoenmakers, J. (2005). *Robust LIBOR modelling and pricing of derivative products*, Chapman & Hall/CRC Financial Mathematics Series (Chapman & Hall/CRC, Boca Raton, FL). ix, 162, 191, 192, 193

Shiryaev, A. (1999). *Essentials of stochastic finance* (World Scientific Publishing Co. Inc., River Edge, NJ). 122

Svensson, L. (1994). Estimating and interpreting forward interest rates: Sweden 1992-1994, National Bureau of Economic Research Working Paper 4871. 98

Vašíček, O. (1977). An equilibrium characterisation of the term structure, *Journal of Financial Economics* **5**, pp. 177–188. 39, 62

Wu, L. (2019). *Interest rate modeling: Theory and practice*, Chapman & Hall/CRC Financial Mathematics Series (CRC Press, Boca Raton, FL), second edition. ix

Wu, X. (2000). A new stochastic duration based on the Vasicek and CIR term structure theories, *Journal of Business Finance and Accounting* **27**. 95

Yor, M. (1985). Grossissements de filtration et absolue continuité de noyaux, in T. Jeulin and M. Yor (eds.), *Grossissements de filtrations: exemples et applications, Lecture Notes in Mathematics*, Vol. 1118 (Springer-Verlag, Berlin), pp. 6–14. 218

Index

Author Index

Printed in the United States
by Baker & Taylor Publisher Services